Geographies of the Internet

This book offers a comprehensive overview of recent research on the internet, emphasizing its spatial dimensions, geospatial applications, and the numerous social and geographic implications such as the digital divide and the mobile internet.

Written by leading scholars in the field, the book sheds light on the origins and the multiple facets of the internet. It addresses the various definitions of cyberspace and the rise of the World Wide Web, draws upon media theory, as well as explores the physical infrastructure such as the global skein of fibre optics networks and broadband connectivity. Several economic dimensions, such as e-commerce, e-tailing, e-finance, e-government, and e-tourism, are also explored. Apart from its most common uses such as Google Earth, social media like Twitter, and neogeography, this volume also presents the internet's novel uses for ethnographic research and the study of digital diasporas.

Illustrated with numerous graphics, maps, and charts, the book will best serve as supplementary reading for academics, students, researchers, and as a professional handbook for policy makers involved in communications, media, retailing, and economic development.

Barney Warf is a Professor of Geography at the University of Kansas. His research and teaching interests lie within the broad domain of human geography, particularly telecommunications and the internet.

Routledge Studies in Human Geography

This series provides a forum for innovative, vibrant, and critical debate within Human Geography. Titles will reflect the wealth of research which is taking place in this diverse and ever-expanding field. Contributions will be drawn from the main sub-disciplines and from innovative areas of work which have no particular sub-disciplinary allegiances.

Ageing and Place
Edited by Gavin J. Andrews and David R. Phillips

Geographies of Commodity Chains
Edited by Alex Hughes and Suzanne Reimer

Queering Tourism
Paradoxical Performances at Gay Pride Parades
Lynda T. Johnston

Cross-Continental Food Chains
Edited by Niels Fold and Bill Pritchard

Private Cities
Edited by Georg Glasze, Chris Webster and Klaus Frantz

Global Geographies of Post-Socialist Transition
Tassilo Herrschel

Urban Development in Post-Reform China
Fulong Wu, Jiang Xu and Anthony Gar-On Yeh

Rural Governance
International Perspectives
Edited by Lynda Cheshire, Vaughan Higgins and Geoffrey Lawrence

For more information about this series, please visit: www.routledge.com/ Routledge-Studies-in-Human-Geography/book-series/SE0514

Geographies of the Internet

Edited by Barney Warf

Routledge
Taylor & Francis Group

LONDON AND NEW YORK

First published 2021
by Routledge
2 Park Square, Milton Park, Abingdon, Oxon OX14 4RN

and by Routledge
52 Vanderbilt Avenue, New York, NY 10017

Routledge is an imprint of the Taylor & Francis Group, an informa business

British Library Cataloguing-in-Publication Data
A catalogue record for this book is available from the British Library

Library of Congress Cataloging-in-Publication Data
Names: Warf, Barney, 1956– editor.
Title: Geographies of the Internet / edited by Barney Warf.
Description: Milton Park, Abingdon, Oxon; New York, NY: Routledge, 2020. |
Series: Routledge studies in human geography |
Includes bibliographical references and index.
Identifiers: LCCN 2019050042 (print) | LCCN 2019050043 (ebook)
Subjects: LCSH: Internet – Social aspects. | Internet – Economic aspects.
Classification: LCC HM851.G467 2020 (print) |
LCC HM851 (ebook) | DDC 302.23/1 – dc23
LC record available at https://lccn.loc.gov/2019050042
LC ebook record available at https://lccn.loc.gov/2019050043

ISBN: 978-0-367-42042-0 (hbk)

ISBN: 978- 0- 367-5 0255- 3 (pbk)

ISBN: 978-0-367-81753-4 (ebk)

Typeset in Times New Roman
by Newgen Publishing UK

Contents

Figures

Tables

Contributors

Anurag Agarwal is currently a Professor in the College of Business at Florida Gulf Coast University. He has a Ph.D. in Information Systems from The Ohio State University. His research interests include Operations, Internet of Things, Optimization, Metaheuristics, Neural Networks, and Genetic Algorithms. His teaching interest include Analytics, Information Systems and Operations. He has published in journals such as *INFORMS Journal of Computing, Omega, Naval Research Logistics, Computers and OR, Annals of OR* and *European Journal of Operational Research.*

Mark Billinghurst is Professor at the University of South Australia in Adelaide, Australia, and at the University of Auckland in New Zealand, directing the Empathic Computing Laboratory in both places. He earned a Ph.D. in 2002 from the University of Washington and researches how virtual and real worlds can be merged, publishing over 400 papers on Augmented Reality, remote collaboration, Empathic Computing, and related topics. In 2013 he was elected as a Fellow of the Royal Society of New Zealand.

Michael L. Black is an Assistant Professor of English at the University of Massachusetts Lowell. His research blends archival reading and text analysis techniques to explore questions around digital authorship and the politics of usability. His work has appeared in *Digital Humanities Quarterly,* the *International Journal of Humanities and Arts Computing,* and *Science, Technology, and Human Values.*

Irene Cheng Chu Chan holds a Ph.D. in Hotel and Tourism Management. She teaches at Hong Kong Polytechnic University. Her research interests include technology management, mobile communication, and customer relationship management in hospitality and tourism.

Ramakrishnan Durairajan is an Assistant Professor in the Department of Computer and Information Science at the University of Oregon, where he co-leads the Oregon Networking Research Group (ONRG). Ram earned his Ph.D. and M.S. degrees in Computer Sciences from the University of Wisconsin – Madison and his B.Tech. in Information Technology from the

College of Engineering, Guindy (CEG), Anna University. Ram's research has been recognized with NSF CRII award, UO faculty research award, best paper Awards from ACM CoNEXT and ACM SIGCOMM GAIA, and has been covered in several fora including the *New York Times*, *MIT Technology Review*, *Popular Science*, *Boston Globe*, Gizmodo, Mashable, among others. Recently, his research on internet topology was named "One of the 100 Greatest Innovations" and has won "Best of What's New" award, in security category, from *Popular Science* magazine.

Emily Fekete is the Social Media and Engagement Coordinator at the American Association of Geographers. She obtained her Ph.D. in Geography from the University of Kansas, where she focused on geographies of consumption, retail, and social media. Her current research interests include geographies of media and communication, particularly social media, cyberterrorism, and online spaces of retailing.

Jayson J. Funke is Adjunct Assistant Professor at Medgar Evers College, City University of New York. He is a radical economic geographer with a concentration in financial geography, international political economy, and geopolitics. He received his Ph.D. in Geography from Clark University, his M.A. in History from Northeastern University, and B.A. in History from the University of Minnesota.

Matthew Haffner is an Assistant Professor in the Department of Geography and Anthropology at the University of Wisconsin – Eau Claire. His interests are in GIScience, digital geographies, and urban geography. He is particularly interested in creating transparent and reproducible solutions to the analysis of spatial data, especially those from crowdsourced datasets like location-based social media.

Jeffrey James is currently Emeritus Professor of Development at Tilburg University, where he was also for a time Director of Graduate Studies in economics at CenTER in Tilburg. Before that he was assistant professor at Boston University and Research Fellow at Queen Elisabeth House, Oxford. He has also worked at the International Labour Organization in Geneva. James has written extensively on issues related to technology and development, with a particular focus on Sub-Sararan Africa. He holds a D.Phil from Oxford.

Aharon Kellerman, Ph.D. Boston University, 1977, is Professor Emeritus at the Department of Geography, University of Haifa, Israel. He further serves as President, Zefat Academic College, Israel. His specialties include the geography of the internet, personal mobilities and globalization.

Matthew Kelley joined the Urban Studies faculty at the University of Washington Tacoma in 2008. He earned a Ph.D. in Geography from Pennsylvania State University in 2007 and then spent a year teaching at

Bucknell University prior to arriving in Tacoma. At UWT, he directs and teaches the Geographic Information Systems (GIS) Certificate Program.

Michel S. Laguerre is Professor and Director of the Institute of Government Studies at the University of California Berkeley. His research interests include contemporary social theory, information technology, Transnational diaspora politics, globalization, and global metropolitan studies. He is the author of numerous books on diasporas, the Caribbean, and globalization.

Rob Law, Ph.D., is a Professor of Technology Management in the School of Hotel and Tourism Management, The Hong Kong Polytechnic University. His primary research interests are technology applications. Professor Law is recognized as one of the world's most prolific and influential scholars in tourism and hospitality management. He has published 250+ articles, attained h-index/i10-index of 81/341, received HKD 25+ million in research grants, and supervised 120 post-doctoral fellows and graduate students.

Wen Lin is a Lecturer in Human Geography at Newcastle University in the UK. Her research and teaching interests include geographic information science (GIS) and urban geography, and her research agenda centers on the intersections between geospatial technologies and the social and political conditions in which these practices are situated. She utilizes combined frameworks from critical GIS research and urban/political geography to investigate processes of geospatial technology constructions within a variety of urban contexts, and theorize their socio-political implications for citizen participation and urban governance.

Elizabeth Mack is an Associate Professor in the Department of Geography, the Environment, and Spatial Sciences at Michigan State University where she teaches courses in economic geography. Dr. Mack's research program evaluates the impact of information and communications technologies (ICTs) on the development trajectory of regional economies. This research program includes broadband infrastructure deployment policy and the impact of broadband on business location and competitiveness. Her work has been funded by the National Science Foundation and the Kauffman Foundation.

Bruno Moriset is an Associate Professor in the Centre for Research in Geography and Planning at the Université Jean Moulin Lyon in Lyon, France. His research centers on the dynamics of e-commerce and digital economies. He authored, among others, *The Digital Economy* with Edward Malecki.

Monica Murero works at the Department of Social Sciences, University of Naples Federico II, Italy. She is the Director and Founder of the E-Life International Institute. Her research concerns innovative technologies for health, such as wearable devices, robotics, and the Internet of Things, and their effects on society. Her theoretical work is internationally known for the interdigital theory.

Todd Patterson is Professor of Geography at Northampton Community College in Bethlehem, Pennsylvania. Much of his research concerns Google Earth.

James B. Pick is Professor of Business at the University of Redlands. He is author or co-author of 165 journal articles, book chapters, and peer-reviewed papers in MIS, GIS, and urban studies, and has authored 13 books, including *Exploring the Urban Community: A GIS Approach* (2012), and *The Global Digital Divides* (2015). He is associate editor of *European Journal of Information Systems* and *Information Technology for Development*. He holds a B.A. from Northwestern University and Ph.D. from the University of California Irvine.

Avijit Sarkar is a Professor of Business Analytics and Operations Research at the University of Redlands School of Business. His primary research interests include examination of global digital divides, spatial patterns of the sharing economy, location-based analytics, and spatial maturity models.

Tyler Sonnichsen is a cultural geographer at the University of Tennessee with interests in media, tourism, gender, and urbanism. Much of my work revolves around ways through which cities change over time, with special consideration for public transit infrastructure, healthcare, and arts funding.

Bhuvan Unhelkar is a Professor of Information Technology at the University of South Florida. He is also a Founding Consultant at Method Science and a Co-Founder/Director at PlatiFi. He has authored 20 books, including *Big Data Strategies for Agile Business* and *The Art of Agile*. His research and teaching interests include Business Analysis & Requirements Modeling, Software Engineering, Big Data Strategies, Agile Processes, Mobile Business, and Green IT. He is also a Fellow of the Australian Computer Society.

Barney Warf is Professor of Geography at the University of Kansas. His research and teaching interests lie within the broad domain of human geography. Much of his research concerns telecommunications, particularly the geographies of the internet, including the digital divide, e-government, and internet censorship. He views these topics through the lens of political economy and social theory. He has authored and edited numerous books and more than 110 journal articles.

Xiang Zhang, Ph.D., University of Kansas, 2018, is a Lecturer in International Development at Nottingham Trent University in the United Kingdom. His research concerns e-commerce in China.

1 Introduction

Barney Warf

Like the wheel and the printing press, the internet has had, and continues to have, profound impacts on the world. Roughly 55% of the planet used the internet in 2019, and for users it has deeply affected communications, consumption, entertainment, politics, and culture. In ways ranging from electronic banking to online education, internet gambling and videogames, e-government and e-commerce, YouTube, Twitter, Facebook, and Google, the internet has become woven into the world economy and everyday life. Never have so many people been able to contact one another so easily, obtain news, purchase goods, file complaints, and save time than today. For large numbers of users, the real and virtual worlds have become inextricably intertwined; for them, the internet is a necessity, not a luxury. Seen this way, the dichotomies off-line/on-line does not do justice to the diverse ways in which the "real" and virtual worlds are interpenetrated. However, for those without access to the information highway, the internet may represent a new source of inequality.

The internet may be defined as a global system of interconnected computer networks, that is, a network of networks, which billions of computers across the planet. Its decentralized architecture allows any computer to communicate with almost any other using the Transmission Control Protocol and Internet Protocol (TCP/IP), the language that allows routers and servers to exchange information with one another. The speeds of connection vary considerably, but have become ever-faster with the widespread adoption of fiber optic cables. In the 21st century, as cellular phones have become ubiquitous, wireless connectivity has become widespread. The World Wide Web is part of the internet, consisting of documents written in the hypertext markup language (HTML), which enables them to be accessed by web browsers. Users find documents on websites, each of which may have several webpages, using their unique Universal Resource Locator (URL) codes. The WWW contains countless billions of hypertext documents, including not only text but figures, music, and video. Finally, cyberspace, a term coined by science-fiction writer William Gibson (1984), refers to the digitized, virtual "world in the wires," that is, computer-mediated communications, augmented reality, and virtual reality. Although cyberspace is often conceived as a domain independent of the "real" or physical world, in reality the virtual and real worlds shape each other continuously.

A brief history of the internet

The internet has a relatively short but fascinating history. (For more thorough treatments, see Hafner and Lyon 1996; Banks 2008; Ryan 2013). It originated in the 1960s under the U.S. Defense Department's Agency Research Projects Administration (ARPA), which designed it to allow computers to communicate with one another in the event of a nuclear attack. ARPA gave rise to innovations such as packet switching, neural networks, queuing theory, adaptive routing, and file transfer protocols. In the process, it created a network quite different from the centralized system of the telephone company (i.e., the monopoly once held by AT&T), which relied on analogue information: rather, the process of digitization facilitated a decentralized, then distributed network. The earliest signs of what eventually became ARPANET connected universities such as Stanford, UCLA, the University of California at Santa Barbara, and the University of Utah. Over time, the military goals were soon supplemented by civilian ones. In 1972, Ray Tomlinson created the first computer messages for personal use, inventing email.

Administration of the internet has varied over time. Between 1984 and 1995, it fell to the National Science Foundation, which transformed ARPANET into NSFNET, a largely academic network of supercomputers in a few select universities. Meanwhile, the world's first cybercommunities arose, such as the WELL (Whole Earth 'Lectronic Link) in San Francisco. On the border of Switzerland and France, the European Particle Physics Lab (CERN) developed hypertext and Universal Resource Locators (URLs), making possible the World Wide Web and user-friendly browsers. Tim Berners-Lee, often called the "father of the World Wide Web," played a key role in this process. In the 1990s, amidst a wave of neoliberal privatization, the U.S. government shifted control to a consortium of telecommunications corporations. Meanwhile, the system began to explode in use on an international scale. The number of websites grew exponentially, from roughly one million in 1990 to almost eight billion in 2019. Next to the mobile or cellular phone, the internet is the most rapidly diffusing technology in world history. Global access deeply shaped by the density, reliability, and affordability of fiber optics lines, which were being laid down in large numbers.

In the late 1990s and the 21st century, two developments greatly affected the internet: Web 2.0 and the rise of the mobile internet. Web 2.0 allows users to contribute material to webpages rather than simply consume their information passively, allowing users to upload content and enjoy instantaneous interactions. Thus were born Amazon, Facebook, YouTube, Wikipedia, and location-based services. The invention of smartphones, or phones that can access the internet, gave rise to the mobile or wireless internet (Arminen 2007; Kellerman 2010). Rapid decreases in the cost of mobile phones made them affordable for vast numbers of people; today more than 90% of the planet owns one. Mobile internet access is particularly important in the developing world. The mobile internet greatly enhanced its accessibility, adding flexibility and convenience. Text messages and Twitter have become common for vast numbers of people.

As the internet became more user-friendly and computer costs declined, the number of the world's netizens world skyrocketed (Figure 1.1). In December, 2019, more than 4.5 billion people used the internet, approximately 58% of the world's population. However, the geographical distribution of internet users is highly uneven: internet penetration rates are far higher in the developed world than the developing one, although they are rising rapidly everywhere. internet growth, however, has still excluded almost half the planet, a phenomenon known as the digital divide, or unequal social and spatial access. This phenomenon takes many forms, including class, education, gender, ethnicity, and age. Everywhere, the young are most likely to use digital technologies; many young people are digital natives, never knowing what the world was like before the internet. Readers of a certain age who recall pre-internet days are in a sense similar to the last generation to experience the world before the automobile became widespread in the early 20th century.

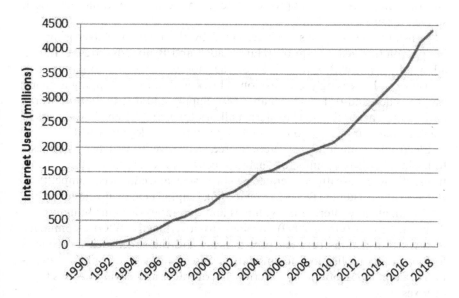

Figure 1.1 Growth in global internet users, 1990–2018.
Source: Author, using data from Internet World Stats (Internetworldstats.com).

The impacts of the internet

The internet has such a large number of effects on peoples, societies, and places that it is difficult to list them all. Some of the major consequences include the economy, social and behavioral effects, and political ones that include governance.

Economic impacts

The internet's economic effects extend into every part of the economy. Farmers can use the internet to manage supply chains, logistics, and output, and control irrigation systems remotely (McKinion et al. 2004). Telecenters in countries such as India allow farmers to acquire information about crop prices and best practices, acquire land titles, and bypass corrupt and intermediaries (Pick et al. 2014). In manufacturing, the internet has given new power to machinery, including robots and artificial intelligence (Caputo et al. 2016). Producer services, too, have been markedly altered, such as with the movement of many clerical functions, call centers, tax preparation, and radiology jobs to the developing world. In finance, enormous sums of funds move at the speed of light through the world's fiber optics networks. Internet banking has reduced the need for labor at the retail end, lowering costs, improving efficiency and convenience (Hanafizadeh et al. 2014). Digital currencies such as Bitcoin are profoundly affecting money supplies and currency markets, while crowdfunding has let many projects emerge that otherwise would have been starved for investment funds.

Consumption, too, has been reshaped by the internet. Advertising has been revolutionized by adware (Hanafizadeh and Behboudi 2012). e-Tailing allows shopping with a few clicks of the mouse, giving rise to giants such as Amazon. Increasingly, old-fashioned brick-and-mortar shopping is giving way to "click-and-order" (Forman 2009). Supply and distribution networks have evolved accordingly. The internet allows, for example, small producers to reach global markets, a boon to small firms in the developing world.

Electronic commerce, or e-commerce, has several varieties. Business-to-business (B2B) ecommerce includes electronic data interchange (EDI) transactions and greatly accelerated product cycles and enhanced competition. Business-to-consumer (B2) e-commerce links suppliers and customers via the web, dramatically altering the volumes and quantities of items bought (e.g., Amazon Prime) (Gong 2009). Business-to-government (B2G) e-commerce involves transactions between private firms and the state, such as requests for proposals; applications for permits, licenses, and patents; online registration of companies; obtaining government contracts; and paying various types of taxes.

In transportation, the internet has unleashed waves of change (Thomopoulos et al. 2015). Smartphones and laptops allow people to work while travelling or at home, for better or worse. Global positioning systems (GPS) have made driving and finding locations far easier. With telework or telecommuting, many employees can work at home. Smart traffic systems have made roads faster and safer. Firms such as Uber and Lyft have created enormous competition for traditional models of transportation. Tourists can use the internet to obtain information about prospective destinations, book hotels, buy tickets, and reserve cars (Condratov 2013; Standing et al. 2014). Taxi drivers and travel agents represent victims of the internet's "creative destruction."

Energy production and consumption have also been greatly affected by the internet. For example, parallel and distributed computing has raised the efficiency of energy suppliers in the face of constraints such as security and environmental limitations. Smart grids – computerized controls and sensor networks – allow for more efficient ways of managing electricity supply and demand, improving the generation, transmission, distribution, and applications. Smart cities, which integrate information technology in numerous ways, including smart traffic systems and smart homes, are a viable means of promoting sustainable development, fostering renewable energy sources, and reducing carbon use (Strengers 2013; Komninos 2014).

Social, behavioral, and psychological impacts

There are innumerable social and behavioral impacts of the internet. Today, more people are more connected technologically to one another than at any other time in human existence. More than one-half of the world's population uses the internet, and social media platforms such as Facebook and Twitter have become thoroughly intertwined with daily existence as a means of communication, news sharing, organizing, and protest. Email is indispensable for almost all internet users: the world sends 145 billion email messages each day. Smartphones and texting have become ubiquitous: the planet sends more than 250,000 text messages every second. Blogs and YouTube have given unprecedented numbers of people the ability to make their views heard.

Digital communications and social media have played significant roles in the formation of individual and social identities, allow people to stay in touch with like-minded others to form classic "communities without propinquity." Some observers argue that meaningful interactions are invariably preceded by earlier face-to-face ties and that digitally mediated communities are ephemeral and lack emotional depth. Digital social media may actually make people less, not more, sociable (Turkle 2011). The internet may even generate subtle changes in brain structure, such as shortened attention spans (Carr 2010). The enormous enhancement of human extensibility that the internet allows has shifted many social interactions from a series of one-to-one ties to networks of one-to-many connections (Adams 2005). Digital networks allow people to present different sides of themselves to different audiences, allowing for fragmented, multifaceted selves to co-exist in one body. Digital media also blur the borders between public and private life. For many people, the digital, on-line self and the non-digital, analogue self have become hopelessly entangled. The internet has thoroughly reshaped many forms of entertainment, including on-line television (e.g., Netflix), videogames, music, and publishing. Online dating sites and pornography have affected sexuality, as have sexting and cyberstalking.

These issues raise serious concerns about digital privacy. Corporations can use data mining to targeted online advertisements and collect propitious volumes of information about individuals, often reselling this data to other vendors. Search engines such as Google or the U.S. National Security Agency collect unimaginable quantities of information about people. The public digital data about private individuals reflects the panopticonic tendencies of contemporary capitalism to monitor and control citizens' behavior.

Education and health care have been reconfigured by cyberspace. The rise of online courses has been steady and inexorable. Libraries have changed their purpose to become repositories of digital data (Herring 2014). Health care has witnessed the rise of telemedicine, telesurgery, and e-health, in which patients can look up diseases, disorders, symptoms, drugs, and solutions (Ball and Lillis 2001).

Political and governmental impacts

The political world has been drastically transformed by the internet. Whereas journalism has been digitized, and traditional newspapers have had to adjust, social media also allows for the propagation of fake news. Blogs have become an important part of political campaigns and commentary. Internet law, or cyberlaw, is another field in which these effects are being felt, including issues like e-contracts, electronic signatures, digital intellectual property rights, and copyright and trademarks (Rustad 2014; Lipton 2015). Other issues concerns pornography, hate speech, online defamation, cybercrimes, internet privacy, and the Right to be Forgotten. Yet other topic concerns the regulation of the internet itself, including issues such as net neutrality. Cybercrimes take a variety of forms, including hacking, identity theft, credit card fraud, and the use of malware such as ransomware, spyware, and phishing.

Electronic government, or e-government, is another example of how the internet is altering the social fabric of societies (Fountain 2001; Davison et al. 2005). E-government includes online contract proposals and submissions of bids, bills, and payments, digital payments of taxes, electronic voting, payment of utility bills, fines, and dues, applications for permits, and licenses, online registration of companies and automobiles, and access to census and other public data. E-government is often held to improve the efficiency of the public sector, raise transparency, and reduce corruption. However, e-government often fails in developing countries (Dada 2006). Finally, internet censorship reflects how some states fear the internet, which allows groups to bypass government monopolies over information access (Deibert 2009). Governments face a choice in the degree of censorship, leading to wide geographic variations (Warf 2010) in its *scope* (or range of topics) and *depth* (or degree of intervention).

Political impacts of the internet also include cyberterrorism and cyberwarfare. Cyberterrorism involves the use of the internet to damage computers and information systems, including the infrastructure, as a result of political motivations (Chen et al. 2014). Cyberterrorism is not simply an attack on human beings, but a way to physically cause damage to the infrastructure. Cyberwarfare attacks can disrupt a country's infrastructure and communications systems, including financial markets, through the use of malicious code. In dissolving the boundaries between the civilian and military spheres, state and non-state actors, the foreign and the domestic, war and peace, cyberwar shifts the location of the battlefront. Given its rising importance, in 2009 the U.S. military established the Cyber Command expressly to wage cyberwar and defend against cyberattacks (Harris 2014).

Misconceptions about the internet

There are numerous myths that swirl around the internet that stem from simplistic readings of the relations among society, information technology, and space. One of the most common is technological determinism, a view that portrays technologies as the driving force behind social changes and views society as the passive recipient of them. This line of thought, which is surprisingly common, often portrays cyberspace acquires the aura of some omnipotent, external actor that drives all other changes, ignoring or marginalizing the social, political, and cultural contexts in which the internet is invariably situated. Technological determinism denies the historical and geographical contingency that characterizes the origins and growth of the internet, and how societies shape cyberspace as much as they are shaped by it. Rather than this simple determinist view, it is more productive to view the internet and its social contexts as simultaneously determinant, shaping one another in multiple, contingent, and unpredictable ways.

Another misconception holds that the internet is only a force for positive changes, as noted in discourses that proclaim it to be inherently emancipatory. Undoubtedly the internet has made obtaining vast amounts of information easier and more convenient, helped to make governments more transparent and efficient, raised productivity, made travel safer, faster, and more convenient, allowed easier shopping, banking, and payment of bills, and revolutionized entertainment and education. Yet inevitably there is the ominous "dark side" of cyberspace, including spam, cybercrime, identity theft, viruses, cyberattacks, bank and credit card fraud, and more. The internet can be used against people as well as for them. Many totalitarian governments have deployed it to great effect to monitor and control citizens and spread propaganda. Ignoring this dark side leads to unrealistic and overly optimistic interpretations of cyberspace.

A significant concern about cyberspace is the "digital divide," the social and spatial inequalities in access that threaten to reproduce social inequalities. Almost half of the world consists of information have-nots, typically those in poor countries, the socially and politically marginalized, the poor and uneducated, and, disproportionately, women. Even in highly developed countries there are pockets of "off-line" citizens. Early expectations that the internet would eliminate inequality and poverty proved to be unfounded. Rather, the digital divide has persisted over time, assuming different forms in different contexts. Thus, as broadband has become the most important medium to send large files over the internet, broadband access is central to the digital divide. This division mirrors the increasing polarization of Western societies in general in an age of unchecked neoliberalism. In an historical moment in which social life is increasingly mediated through computer networks, discrepancies in internet access threaten to sustain and reproduce class, ethnic, and gender inequalities.

A third myth pertaining to the internet is that it annihilates space, rendering geography meaningless. Numerous authors proclaimed the ostensible "death of distance," the "end of geography," and a "flat world" (Friedman 2005). Combatting these simplistic assertions, which typically arise from technological determinism, geographers have illustrate charted the spatial dimensions of cyberspace, its uneven social and spatial diffusion, and how its impacts assume different forms in different contexts (Kellerman 2002, 2016; Zook 2005; Warf 2012). This literature demonstrates the rootedness of cyberspace in social relations and serves as a necessary antidote to many prevailing utopian and technocratic interpretations.

Outline of this volume

This volume consists of three intertwined sections that address different facets of the spatiality of the internet. Part I concerns the history, technology, and geographies of the internet, that is, how its various components came into being, the various modalities that give it concrete form, such as fiber optics, and its uneven spatial distribution. Part II focuses on the political economy of the internet, that is, its impacts on commerce, retail trade, and tourism, as well as e-government; it also addresses the sticky issue of the Chinese web, where politics and economics are seamlessly fused. The third part addresses the internet in everyday life, where applications such as Google Earth, Twitter, and augmented reality have increasingly blurred the boundaries between the virtual and physical worlds, as well as its implications for neogeography, ethnography, and diaspora studies. It concludes with the Internet of Things, in which billions of devices have become web-enabled.

Initiating Part I with Chapter 2, Aharon Kellerman explores the definitions and changing meanings of cyberspace. He delves into the relations between cyberspace – the world in the wires – and virtual and internet spaces. He then turns to the inter-relations between cyberspace and the physical, analogue world, and concludes by examining cyberspace as a hybrid space.

"The World Wide Web as Media Ecology," the title of Chapter 3 by Michael Black, offers a media studies perspective on the history and geography of the World Wide Web. He discusses several technologies that shape the topology of the web, particularly hyperlinks. With roots stretching back to the 1940s, hyperlinks evolved over time to create a fundamentally new form of knowledge creation and storage, one widely used today. In the 1990s, Tim Berner-Lee's new web-based platform rapidly grew into today's internet using an alphabet soup of acronyms such as HTML, HTTP, and URLs. The essay takes the reader through the rise of numerous means of surfing the web and the browser wars of the 2000s. Today's Web 2.0 therefore, is the contingent, path-dependent outcome of a long series of experiments involving programmers, corporations, and users.

Ramakrishnan Durairajan, in Chapter 4, discusses the robustness and/or fragility of the internet in light of its geographically grounded infrastructure. In presenting the topological structure of the internet, particularly its fiber optics routes, he touches on a series of inter-related topics: dark fiber; data centers; mishaps that disturb network functionality; and the poorly understood complexity of the entire system. He illustrates this topic with a series of long-haul fiber maps.

Chapter 5, by Elizabeth Mack, concerns the history of broadband. From its origins in the Cold War to its diffusion among the public in the 2000s, broadband vastly accelerated the speed of internet connections. She discusses the variety of platforms used, including wireless broadband, and the numerous innovations that sprang forth from this technology, such as web browsers, Web 2.0 and smartphones. Broadband has been extensively studied in terms of its economic impacts and social benefits. Mack concludes with comments on the broadband digital divide and the future of the technology.

The mobile internet is among the most important features of cyberspace today. For many people, it is the only means of logging onto the information highway. Matthew Kelly, in Chapter 6, looks at the history of the mobile internet in light of the smartphone, and the associated growth of Webs 1.0, 2.0, and 3.0. The chapter addresses the impacts on everyday life, such as locative technologies, contextualized data and algorithmic searches, and mobile artificial intelligence. The chapter also addresses the seen and unseen consequences of mobile internet use, including the mobile digital divide and privacy concerns.

Rural areas throughout the world have been frequently left behind by the internet revolution. Jeffrey James takes up this topic in Chapter 7, in which he discusses the reasons behind the lower internet penetration rates there. The urban–rural divide is one of the foremost characteristics of the geography of the internet. James notes the potential benefits the internet offers to rural denizens, including telemedicine and educational opportunities, and the possibilities opened up by mobile internet access there. Constraints include the costs of access and thus affordability and the lack of user capabilities (e.g., literacy). He concludes by examining policy options.

Digital divides at multiple spatial scales are a significant feature of inequality in cyberspace. At the global level, James Pick and Avijit Sarkar, in Chapter 8, summarize the global digital divide, which, as they note, is really more of a continuum than a dichotomy. Their analysis reveals how the divide has changed over time, various measures of the phenomenon, and how it can be fruitfully theorized. They then turn to factors that influence digital divides, such as demography, geographic location, employment status, policy measures, and social capital. They emphasize the spatial unevenness of divides, and discuss its status in the United States, China, India, Japan, and Africa.

Part II, on the political economy of the internet, opens with Chapter 9, by Bruno Moriset, who focuses on e-commerce, which has exploded in size and significance. All over the world the internet is reshaping supply chains, product cycles, and corporate strategies, and has given rise to behemoths such as Amazon. Moriset explores various forms and business models of e-commerce, its links to the gig economy (e.g., Uber) and the rise of the "winner-take-all" economy, in which large firms gain via their economy of scale. But e-commerce has diffused unevenly across the planet's surface; in the U.S. its role is limited compared to that in China. The chapter also focuses on the role of tax regimes in shaping this geography as well as the logistics involved. Moriset then turns to how e-commerce has affected local economic landscapes, including retail trade. He concludes by looking at the impacts on rural areas.

One particularly important part of e-commerce is electronic retailing, or e-tailing, the subject of Chapter 10, by Emily Fekete. Electronic sales constitute a growing part of retail sale revenues, and the phenomenon has changed how people conduct shopping. The chapter traces the origins and growth of online shopping. Fekete shows that far from simply displacing "brick and mortar" shopping, e-tailing often complements it. She concludes by examining alternative online places of shopping.

In Chapter 11, Jayson Funke delves into the world of fintech, where finance and technology intersect. He notes that "Financial systems are inherently information systems," and explores how territory, finance, information, and technology have become seamlessly integrated. He embeds this topic within wider understandings of neoliberalism and the financialization of the economy, including global debt, which gave rise to global financial flows that cross borders with ease. Along the way, fintech gave rise to spin-offs such as NASDAQ, offshore financial centers, big data analytics, and cryptocurrencies such as Bitcoin. The chapter concludes with a warning that practices such as credit scoring are integral to the panopticonic operation of contemporary capitalism.

Tourism, the world's largest industry in terms of employment, has also felt the repercussions of the internet revolution. One out of six people in the world is a tourist annually. In Chapter 12, Irene Cheng Chu Chan and Rob Law analyze the origin and growth of e-tourism. They look at the technologies that have reshaped the industry, giving rise to a new business model, online travel agencies, as well as the adoption of artificial intelligence. They also address key issues and debates, such as the impacts on consumer behavior and demand, business functions such as intermediaries, and marketing. They conclude by turning to the spatial implications of this transformation.

Barney Warf explores the spatiality of e-government in Chapter 13. All over the world, the internet has reshaped how states interact with their citizens. Warf opens by noting the varieties of forms of e-government. He then proceeds to theoretical perspectives on the topic, including various stages models and the widely used technology acceptance model. The next part concerns obstacles to the successful implementation of e-government, such as gender roles, resentful bureaucrats, and the digital divide. The chapter concludes with an examination of e-government in practice in three East Asian cities, Shanghai, Seoul, and Singapore.

China has the world's largest population of netizens (roughly 850 million in 2020). In Chapter 14, Xiang Zhang summarizes the multiple dimensions of the Chinese internet. He examines the rapid growth of cyberspace there and the persistence of a digital divide in the country. He describes the geography of China's internet in light of its profoundly uneven spatial development. The chapter then describes the growth and geography of e-commerce in China. The chapter concludes by noting the controversies pertaining to the internet in China and challenges to the government's strict censorship, such as the Great Firewall.

Todd Patterson's examination of Google Earth comprises Chapter 15, which initiates Part III. The most widely used virtual globe, Google Earth and its by-products (e.g., Google Street View) have had profound impacts of geographical imaginations across the planet, enhancing, as he notes, spatial thinking and education. It has also found its way into a variety of personal applications, such as virtual tours of parks, and in government, such as national security and public service provision. Academics have used Google Earth to study the structure of urban land use and in health geographies. Commercial applications include land cover studies and engineering services. Patterson also notes how it has been incorporated into consumer technologies such as location-based services.

The internet has opened up new worlds of human experience in which the physical and the virtual become hopelessly entangled. Mark Billinghurst, in Chapter 16, writes about augmented reality (AR), in which digital technologies complement the senses in real time (unlike virtual reality, which replaces the real world with a digitized immersive environment). The chapter illustrates the characteristics of augmented technologies in detail and their evolution over time, using copious illustrations. From head-mounted displays to smartphones, today vast numbers of people have access to this technology. The chapter also notes the applications of AR in multiple contexts, including architecture, health care, marketing, and entertainment.

As social media has exploded in its number of users and applications, Twitter has emerged as one of the dominant forms of communications over the internet. More than 335 million people worldwide tweet today. Matthew Haffner, in Chapter 17, begins with the history of location tagging or geotagging and then the rise of crowdsourced geographic information, which raises a number of issues and questions. He also studies how Twitter is used, such as continually streamed, spatially differentiated content production. Twitter has been used to analyze geographic trends in word usage, trip generation patterns, and hazards and disaster research. The chapter concludes by examining challenges in the use of Twitter data, such as gender and racial biases, computational dilemmas, and privacy concerns.

In Chapter 18, Wen Lin takes up the topic of neogeography, the production of spatial knowledge by non-experts. User-generated geographic knowledge has grown rapidly, and its implications are unclear. Lin discloses the origins of neogeography, then moves to on-going debates in the field, discussions about how spatial data is collected, and the social and political ramifications. For example, there exists a widespread impression that neogeography is inherently empowering and facilitates participant mapping. Finally, she addresses the subjective experiences of neogeography and its potential to introduce affect and emotion into cartography.

The internet has become indispensable to research of multiple kinds. In Chapter 19, Tyler Sonnichsen focuses on its use in ethnography. As ever-larger domains of human interaction move online, the web offers a wealth of ethnographic data, notably on social media. The study of internet-mediated communication has become a discipline unto itself. Much more than simply text, the material for such research includes images and videos, as YouTube, Snapchat, Tumblr, Skype, and Instagram attest. Smartphones offer yet another avenue to pursue this line of work. Online ethnographic works also allows insights into the role played by affect and emotion, the pre-linguistic landscape that underlies much of human behavior that is typically understood through non-representational theory. Sonnichsen concludes by turning to the ethics of such work, which often turns many people into inadvertent subjects.

Digital diasporas are another burgeoning field of internet-related research. In Chapter 20, Michel Laguerre elaborates on how different generations of immigrants use social media to forge linkages between their home and host countries. These cross-border online linkages form a cosmonational space. Laguerre examines the research dimensions of this phenomenon, the politics of foreign policy that are frequently involved, and how race and gender enter into this issue in complex ways. Social media, in this reading, leads to online performances of identity. He concludes with comments on the cyber-cartography of digital diasporic communities.

Chapter 21, by Monica Murero, concerns a new development in the geography of cyberspace, the wearable internet. The miniaturization of digital technologies has led to a proliferation of internet-connected watches, rings, headsets, glasses and health care devices. The data produced by these items – part of the broader Internet of Things – serve multiple purposes: users can track their own health; marketing companies utilize them; they serve as navigational tools; and they are useful for location-based services. Murero traces the surprisingly long history of this technology and the new territories they open up, such as implanted devices. She describes how they work, such as the sensors, monitors, Bluetooth connections, and the rapid proliferation of apps that service them. She examines controversies surrounding the wearable internet, including surveillance, the quantified self, and artificial intelligence. The chapter concludes by stressing the need for regulation of this industry.

Anurag Agarwal and Bhuvan Unhelkar, in the last chapter (22), conclude the volume by describing the Internet of Things (IoT). Today more devices are internet-enabled than there are people in the world. They start by outlining the history of the IoT, and move on to its industrial counterpart, the Industrial Internet of Things. They summarize the innumerable applications of this innovation, notably in the realm of geospatial technologies, the challenges involved in its use, and note future research directions. In an age of ubiquitous, mobile, and invisible computing, the IoT may well define the future of the internet.

References

Adams, P. 2005. *The Boundless Self: Communication in Physical and Virtual Spaces.* Syracuse, NY: Syracuse University Press.

Arminen, I. 2007. Review essay: Mobile communication society. *Acta Sociologica* *50*:431–437.

Ball, M., and Lillis, J. 2001. E-health: Transforming the physician/patient relationship. *International Journal of Medical Informatics 61*(1):1–10.

Banks, M. 2008. *On the Way to the Web: The Secret History of the Internet and its Founders.* New York: Springer-Verlag.

Caputo, A., Marzi, G., and Pellegrini, M. 2016. The Internet of Things in manufacturing innovation processes: development and application of a conceptual framework. *Business Process Management Journal 22*(2):383–402.

Carr, N. 2010. *The Shallows: What the Internet is Doing to our Brains.* New York: W.W. Norton.

Chen, T., L. Jarvis, and S. Macdonald (eds.) 2014. *Cyberterrorism: Understanding, Assessment, and Response.* New York: Springer.

Condratov, I. 2013. e-Tourism: Concept and evolution. *Ecoforum 2*(1–2):58–61.

Dada, D. 2006. The failure of e-government in developing countries: a literature review. *Electronic Journal of Information Systems in Developing Countries 2697*:1–10.

Davison, R., Wagner, C. and Ma, L.. 2005. From government to e-government: A transition model. *Information Technology & People 18*(3):280–299.

Deibert, R. 2009. The geopolitics of internet control: censorship, sovereignty, and cyberspace. In H. Andrew and P. Chadwick (eds.) *The Routledge Handbook of Internet Politics*, pp. 212–226. London: Routledge.

Forman, C., Ghose, A., and Goldfarb, A. 2009. Competition between local and electronic markets: how the benefit of buying online depends on where you live. *Management Science 55*:47–57.

Fountain, J. 2001. *Building the Virtual State: Information Technology and Institutional Change.* Washington, DC: Brookings Institution Press.

Friedman, T. 2005. *The World is Flat: A Brief History of the 21st Century.* New York: Picador.

Gibson, W. 1984. *Neuromancer.* London: Gollancz.

Gong, W. 2009. National culture and global diffusion of business-to-consumer e-commerce. *Cross-cultural Management 16*(1):83–101.

Hafner, K. and Lyon, M. 1996. *Where Wizards Stay Up Late: The Origins of the Internet.* New York: Simon and Schuster.

Hanafizadeh, P. and M. Behboudi (eds.) 2012. *Online Advertising and Promotion: Modern Technologies for Marketing.* Hershey, PA: IGI Global.

Hanafizadeh, P., Keating, B. and Khedmatgozar, H. 2014. A systematic review of internet banking adoption. *Telematics and Informatics 31*(3):492–510.

Harris, S. 2014. *@War: The Rise of the Military-Internet Complex.* New York: Houghton Mifflin.

Herring, M. 2014. *Are Libraries Obsolete? An Argument for Relevance in the Digital Age.* Jefferson, NC: McFarland & Company.

Kellerman, A. 2002. *The Internet on Earth: A Geography of Information.* London: Wiley.

Kellerman, A. 2010. Mobile broadband services and the availability of instant access to cyberspace. *Environment and Planning A 42*:2990–3005.

Kellerman, A. 2016. *Geographic Interpretations of the Internet.* Dordrecht: Springer.

Komninos, N. 2014. *The Age of Intelligent Cities: Smart Environments and Innovation-for-all Strategies.* London and New York: Routledge.

Lipton, J. 2015. *Rethinking Cyberlaw: A New Vision for Internet Law.* Cheltenham: Edward Elgar.

McKinion, J., Turner, S., Willers, J., Read, J., Jenkns, J., and McDade, J. 2004. Wireless technology and satellite internet access for high-speed whole farm connectivity in precision agriculture. *Agricultural Systems 81*(3):201–212.

Pick, J., Gollakota, K. and Singh, M. 2014. Technology for development: Understanding influences on use of rural telecenters in India. *Information Technology for Development 20*(4):296–323.

Rustad, M. 2014. *Global Internet Law.* St. Paul, MN: LEG.

Ryan, J. 2013. *The History of the Internet and the Digital Future.* New York: Reaktion Books.

Standing, C., Tang-Taye, J.-P., and Boyer, M. 2014. The impact of the internet in travel and tourism: A research review 2001–2010. *Journal of Travel & Tourism Marketing 31*(1):82–113.

Strengers, Y. 2013. *Smart Energy Technologies in Everyday Life: Smart Utopia?* New York: Palgrave Macmillan.

Thomopoulos, N., M. Givoni and P. Rietveld (eds.) 2015. *ICT for Transport: Opportunities and Threats.* Northampton, MA: Edward Elgar.

Turkle, S. 2011. *Alone Together: Why We Expect More from Technology and Less from Each Other.* New York: Basic Books.

Warf, B. 2010. Geographies of global internet censorship. *GeoJournal* 76(1):1–23.

Warf, B. 2012. *Global Geographies of the Internet.* Dordrecht: Springer.

Zook, M. 2005. The geography of the internet. *Annual Review of Information Science and Technology* 40:53–78.

Part I

Conceiving the history, technology, and geography of the internet

2   Is cyberspace there after all?

Aharon Kellerman

Gibson (1984) originally proposed the term "cyberspace" in his novel *Neuromancer*, as a science-fiction notion. The term acquired its connotation for a rather metaphorical space within which the internet operates, as of the early 1990s, towards the commercial introduction of the internet in 1995. The use of the term cyberspace for a class of virtual space was simultaneously coupled with the adoption of numerous additional terms, borrowed from words and terms originally coined for the use of physical space. These terms included, for instance, the words home, site, link, browsing, moving, and more. The adoption of a geographical language for the emerging uses of virtual entities, at the time, reflected the primal role of physical space and its uses in human life, thus making it easy for individuals to operate within virtual entities, notably the internet (Schlottmann and Miggelbrink 2009; Kellerman 2016).

This chapter will present and discuss four dimensions concerning the notion of cyberspace, and its adoption for virtual space in general, and for the internet in particular. First, it will present the meanings, uses, and interpretations of cyberspace, proposed notably from the mid-1990s until the early 2000s. Second, it will describe the relationships between cyberspace and related image spaces, notably virtual space and internet space. Third, it will outline some of the relationships between cyberspace and physical space, and fourth, it will discuss two notions, which have been proposed following the wide dissemination of the metaphorical cyberspace. The first among these two notions will be hybrid space, emerging in the 2000s, and it will be followed by the notion of spatial media, developing in the 2010s. Thus, the following discussions will attempt to cope with the question posed soon after the beginning of the geographical exploration of cyberspace: "Is cyberspace a kind of space?" (Adams and Warf 1997, 141).

Meanings of cyberspace

Numerous meanings and features of cyberspace as a geographical concept and entity have been proposed, notably through the 1990s. From a spatial perspective, cyberspace was widely viewed as synonymous with information space (Thrift 1996), being "invisible to our senses" (Batty 1993, 615), and, thus, constituting a geographic metaphor for disembodiment (Adams 1997; Tranos and Nijkamp 2013). From a digital perspective, cyberspace was related to as "a multi-media skein of digital networks" (Graham 1998, 165).

In the early phase of cyberspace study, cyberspace was viewed as constituting a space for itself, being neither absolute nor relative (Wang, Lai, and Sui 2003). Thrift argued for information spaces in general to "signal new spatial logics which respect none of the apparently Newtonian constructs of space. ... They are connected to the rise of images and signs as the means by which our society makes sense of itself" (Thrift 1996, 1467). Side by side with the rather virtual views of cyberspace, it was still viewed as being spatially and materially based through its real space infrastructure (Zook et al. 2004), and as interacting with real environments (Graham 1998; Light 1999).

More specifically, cyberspace received several spatial definitions since the early 1990s, reflecting four of its dimensions:

1. *Artificial reality*: "Cyberspace is a globally networked, computer-sustained, computer-accessed, and computer-generated, multidimensional, artificial, or 'virtual', reality" (Benedikt 1991, 122; see also Kitchin 1998a).
2. *Interactivity space*: "interactivity between remote computers defines cyberspace...cyberspace is not necessarily imagined space – it is real enough in that it is the space set up by those who use remote computers to communicate" (Batty 1997, 343–344).
3. *Conceptual space*: "the *conceptual space* within ICTs (information and communication technologies), rather than the technology itself" (Dodge and Kitchin 2001, 1).
4. *Metaphorical space*: "the idea of 'cyberspace' is deployed as an inherently geographic metaphor" (Graham 2013, 178).

All of these four definitions approach cyberspace from the perspective of its users' experiences, with cyberspace viewed as an artificial reality, as a communications platform, or as conceptual or rather as metaphorical spaces. Out of these four spatial definitions for cyberspace, the first three ones place it within the wider spectrum of information technologies, so that cyberspace is viewed as including also communications media, that is, radio, television, and fixed and mobile telephones. The common thread for all of these communications media is that they were originally invented prior to the invention of computers in the late 1940s, whereas the internet is a distinct medium in this regard, since it has been computer-based since its innovation in the 1960s (Kellerman 2019).

The four definitions may be regarded as being complementary, rather than contradictory, to each other. Therefore, cyberspace can be viewed as constituting simultaneously a virtual, interactive, conceptual, and metaphorical spatial entity. Such a multiple approach to the nature of cyberspace was proposed more generally by Strate (1999, 383), in his noting of cyberspace as being "better understood as a plurality rather than a singularity." Strate (1999) further proposed to rank the meanings of cyberspace: zero order cyberspace being the ontological nature of cyberspace as a virtual reality; first order cyberspace constituting the physical space of cyberspace hardware, being simultaneously also a conceptual space mediating between cyberspatial ontological and physical dimensions; and second order cyberspace constituting a synthesis between orders zero and one.

Our discussion so far, has viewed cyberspace as a space by itself, however cyberspace can be looked upon from additional perspectives as well, spatial and non-spatial alike. For instance, cyberspace may be regarded as exhibiting some representations of physical space through digital maps, pictures, and graphs, which may be used for the study of physical space, as well as for navigation within it (Zook, Dodge, Aoyama, and Townsend 2004; Zook and Graham 2007).

Furthermore, and again from a spatial perspective, I mentioned previously the metaphorical adoption of spatial daily words and terms for the construction and naming of cyberspace, as well as for its use, notably for the operation of the internet. This latter adoption of geographical terms for the use of the internet has been universal, stemming from the everyday familiarity of people with physical space, expressed in internet notions such as site, browsing, and moving (Wilken 2007; Graham 2013). Thus, the adoption of spatial metaphors for cyberspace was claimed to be founded on the human experience since "early in life and is essential for survival" (Tversky 2000, 76; see also Couclelis 1998). Spatial metaphors have turned out to be convenient for additional dimensions of information use, as well, such as organization, access, integration, and operation (Tversky 2001).

The wide adoption of spatial language for the routine manipulation of cyberspace through the internet presents a process of spatialization for cyberspace (Kellerman 2007). Couclelis (1998) commented on this adoption of metaphors as involving

> the mapping of one domain of experience into another, more coherent, powerful, or familiar one ... the metaphor performs a cognitive fusion between the two, so that the things in the source domain are viewed as if they really belonged in the target domain.
>
> (214–215)

Wide-ranging metaphors were termed by Lakoff and Johnson (1980, 14) as *orientational metaphors*, and the rather wide-ranging spatial metaphors used for cyberspace and its operation seem like an obvious case of orientational metaphors.

Cyberspace was defined also from some non-spatial user-oriented perspectives. For example, for Mitchell (1995, 8), "cyberspace is profoundly *antispatial*," and for Mizrach (1996) cyberspace constitutes a "consensual hallucination." Hence, "under the right conditions, cyberspace becomes a dream world, not unlike the world which emerges when we sink to sleep" (Suler 1999). However, internet cyberspace differs from dream cyberspace, in that internet users consciously navigate within the publicly available Web, as opposed to dreamers" rather unconscious navigations within their personal dream-cyberspace.

Cyberspace and image spaces

Following the exploration of the notion of cyberspace per se, I will now place it within the wider family of image spaces. I begin with a presentation of the general notion of image space, followed by the sketch of a scalar model, differentiating among four visual classes, from the wider to the specific. First, and being the widest image space, is virtual space, constituting the visual presentations of physical space and material artifacts. The second class is cyberspace, being a sub-class of virtual space. It refers to digital communications and information media. The third class is internet space, being a sub-class of cyberspace, and referring to digital communications and informational internet spaces. Finally, the fourth class of internet screen space constitutes a sub-class of internet space, and it relates to the visual interface between internet information and communications spaces and their users.

The common denominator among the three classes of virtual, cyber, and internet spaces is that they all constitute image spaces. Images are usually viewed as visual representations of material entities, but they possess even wider connotations, as Jay (1994, 8–9) commented: "There is [therefore] something revealing in the ambiguities surrounding the word 'image,' which can signify graphic, optical, perceptual, mental or verbal phenomena." Hence, Aumont (1997) identified three channels for image space expressions: spectators" perceptions, image transmission apparatuses, and the images themselves. For the latter class of the images themselves, Aumont (1997) focused on painting, film, and photography, noting generally on space representations, "that space is a much more complex category than its iconic representation" (160), thus calling for adjustments pertaining to its image presentations, notably through the use of perspectives in its painting.

Initially, geographers attributed images to mental spaces, that is, imagined spaces, and to their visual expression via mental maps (Phillips 1993). This was followed by attention paid to space in pre-cinematic and cinematic film technologies (e.g., Doel and Clarke 2005), as well as to micro-spaces such as slides (Rose 2003) and diagrams (Petersson 2005). The common thread among these explorations is their engagement mainly with expressions of physical space within certain media, rather than on the media themselves as constituting spaces.

It was Ash (2009 who paved an initial road for the geographical study of images as space classes by themselves. In his study of video game screens as spaces, he asserted that space in visual images "can be considered as a surface, a flat image presented on the screen" (Ash 2009, 211). He further developed a number of ideas pertaining to visual image spaces, attempting to tie together the distinct classes proposed by Aumont (1997), which we mentioned before (spectators' perceptions, image transmission apparatuses), and the images themselves. First among these ideas is the relationship of image spaces with physical space. Thus, visual images simultaneously represent the real world, even if in skewed, distorted, or imagined forms, as well as newly created image spaces. Second, "the 'being' of images consists of both a *materiality* and a *phenomenality*, acting in concert with each other, in their constitution of the conditions for being able to 'see' or experience the image at all" (Ash 2009, 2107–2108). Third, image spatiality constitutes an existential spatiality, in its being constructed by activities, as well as by engagement of image users in them. Image spaces constitute, therefore, two things simultaneously: personally imagined spaces as perceived by image users, and material or visual images representing physical space.

Image spaces possess an even wider connotation in their inclusion also of metaphorical spaces, presented and verbally described in non-visual literal texts, mainly in prose and poetry writings. Contemporarily, metaphorical image spaces include also digital visual images and representations, notably those transmitted via the internet, for instance the wide application of spatial notions for internet operations, which we noted already before. In addition to metaphorical spatial expressions adopted for internet operations, the visual interface between the internet and its users through internet screens may also be considered as image spaces.

Following the discussion so far of the general category of image space, I now move to the examination of virtual space, within which cyberspace nests, followed by internet space, nesting within cyberspace, and, finally, internet screen spaces, expressing internet images. I focus mainly on the specific qualities of each of these image space sub-classes, as well as on geographical notions relevant to their understanding. Kinsley (2014, 365) reviewed virtuality and noted in this regard the nuanced range of interpretations for virtual space as a digital entity. Thus, "the 'virtual' of 'virtual geographies' tends to mean simulation of a kind of digital liminality, akin to a space 'between' screen and body, data and machine" (Crang, Crang, and May 1999, 6).

Following Ettlinger (2008, and see the discussion by Grosz 2001), I propose a distinct connotation for virtual space, so that digital cyberspace constitutes its subset. Ettlinger (2008) claimed that "virtual space is the visible world of pictorial images: paintings, films, photographs, TV programs, video games, or any other pictorial medium – that is, physical devices that allow us to experience through them something that is not physically there" (xi). Hence, "virtual space is not the world of dreams" and "virtual space is not a hallucination" (31), whereas "referring to the internet in terms of a space, [therefore,] is valid only metaphorically – as a conceptual type of space" (27), and "cyberspace with all its complexity and elaboration is only a specifically-defined subset of virtual space" (33).

The nature of the virtual realm and its geography are complex by their very virtuality, turning it difficult to interpret them along the classical and basic differentiation between abstract and relative spaces (see Curry 1998), with virtual space possibly constituting a merge between these two classes of space (Hillis 1999, 77). The experiencing of virtual space might be close to but will never be identical to that of physical space (Crang, Crang, and May 1999). The interpretation of the virtual as something "which is not physical but emulates the physical" was attributed by Farman (2012, 37) to 17th-century Christianity.

The very existence of virtual space implies also a process of virtualization. Lévy (1998) studied this process at the time, and noted that "when a person, community, act, or piece of information are virtualized, they are 'not-there', they deterritorialize themselves" (29), and "if cyberspace results from the virtualization of computers, the electronic highway reifies this virtual world" (160). Hence, virtualization amounts to a process of turning material things into virtual presentations, not necessarily cyber ones. In other words, the virtualization process implies a process of transformation of things into virtual presentations, whereas cyberspace denotes the digital condition of visual exhibition of virtual things, expressed mainly through television and the internet. The turning of something into a virtual condition does not necessarily imply its being presented over cyberspace, since it can be expressed through painting, for example. However, the opposite is true: things presented in cyberspace are always virtual.

I discussed earlier the nature of cyberspace as a virtual entity, so I move now to a brief presentation of the internet as a sub-class of cyberspace. The internet constitutes foremost a cyberspatial communications and informational technology, which facilitates visual presentations of information to its users. Concerning its status *vis-à-vis* the physical space world, it was suggested that "the Internet can be thought of as a space attached to the earth" (Wang, Lai, and Sui 2003, 383). The history and development of the internet have been presented elsewhere (Kellerman 2002, 2016), so that it will suffice here to note its internal structure. The internet includes two major functions or components, both of which constitute metaphorical spaces: information space consisting of the Web and its websites, and communications space, which includes e-mail platforms and Web 2.0 social networking applications, led by Facebook and Twitter (Kellerman 2007). Both of these two metaphorical spaces become visually reified to internet users through internet screen spaces.

The two internet categories of information and communications spaces become frequently interfolded, for instance when internet users send e-mails through a website to the website owner rather than separately through e-mail systems, or when e-mails and network messages include links to pictures, websites, and/or data. This interfolding and frequently even fusing of the internet information and communications spaces attest to the oneness of the double-function internet, at least from its users' perspective. However, each of the two internet cyberspace classes may frequently fuction independently of the other: for instance audial personal communications normally do not involve simultaneous transmissions of textual datasets.

Another classification for the internet is its internal division into domain names, signified by organizational and national codes, which comprise an integral component for website addresses, as well as for personal e-mail addresses (Wilson 2001).

The internet has been regarded as a unique social landscape, comprised of spatial elements. For instance, "the internet, as a platform for *virtual interactions* among individuals and organizations, has necessarily a geographical component" (Tranos and Nijkamp 2013, 855), and "the only communication medium that rivals the topological flexibility of computer networks is place itself" (Adams 1998, 93). A significant element of internet communications is their ability to communicate anonymously and in most egalitarian ways. As was noted already by Lévy (1998) before the construction of Web 2.0, "here we no longer encounter people exclusively by their name, geographical location, or social rank, but in the context of centers of interest, within a shared landscape of meaning and knowledge" (Lévy 1998, 141). Moreover, "cyberspace provides social spaces that are purportedly free of the constraints of the body; you are accepted on the basis of your written words, not what you look like or sound like" (Kitchin 1998b, 386; see also Mizrach 1996).

Internet screen spaces constitute a sub-class of internet spaces, providing for a visual interface between internet information and communications spaces, on the one hand, and their users, on the other. Computer screens in general have been explored from phenomenological perspectives (Introna and Ilharco 2006), as well as from ethological ones (Ash 2009), and here we add a geographical framework for the specific internet screens. "Online interaction is currently dominated by visual interfaces, rather than aural, tactile, or olfactory interfaces," and such digital spaces facilitate the spatialization of non-spatial data (Fabrikant 2000; Zook, Dodge, Aoyama, and Townsend 2004). The nature of the internet as constituting both an informational and a communications system implies that screen-spaces may present all possible visual modes: texts, pictures, maps, and landscapes, as well as combinations among these modes.

Internet screen spaces, by their nature as consisting of computer codes, are not constant and stable spaces like printed or painted virtual ones and they may disappear by pre-programmed codes, or in response to instant actions executed by internet users. Internet subscribers may make routine and repetitive use of specific screen-spaces, such as their homepages, news services, and banking and shopping websites and such repetitive used screens may present to the users pages with fixed structures and colors, but with some contents changes. Hence, internet users may find it difficult to cognize and eventually draw cognitive maps for instantly appearing and disappearing virtual landscapes and informational screens (Kellerman 2007). Kwan (2001) noted, therefore, that in the analogue world, space and its maps are two completely separate entities, whereas for internet screen spaces, space and its maps may converge. Hence, "cognitive communications cyberspaces are personally unique, and cannot be aggregated, whereas cognitive maps relating to a specific area may be compared and conclusions on a wider societal knowledge of an area drawn" (Kellerman 2014, 9). In telephone calls, notably video ones,

> the virtual is imagined as a "space" between participants, a computer-generated common ground which is neither actual in its location or coordinates, nor is it merely a conceptual abstraction, for it may be experienced "as if" lived for given purposes.
>
> (Shields 2003, 49)

The use of the internet implies the visual exposure of individuals to cyberspace, side by side with their presence in physical space, thus amounting to their co-presence or telepresence in the two spaces (Lévy 1998; Urry 2000; Kaufmann 2002; Kellerman 2016). Graham (2013) mentioned in this context that cyberspace has been variously "conceived of as both an ethereal dimension which is simultaneously infinite and everywhere… and as fixed in a distinct location" (179), but he objected this view.

Relations between cyberspace and physical space

Following the discussion of cyberspace, and its position as a subclass within the wider class of image space, I move now to an exploration of some of the relations between the metaphorical cyberspace, on the one hand, and physical space, on the other. The interpretation of cyberspace as being somehow similar to physical space in its nature, as well as in its experiencing by its users, is not novel. For instance, "virtual environments contain much of the essential spatial information that is utilized by people in real environments" (Péruch et al. 2000, 115), and "human behavior in cyberspace bears certain similarities with spatial behavior in the physical world" (Kwan 2001, 33). However, a major difference between the two spaces involves the comprehension of distance: "what is near in physical space is often far in cyberspace, and vice versa" (Adams 1998, 93; see also Pickles 2004).

Metaphorical cyberspace has not developed apart from physical space, since, as we mentioned already, its hardware, as well as its users, are located in physical space. Thus, "cyberspace is hardly immaterial in that it is very much an embodied space" (Dodge 2001, 1), and from yet another perspective, "information systems redefine and do not eliminate geography," and even more so, "electronic space is embedded in, and often intertwines with, the physical space and place" (Li, Whalley, and Williams 2001, 701). Thus, the internet within cyberspace "is shaped by, and reflects, the place-routed cultures in which it is produced and consumed" (Holloway and Valentine 2001, 153). Still, however, the internet constitutes a "different human experience of dwelling in the world; new articulations of near and far, present and absent, body and technology, self and environment" (Crang, Crang, and May 1999, 1). Thus, cyberspace enjoys its own geography, it is symbol-sustained (Benedikt 1991; Batty 1997), and it possesses its own materiality as well (Kinsley 2014).

The use of the Web and its experiences involve some significant resemblance to physical space: "space isn't a mere metaphor. The rhetoric and semantics of the Web are those of space. More important, our *experience* of the Web is fundamentally spatial" (Weinberger 2002, 35). Hence, "cyberspace itself is deeply structured geographically" (Warf 2006, xxvii).

The variety of differences between physical and virtual spaces have been presented and discussed elsewhere (see e.g., Kellerman 2014), and the relationships between virtual space and physical (absolute and relative) spaces were further elaborated by Wang, Lai, and Sui (2003). Despite of the rather illusive and metaphorical nature of cyberspace, it may be considered as an ontic entity, which involves geographical experiencing by internet users. This experiencing of cyberspace differs from that of physical space, and in numerous ways. First, the cyberspace experienced by internet users is usually more extensive in its spatial extent as compared to that of users of physical space, since internet users may contact websites and people remotely from their location in physical space. Second, cyberspace browsing may be much more intensive in time than travels in physical space. Third, and in contrast to the first two points, cyberspace experiencing constitutes a rather shallow experience, as compared to that of physical space, in its perceptional imprint on users (Kellerman 2007). Fourth, cyberspace experiencing lacks almost any bodily involvement by internet users, and this may contribute to the lower experiential imprint of cyberspace on internet users.

Hybrid space: cyberspace and physical space

In the first phase of thought on cyberspace, which we have discussed so far, metaphorical cyberspace has been considered as an entity by itself, enjoying some anchoring in and relationships with physical space. The second phase of thought regarding the nature of cyberspace, opposed the separation between physical space and cyberspace (Kellerman 2019). Already back in 1998, internet space was viewed as interacting with physical space, in that the two spaces "stand in a state of *recursive interaction*, shaping *each other* in complex ways" (Graham 1998, 174). This view matured later into the idea of hybrid space, proposing a connection between physical and digital spaces. The notion of spatial hybridity emerged, at the time, in light of the development of the internet, as of the 2000s, into a most versatile, comprehensive, and powerful information system. The internet was enhanced by then through its integration with new technologies, notably GPS (global positioning systems), which have imported physical space images and locations into internet space, through applications such as Google Maps and Google Earth for satellite images, and Google Street for physical space pictures (Zook and Graham 2007a, 2007b, 2007c; Crutcher and Zook 2009).

Another development which may have contributed to the view of convergence between physical and internet spaces was the introduction and wide adoption of smartphones (as of 1993), broadband (as of the late 1990s), mobile broadband connectivity (as of 2001), and wi-fi (as of 1997). These technologies have facilitated a wider range of uses for the internet, coupled with the facilitation of users' permanent and mobile access to it through smartphones. Thus, a practical convergence between physical and internet spaces for the performance of several routine activities has emerged, such as for shopping and social networking, activities in which physical and internet spaces may complement each other (Kellerman 2014, 2019).

It was Kluitenberg (2006) who first put forward the idea of convergence and interfolding of physical and virtual spaces, notably within mobile contexts, into a joint "hybrid space," and this notion was further developed by de Souza e Silva (2006). The term hybrid space, and the duality which it involves, have received numerous nuances, such as "doubling-of-space" (Moores 2012), and "more-than-real" space (McLean 2016; McLean et al. 2016), side-by-side with additional phrases relating to relations between the spatial and the digital in general (see Ash et al. 2018; Kinsley 2014).

The notion of hybrid space, as well as its derivatives, has been largely developed through feminist interpretations of the digital. These interpretations criticized, for instance, the "sweeping erosion of locational privacy" involved in the mediation of everyday life by digital media" (Leszczynski and Elwood 2015, 22). Thus, the idea of hybrid space refers to "a world where the digital and the material are not separate but entangled elements of the same processes, activities and intentionalities" (Pink et al. 2016, 1; see also Wilson 2001).

The notion of hybrid space does not contradict the conception of internet platforms as metaphorical cyberspace, since the digital/virtual space within hybrid space is still a metaphorical one, even when becoming hybridized with physical space. Spatial hybridity implies an "always-on" connection, while our very moving "transforms our experience of space by enfolding remote contexts inside the present context" (de Souza e Silva 2006, 262). The locations of hybridizations, notably within cities, constitute "net localities" (de Souza e Silva and Frith 2012; see also Farman 2012). Hybrid space constitutes also a networked space, in that the connection between the far and the contiguous is carried out in a rather borderless way, without the feeling of "entering" the internet. Thus, a "hybrid reality" is experienced, in which social practices occur simultaneously in physical and digital spaces within a mobile context. Moreover, "flows through and beyond online spaces produces reality" (McLean 2016, 509), which involves "materialities in which potential becomes actualized and through which digital mediation is afforded" (Kinsley 2014, 365).

Spatial identity for digital media

Based on Crampton (2009, 2014), it was for Leszczynski (2015; see also Timeto 2015; Kellerman 2019) to go one-step further in the search for spatial identities for digital systems, when she focused on the notion of "spatial media." This notion of "spatial media" constitutes the third phase of spatial interpretations for digital systems, following cyberspace and hybrid space. Spatial media "refers to both new technological objects (hardware, software, programming techniques, etc.) with a spatial orientation, as well as to nascent geographic information content forms produced via attendant practices with, through, and around these technologies" (729). Thus, the spatial identity of digital systems, notably the internet, has moved from spatial metaphors attached to them, and, subsequently, also from their hybridity with physical space, to the very possible possession of spatial orientation and/or spatial content by digital systems of codes. This interpretation of the spatial within the digital applies not just to the internet, but to much wider digital systems as well. Furthermore, through the notion of spatial media we may view contemporary human life as a product of mediations among technology, space, and society, and our spatial experiences in physical space as supplemented by spatial information as provided by digital systems.

The notion of spatial media/tion, as offered by Leszczynski (2015) for digital systems, makes it possible to identify some "true" spatial elements within digital systems, such as their orientation and content, rather than metaphorical spatial notions attributed to them. However, the nexus technology–space–society, which stands at the basis of the spatial media/tion thesis, remains rather theoretical or conceptual, focusing on the very nature of digital systems, without reference to human agency, or the everyday uses of digital systems, notably the internet, by individuals. For internet users, the internet has come to constitute a metaphorical space in general, and notably so in their use of the internet as an action space or arena, in which users can perform activities that they traditionally used to carry out in physical space (Kellerman 2014, 2019).

Conclusion

I reviewed in this chapter three options for cyberspace in general, and for internet space in particular, and these options emerged in a historical sequence. First, cyberspace was proposed in the 1990s as a kind of stand-alone metaphorical spatial entity within the wider metaphorical virtual space, enjoying some anchoring and relations with physical space. Second, as of the 2000s, cyberspace was viewed as a metaphorical space existing jointly with physical space, with internet users being co-present in this hybrid space. Third, the metaphorical interpretations attributed to digital systems have been ignored as of the 2010s, and, alternatively the notion of spatial media was proposed, referring to code systems that enjoy some geographical orientation or contents.

The latter notion of digital media is a conceptual one, whereas hybrid space relates to human agency. From the perspective of its users, the internet has turned into a space-like entity that permits its users to perform physical-space-like human actions within and through it. Thus, when interpreting human action vis-à-vis the internet per se, rather than theorizing the spatial meanings of the digital realm in general, the use of the term "internet space" is called for. The dual-space society, described elsewhere (Kellerman 2019) focuses on internet users who function within a hybridized space that mediates between physical space and digital media, with the latter of which being meta-phorically space-like. Thus, internet users operate contemporarily within a dual-space society, side by side with the wider existence of spatial media.

References

Adams, P. 1997. Cyberspace and virtual places. *Geographical Review* 87:155–171.

Adams, P. and Warf, B. 1997. Introduction: cyberspace and geographical space. *Geographical Review* 87:139–145.

Adams, P. 1998. Network topologies and virtual place. *Annals of the Association of American Geographers* 88:88–106.

Ash, J. 2009. Emerging spatialities of the screen: Video games and the reconfiguration of spatial awareness. *Environment and Planning A* 41:2105–2124.

Ash, J., Kitchin, R., and Leszczynski, A. 2018. Digital turn, digital geographies? *Progress in Human Geography* 42:25–43.

Aumont, J. 1997. *The Image.* C. Pajackowska (trans). London: British Film Institute.

Batty, M. 1993. The geography of cyberspace. *Environment and Planning B* 20:615–616.

Batty, M. 1997. Virtual geography. *Futures* 29:337–352.

Benedikt, M. 1991. Cyberspace: some proposals. In M. Benedikt (ed.) *Cyberspace: First Steps*, pp. 119–224. Cambridge, MA: MIT Press.

Couclelis, H. 1998. Worlds of information: The geographic metaphor in the visualization of complex information. *Cartography and Geographic Information Systems* 25:209–220.

Crampton, J. 2009. Cartography: Maps 2.0. *Progress in Human Geography* 33:91–100.

Crampton, J. 2017. New spatial media. *Open Geography.* https://opengeography. wordpress.com/2014/06/06/new-spatial-media/.

Crang, M., Crang, P., and May, J. 1999. Introduction. In M. Crang, P. Crang, and J. May (eds.) *Virtual Geographies: Bodies, Space and Relations.* pp. 1–23. London: Routledge.

Crutcher M. and Zook, M. 2009. Placemarks and waterlines: Racialized cyberscapes in post-Katrina Google Earth. *Geoforum* 40:523–534.

Curry, M. 1998. *Digital Places: Living with Geographic Information Technologies.* London and New York: Routledge.

de Souza e Silva, A. 2006. From cyber to hybrid: Mobile technologies as interfaces of hybrid systems. *Space and Culture* 9:261–278.

de Souza e Silva, A. and Frith, J. 2012. *Mobile Interfaces in Public Spaces: Locational Privacy, Control, and Urban Sociability.* London: Routledge.

Dodge, M. 2001. Guest editorial. *Environment and Planning B: Planning and Design* 28:1–2.

Dodge, M. and Kitchin, R. 2001. *Mapping Cyberspace*. London: Routledge.

Doel, M., and Clarke, D. 2005. Emerging space and time: Moving pictures and motionless trips. *Journal of Historical Geography* 31:41–60.

Ettlinger, O. 2008. *The Architecture of Virtual Space*. Ljubljana: University of Ljubljana.

Fabrikant, S. 2000. Spatialized browsing in large scale data archives. *Transactions in GIS* 4:65–78.

Farman, J. 2012. *Mobile Interface Theory: Embodied Space and Locative Media*. New York: Routledge.

Gibson, W. 1984. *Neuromancer*. London: Gollancz.

Graham, S. 1998. The end of geography or the explosion of place? Conceptualizing space, place and information technology. *Progress in Human Geography* 22:165–185.

Graham, M. 2013. Geography/internet: Ethereal alternate dimensions of cyberspace or grounded augmented realities? *Geographical Journal* 179:177–182.

Grosz, E. 2001. *Architecture from the Outside: Essays on Virtual and Real Space*. Cambridge, MA: MIT Press.

Farman, J. 2012. *Mobile Interface Theory: Embodied Space and Locative Media*. New York: Routledge.

Hillis, K. 1999. *Digital Sensations: Space, Identity, and Embodiment in Virtual Reality*. Minneapolis: University of Minnesota Press.

Holloway, S. and Valentine, G. 2001. Placing cyberspace: Processes of Americanization in British children's use of the internet. *Area* 33:153–160.

Introna, L., and Ilharco, F. 2006. On the meaning of screens: Towards a phenomenological account of *Screenness, Human Studies* 29:57–76.

Jay, M. 1994. *Downcast Eyes: The Denigration of Vision in Twentieth Century French Thought*. Berkeley: University of California Press.

Kaufmann, V. 2002. *Re-thinking Mobility: Contemporary Sociology*. Aldershot: Ashgate.

Kellerman, A. 2002. *The Internet on Earth: A Geography of Information*. London: Wiley.

Kellerman, A. 2007. Cyberspace classification and cognition: Information and communications cyberspaces. *Journal of Urban Technology* 14:5–32.

Kellerman, A. 2014. *The Internet as Second Action Space*. London and New York: Routledge.

Kellerman, A. 2016. *Geographic Interpretations of the Internet*. Dordrecht: Springer.

Kellerman, A. 2019. *The Internet City: People, Companies, Systems and Vehicles*. Cheltenham, UK and Northampton, MA: Edward Elgar.

Kinsley, S. 2014. The matter of "virtual" geographies. *Progress in Human Geography* 38:364–384.

Kitchin, R. 1998a. *Cyberspace: The World in the Wires*. Chichester: Wiley.

Kitchin, R. 1998b. Towards geographies of cyberspace. *Progress in Human Geography* 22:385–406.

Kluitenberg, E. 2017. The network of waves: Living and acting in a hybrid space. *Open* 11. http://socialbits.org/_data/papers/Kluitenbergpercent20-percent20 Thepercent20Networkpercent20ofpercent20Waves.pdf.

Kwan, M.-P. 2001. Cyberspatial cognition and individual access to information: The behavioral foundation of cybergeography. *Environment and Planning B* 28:21–37.

Lakoff, G. and Johnson, M., 1980. The metaphorical structure of the human conceptual system. *Cognitive Science* 4(2):195–208.

Leszczynski, A. 2015. Spatial media/tion. *Progress in Human Geography* 39:729–751.
Leszczynski, A. and Elwood, S. 2015. Feminist geographies of new spatial media. *Canadian Geographer* 59:12–28.
Lévy, P. 1998. *Becoming Virtual: Reality in the Digital Age*. R. Bononno (trans). New York and London: Plenum Trade.
Li, F., Whalley, J., and Williams, H. 2001. Between physical and electronic spaces: The implications for organizations in the networked economy. *Environment and Planning A*, 33:699–716.
Light, J. 1999. From city space to cyberspace. In M. Crang, P. Crang, and J. May (eds.) *Virtual Geographies: Bodies, Space and Relations*. pp. 109–130. London: Routledge.
McLean, J. 2016. The contingency of change in the Anthropocene: More-than-real renegotiation of power relations in climate change institutional transformation in Australia. *Environment and Planning D: Society and Space* 34:508–527.
McLean, J., Maalsen, S., and Grech, A. 2016. Learning about feminism in digital spaces: Online methodologies and participatory mapping. *Australian Geographer* 47:157–177.
Mizrach, S. 1996. Lost in Cyberspace: A Cultural Geography of Cyberspace. www2.fiu.edu/~mizrachs/lost-in-cyberspace.html.
Mitchell, W. 1995. *City of Bits: Space, Place and the Infobahn*. Cambridge, MA: MIT Press.
Moores, S. 2012. *Media, Place and Mobility*. Basingstoke: Palgrave Macmillan.
Péruch, P., Gaunet, F., Thinus-Blanc, C., and Loomis, J. 2000. Understanding and learning virtual spaces. In R. Kitchin and S. Freundschuh (eds.) *Cognitive Mapping: Past, Present, and Future*. pp. 108–115. London: Routledge.
Petersson, D. 2005. Time and technology. *Environment and Planning D: Society and Space* 23:207–234.
Phillips, R. 1993. The language of images in geography. *Progress in Human Geography* 17:180–194.
Pickles, J. 2004. *A History of Spaces: Cartographic Reason, Mapping and the Geocoded World*. London and New York: Routledge.
Pink, S., Ardèvol, E., and Landzeni, D. 2016. *Digital Materialities: Design and Anthropology*. London: Bloomsbury.
Rose, G. 2003. On the need to ask how, exactly, is geography "visual"? *Antipode* 35:212–221.
Schlottmann, A. and Miggelbrink, J. 2009. Visual geographies – an editorial. *Social Geography* 4:1–11.
Shields, R. 2003. *The Virtual*. London and New York: Routledge.
Strate, L. 1999. The varieties of cyberspace: Problems in definition and delimitation. *Western Journal of Communication* 63:382–412.
Suler, J. 1999. Cyberspace as a dream world. http://users.rider.edu/~suler/psycyber/cybdream.html.
Thrift, N. 1996. *Spatial Formations*. London: Sage.
Timeto, F. 2015. *Diffractive Technospaces: A Feminist Approach to the Mediations of Space and Representation*. Farnham UK and Brlington VT: Ashgate.
Tranos, E., and Nijkamp, P. 2013. The death of distance revisited: Cyber-place, physical and relational proximities. *Journal of Regional Science* 53:855–873.
Tversky, B. 2000. Some ways that maps and diagrams communicate. In C. Freska, W. Brauer, C. Habel, and J. F. Wender (eds.) *Spatial Cognition II: Integrating Abstract*

Theories, Empirical Studies, Formal Methods, and Practical Applications. pp. 72–79. Berlin: Springer.

Tversky, B. 2001. Spatial schemas in depictions. In M Gattis (ed.) *Spatial Schemas and Abstract Thought.* pp. 79–112. Cambridge, MA: MIT Press.

Urry, J. 2000. *Sociology beyond Societies: Mobilities for the Twenty-first Century.* London: Routledge.

Wang, Y., Lai, P. and Sui, D. 2003. Mapping the internet using GIS: The death of distance hypothesis revisited. *Journal of Geographical Systems* 5:381–405.

Warf, B. 2006. Introduction. In B. Warf (ed.) *Encyclopedia of Human Geography.* Thousand Oaks, CA: Sage.

Weinberger, D. 2002. *Small Pieces Loosely Joined (A Unified Theory of the Web).* Cambridge: Cambridge University Press.

Wilken, R. 2007. The haunting affect of place in the discourse of the virtual. *Ethics, Place and Environment* 10:49–63.

Wilson, M. 2001. Location, location, location: The geography of the Dot.com problem. *Environment and Planning B: Planning and Design* 28:59–71.

Zook, M., Dodge, M. Aoyama, Y., and Townsend, A. 2004. New digital geographies: Information, communication, and place. In S. Brunn, S. Cutter, and J. Harrington (eds.) *Geography and Technology.* pp. 155–176. Dordrecht: Kluwer.

Zook, M. and Graham, M. 2007a. Mapping digiplace: Geocoded internet data and the representation of place. *Environment and Planning B* 34:466–482.

Zook, M. and Graham, M. 2007b. The creative reconstruction of the internet: Google and the privatization of cyberspace and digiplace. *Geoforum* 38:1322–1343.

Zook, M. and Graham, M. 2007c. From cyberspace to digiplace: Visibility in an age of information and mobility. In H. Miller (ed.) *Societies and Cities in the Age of Instant Access.* pp. 241–254. Dordrecht: Springer.

3 The World Wide Web as media ecology

Michael L. Black

When interviewed for the documentary *No Maps for These Territories* (2000), science-fiction author William Gibson commented that widespread use of the internet had led us into a "post-geographic" future. The internet, he suggests, has collapsed all sense of place and space by allowing information to flow freely to everyone and anyone (*No Maps for these Territories* 2000). Like much of Gibson's fiction, his comments in the interview are full of provocation, intended to inspire his audience to reflect on the way that a sense of the "future" has permeated our culture in no small part through the marketing rhetoric that informs our everyday understanding of information technologies (Sterling 1995). Yet if we step outside of the futurist rhetoric of the internet and consider more concretely its data flows, we can begin to see a definitive structure that shapes our access to information in specific ways. Far from resembling some sort of infinitely accessible, rhizomatic dimension, the internet has a definitive topology, albeit one that continues to change as new technologies become integrated into it.

For most internet users, the World Wide Web – or simply "the web" – serves as a primary interface. Our ability to find and produce information across the internet has been structured by the component objects and tools that the web supports. At its most basic level, the web is comprised of hierarchal databases of hypertext documents that are structured through a mixture of markup and scripting languages. These hierarchal databases rest atop the internet's interwoven addressing and routing protocols which link individual databases together into a large network that is at the same time both decentered and highly centralizing. Although some of the basic technologies underlying the web have remained relatively stable, our experience of the web itself has changed in ways both subtle and expansive. Corporate histories of the web characterize these changes in broad terms – before and after the dot-com bubble, Web 1.0 and Web 2.0, etc. – and many of the labels used in popular discourse to describe the history of the web are marketing constructs that elide many of the perceptible changes to the user experience of the web that have occurred across and within these eras. This chapter instead adopts a media studies approach to considering the history of the web with a specific focus on how changes to basic web technologies have changed how we author and read web content.

In adopting a media studies approach to considering the geography of the web, this chapter views the web as a media ecology. In his study of the relationship between analog and digital media, Friedrich Kittler has shown that many of the communicative and cognitive patterns that are often characterized as unique to computing originated in earlier media like the gramophone or typewriter (Kittler 1999). Notably, Kittler builds on Marshall McLuhan's (1964) maxim that the "content of any medium is always another medium," showing how media do not exist in isolation but rather are always borrowing on and referring to one another – both contemporaneously and historically. The same could be said of the web, which is itself an assemblage of technologies: programming languages, images, links, video, sound, and databases. Since the web first came online, new technologies have been incorporated into it and existing ones have changed. With each new addition or change, the structure of the web has also transformed, allowing for new forms of access or restricting existing ones. In the pages that follow, I will discuss several of the technologies that structure the web's topology but with a primary focus on hyperlinks, a defining technology of the web that pre-exists the web itself but that is now rarely found apart from it.

The components of the web

Before discussing the history and the evolution of the web's structure, it is worth taking a moment to consider how the web's various component technologies fit together. According to internet historian Niels Brügger (2018), there are five analytical perspectives one can adopt when discussing the web. First, a discussion of "web elements" includes the basic informational content that appears on the web, regardless of medium. The most common web elements are text, hyperlinks, and images but can also include embedded videos, sound players, interactive modules, forms, buttons, etc. Second, "web pages" are groupings of elements that are accessed at specific universal resource locators (URLs) or web addresses. In the early years of the internet, web pages were static hypertext documents, but now they are often generated dynamically each time users access their respective URL. Third, "websites" are hierarchal groupings of web pages. Websites can exist independently or can be nested within one another. In the interest of establishing a firm technical definition, one could say that a website is any grouping of web pages that are collected together and presented as a singular resource. Four, "web spheres" are groups of websites that share some common theme, or community. The delineation between different spheres is somewhat arbitrary and should be established by researchers to situate the boundaries of a particular study. For example, one might consider all websites that are part of the "gov" top level domain to be part of the United States federal government web sphere. One could also state that while all university websites are part of the .edu top level domain, not all websites within that top level domain are for universities as it could also include primary and secondary schools, liberal arts colleges, or other educational institutions. Finally, the fifth strata is the "web itself." When speaking of the web itself, we typically refer to web as a technology or to the various technologies that support the web, each of which has its own individual history.

Much like the structure of sites and pages, Brügger's model suggests a hierarchal relationship among the various components of the web. Our basic conceptualization and understanding of the web itself is, in other words, predicated upon the way that the most basic elements are organized across pages, sites, and spheres. Yet, it is important to recognize that changes can occur at any level of this model, reconfiguring the form and function of components at other levels. In popular culture, the labels Web 1.0 and Web 2.0, for example, often refer to the rise of "participatory culture" on the web via the emergence and widespread adoption of social media and open content sharing platforms like YouTube (O'Reilly 2009). From a media ecology perspective, however, a primary difference between these web eras is in the way that hyperlinks function. By definition, hyperlinks are interactive references in a document that, when engaged, direct a web browser or other document viewer to another location. In the earliest days of the web, that location was another document. Now, hyperlinks increasingly lead to documents that do not yet exist but which are generated upon request. When one visits a page containing a YouTube video, for example, the document displayed on the screen does not exist in storage on YouTube's servers. Rather, visiting the linked location generates a request to YouTube's servers to produce a document that allows one to view the video associated with that location, populating the page with different suggestions and loading the latest set of comments associated with that video into the space below the embedded media player. The next time one visits that location, the page will look similar but much of the content loaded alongside the video will be different. This change to how we use hyperlinks is the result of differences in the web itself across the two eras: the increased computing power of servers, the development of content management software, and ready availability of broadband internet service.

On the origins of hyperlinking

Although the web has become the most common public interface for the internet, it was not the first attempt to develop a hypertext system. The concept of hyperlinks can be traced back at least as far as 1945, to an essay titled "As We May Think," written by Vannevar Bush (1945), then the Director of the United States Office of Scientific Research and Development. During World War II, Bush had been instrumental in steering the United States' nuclear energy and weapons programs. After the war, he argued that American scientists needed to find peacetime applications for the advanced technologies they were developing. Among them were early computers. In his essay, Bush imagined repurposing computers from calculating machines into systems for information retrieval that could act as extensions of the human mind. Bush called his system the "memex" and described it as a tool for traversing libraries of documents that referenced one another through arbitrary associations made by readers.

Bush's lengthy explanations of how the memex would function mechanically are now forgotten, but his descriptions of the way that associations could be used as a navigational mechanism influenced the development of early hypertext software. The essential feature of the memex, he explains, would be reader generated "associations ... whereby any item may be caused at will to select immediately and automatically another." Although the first time a user engages the memex, they may need to explicitly recall each document through reference to a table of contents or index, while reading they would be able to "tie" documents together by establishing associations between specific sections of them. Reading via the memex would thus take the form of "building a trail" which they or other users would be able to follow in future sessions. Because Bush intended his memex to resemble human thought processes, he suggested that the nature of these associations would be arbitrary – users can create them at any point in any document, leading to any point in any other document or even the same document. The only criteria for establishing an association was that it called to mind some other idea located elsewhere in the memex's library. In this sense, Bush's associations would add another layer of meaning to documents by explicitly marking the casual intertextual references among them that are common to human interpretive behavior but that might not be representable in more formal, bibliographic reference systems.

Although Bush himself never used the term "hypertext," his description of the memex served as an inspiration for many pre-web hypertext software systems. Early commercial hypertext like Guide (1982), Hypercard (1987), Storyspace (1987), and Dynatext (1990) resembled Bush's memex in that users could create hyperlinks among units of text. Yet, many of these early commercial hypertext systems were designed to reproduce the print economics of books in a number of ways. First, the products of these systems were smaller in scale than the library of documents that Bush envisioned. Typically, hypertexts produced through them were closed systems that did not link outside of themselves and were intended to be read as complete works. Second, these software systems structured "author" and "reader" as two distinct roles. Whereas users could create links arbitrarily while authoring, hypertexts published through these software systems were static forms in the sense that users could no longer add links to a hypertext once a work was packaged for distribution. Additionally, these commercial hypertexts were not publically accessible via the internet. Instead, they were distributed on disk media. Many adopters of these early systems were in fact critical of the web, seeing it messier and less user-friendly in comparison to the carefully authored, offline texts they were producing (Barnet 2012).

Other influential, early descriptions of hypertext can be found in the writings of Douglas Engelbart and Ted Nelson. Unlike Bush's memex, Engelbart's and

Nelson's hypertext systems underwent active development; however, neither of them matured into publicly available technologies. Engelbart's hypertext software was built into his oN-Line System, an early prototype for graphic user interfaces. Engelbart developed his hypertext software while serving as the Director of the Augmentation Research Center at the Stanford Research Institute. While Engelbart is popularly remembered for introducing mouse input hardware, the media software he developed to use with it was very much an attempt to reproduce some of Bush's ideas. Like Bush, Engelbart believed that networked document systems could serve as an extension of human cognitive behavior and saw hypertext as a system for "bootstrapping" consciousness to a higher level (Bardini 2000). Rather than seeing hypertext as a system for managing a library of documents, however, Engelbart's workstation framed hypertext as a way to draft single documents, allowing authors to quickly rearrange "concept packets" that he called "cards" in order to make document structure a more integral and dynamic part of the authoring process (Englebart 1962). Although Engelbart's 1968 demonstration of technology fascinated many, his oN-Line System ultimately saw limited adoption both because it was expensive to implement and because many users found it difficult to operate.

Nelson's work on hypertext emerged out of his political writings on early personal computers. Many of the terms we now use to describe hypertext can be found in his work. Nelson's interest in hypertext and in the politics of computing began in 1960, when he was a graduate student in sociology at Harvard University. For Nelson, hypertext was a way for people to assume control over their own learning process. Whereas classrooms forced students to explore topics at the behest of their teacher, Nelson argued, computerized libraries would allow students to follow their own interests, wherever they might lead (Nelson 1987). To realize this liberating vision of self-driven education, libraries would need to not only be digitized but they also needed to be networked and open, accessible to anyone with a computer. Nelson would spend the next few decades attempting to build a software system that would support this vision. He called his proposed hypertext software "Xanadu" and recruited programmers who shared his political views on computing to help him build it. Nelson attempted to steward Xanadu's uneven development throughout the 1970s and 1980s but was not able to produce a working version until returning to the project in the late 1990s. While some accounts of Xanadu's development place the blame on Nelson's own management style, its struggles can also be attributed to the project's private development with an eye towards commercialization during a time before most personal computers were not advanced enough nor networked enough to support regular access to the type of document network Xanadu might have supported (Wolf 1995).

These early hypertext systems are instructive insofar as the alternative models of hypertext they implemented highlight some key structural features of the web. In his review of early hypertext technologies, George Landow identifies at least six different kinds of hyperlink structures. Lexia-to-lexia links join whole documents, typically leading from the end of one to the beginning of another. String-to-lexia links move from a particular moment in one document to the beginning of another. String-to-string links move from a particular moment in one to a particular moment in another. Depending on the software used, Landow notes, each of these types of link structures could be executed in a unidirectional or bi-directional form. In practice, the web has been built around unidirectional links, most commonly string-to-lexia although string to string linking is also possible (Landow 2006). Unidirectional links create associations without regard to whether their destination exists, allowing for the possibly of "broken" links if the destination document is removed after the link is created. Nelson's Xanadu, by comparison, used bidirectional links. In Xanadu, a link could only be formed only if both documents were able to refer to one another. Furthermore, the web's unidirectional links exist within documents themselves, whereas Xanadu's bidirectional links were managed by a centralized database. If one document involved in a bidirectional link were to be deleted, then the central database would be updated to remove any links associated with it. Whereas broken links are a persistent problem on the web, alternatives like that presented by earlier hypertext models like Xanadu's would have made them impossible (Nelson 2000). Although the web does not support bidirectional links explicitly, the incorporation of a "back" button in web browsers has made most links on the web bidirectional in practice.

Decentralization and the structure of Berners-Lee's web

For better or for worse, the looseness of the web's unidirectional links has been an important factor in its long-term success. Although today's web is made possible through several different layers of networking technologies and a combination of client and server-side programming languages, the web's two fundamental technologies are the HyperText Mark-Up language (HTML) and the HyperText Transfer Procotol (HTTP). HTML is a programming language that defines the structure of hypertext documents, and HTTP provides rules that structure access to HTML documents over a network. Both were designed by Tim Berners-Lee during the late 1980s as part of his work at the CERN particle physics laboratory. Independently of one another, several scientists and engineers at CERN had experimented with hypertext systems in order to share their research, but none had ever succeeded in getting their colleagues to adopt their software. Berners-Lee, too, had experimented with hypertext prior to developing HTML and HTTP, concluding both from his own experience and observation of his peers at CERN that hypertext software would only succeed as a document sharing tool if it imposed as few rules as possible on authors (Berners-Lee and Fischetti 1999). Whereas systems like Nelson's Xanadu would, in theory, require the author of two documents to agree with one another to establish bidirectional links between their works,

Berners-Lee's HTML allowed authors to establish hyperlinks on a whim, without any expectation of reciprocity or notification to the author of a destination document that a link had been created.

The avoidance of rules for authors also influenced the decentralized design of the servers that facilitate access to the web's hypertext documents. Berners-Lee believed that any form of centralized management would not only act as a gatekeeping mechanism, discouraging adoption, but would also require someone to actively oversee the expansion of a hypertext network. Instead, Berners-Lee enacted a server–client relationship in HTTP would allow anyone to create their own document library. By design, all one needed to do to join the hypertext network he was creating at CERN was install "httpd," a piece of server software that would handle requests from a specialized client application for called a "browser." Once HTML documents were placed into a directory monitored by httpd, anyone who had access to the server's internet address could view them through their browser. The web takes its name from the first browser application developed by Berners-Lee, "WorldWideWeb."

While Berners-Lee designed the technology that would grow into the web to be decentralized, in practice users required some sort of index to find documents. Although anyone willing to setup a server could be an author, HTTP served hypertext documents as read-only, meaning that browsing and document creation were structured as separate activities. This meant that the ability of readers to navigate a hypertext network relied on authors linking both among documents on their own server but also outward to other servers. In order to facilitate access to HTTP servers across the internet, Berners-Lee established a directory at info.cern.ch and added links to any server upon request. In this sense, the technology of the web was decentralized for authors – in that, anyone could publish their own library of hypertext documents – but has always in practice been centralized for readers. Unless a reader were given the precise address of a server or document, they needed some sort of index to start their browsing from. As we will see below, the centralization of web reading has taken a number of different forms in the decades following the creation of Berners-Lee's first web directory.

Even though the web was designed initially for use within CERN, its loose, decentralized design allowed it to quickly spread off campus. In order to facilitate its spread, Berners-Lee did not claim copyright or patents on HTML or HTTP and was able to convince CERN to similarly waive any claims they might have had to the software. As a result, the httpd server software was soon ported over to other operating systems, most notably UNIX, and development began on web browser clients for home computer operating systems. By late 1991, the web was in theory publicly accessible, but it would not become popularized until personal computer operating systems with graphic user interfaces became the norm in the mid-1990s. On these new operating systems, users would eventually access the web through browsers like NCSA's Mosaic (1993) and Netscape's Navigator (1994).

During its first decade of public adoption, the web was experienced through a process of discovery as users followed links away from nodal centers. Although it does not conform neatly to the post-dot-com bubble transformation implied in Web 2.0 models of internet history, one could argue that a primary difference between our experience of the early web and today's web is the replacement of manually curated link directories with search services powered by automated web-crawlers. Today, search engines like Google represent themselves as all-encompassing, comprehensive tools for finding information on the web; however, one of the most popular search services in the mid to late 1990s, Yahoo!, was structurally similar to Berners-Lee's directory page in that its indices were hand-curated. Yahoo! also effected a "portal" style that tried to persuade users to visit its site first during every browsing session. Even those early automated search services similarly presented themselves as portals, creating directories and categorizing links to mimic Yahoo!'s hand-curation. Portals presented the web as a network of hierarchies that users could move through.

In terms of the web as a network, movement away from nodes like portals promised the discovery of new, unknown information at the periphery. Portals felt largely commercial, and moving outward from them promised the discovery of something novel. Portals were not all encompassing, and it was not uncommon for websites to include a directory of their own: commonly, a page called "Links" that listed related websites that their authors found personally useful or interesting. In addition to these large directory services and lists of links on individual websites, community-based directory forms like "webrings" also emerged that focused on linking together spheres of websites that addressed specific subjects. Each of these less formal directory forms promised access to websites not yet listed in a commercialized portal. As Megan Sapner Ankerson explains, early discussions of web browsing were often characterized by a search for "coolness." Some web portals even emphasized their sites as tools for discovery by including links to a "cool site of the day" that would highlight a notable, new addition to their index (Ankerson 2018). The loose, minimal design of the web's basic technologies may have made it difficult to locate new information during its first decade of operation, but it also ensured that the web continued to grow even as major nodes in its network collapsed following the Dot-Com Bubble collapse in the early 2000s.

Web 2.0 and the myth of decentralization

The technologies of the web may have been designed to promote decentralization, but in practice the web has been a site of tension between competing centers of technological influence. The "Browser Wars" of the 1990s, for example, show that even though HTML is an open standard, meaning that its full specification is publicly and freely accessible, the interpretation of that standard is ultimately decided by the people who write the software that render hypertext documents. In 1994, Berners-Lee founded the World Wide Web Consortium (W3C), a non-profit organization that would manage the HTML language in order to ensure that it would remain an open standard. Although the W3C maintained an official definition of each of HTML's mark-up "tags," both Microsoft and Netscape introduced new tags in an effort to style their browsers as more fully featured than the other – <marquee> and <blink>, respectively. Neither of these tags were a part of the W3C's definitions and were not, until more recently, supported by competing browsers. While the Browser Wars effectively ended with Netscape's transformation into the non-profit Mozilla Foundation, beginning in 1998 with the open sourcing of its browser software, the consequences of the conflict between the two developers extended into the next decade (Cusumano and Yoffie 1998). As HTML has expanded to incorporate other document languages, like Cascading Style Sheets (CSS), Microsoft's and Mozilla's web browsers interpreted many of the new hypertext elements differently. Throughout the late 1990s and early 2000s, it was not uncommon to see a "Best Viewed With" label in the footer or navigation frame of a webpage, indicating which browser the author had written their code towards. Designing a document to look similarly across all browsers, depending on the complexity of its code, required extensive testing and may have even required developers to write separate CSS definitions for each browser.

By the late 2000s, the hand-coded nature of the early web gave way to the automated page generation of content management systems. Whereas during the web's first decade, authoring and reading hypertext documents on the web required two different pieces of software – a text editor and a web browser, respectively – content management systems enabled web authors to add an interactive online interface to their websites, allowing them to add to or change their website through their browser. Instead of having to use a text editor to author new documents in their entirety, content management systems provided a template shell and database backend that would store the media content of all pages and ensure that they were displayed with a uniform look and feel across a website. When using a content management system, authors visit a designated interface page on their website, enter their login credentials, and begin writing text, styling their documents, and uploading other media content through a series of forms. Once the document is ready to be made available on their site, authors select to "publish" it to a URL. Whenever a reader visits that URL, the content management system processes the document's contents through the website's template and produces a page populated with the new content that matches the overall look and feel of the rest of the site's pages. Today, there are a variety of content management systems available, including a variety of open source packages such as Drupal, Joomla, Jekyl, and Wordpress.

Content management systems helped to alleviate many of the difficulties associated with hand-coding hypertext web documents, including accounting for differences in browsers. At the same time, however, they have also contributed to a homogenization of design across the web. Although Berners-Lee envisioned a web on which everyone connected server machines that they themselves owned, this structure of the web existed only very briefly. Instead, many web authors established websites through "hosting services." Prior to the rise of content management systems, many hosting services – both free and paid alike – provided their users with nothing more than a URL for their website and a directory to which they could upload their hand-written HTML documents using file transfer protocol software. Popular early hosts like Angelfire, Geocities, and Tripod imposed few, if any, restrictions on website design initially, apart from inserting advertisement frames in lieu of charging authors a hosting fee. Many modern hosts, however, utilize content management systems; and while content management systems allow authors to customize the template shell of their websites, some hosts may restrict customization or impose a fee to adjust certain features of the template. Even in cases when users have relative freedom to make adjustments to a content management system's template, invariably all websites created with a specific content management system will share some base level of design features that are a product of the algorithms that compile a website's documents from the site's content database.

Just as content management systems have automated many aspects of web authoring, today's search engines are made possible through automated web reading. Google's current dominance of web search is predicated upon its web crawling software, an algorithmic system for following links across websites and storing information about them to populate its search database. Google did not invent web crawler software or automated search engines. Before Google was made publicly available in 1997, services like AltaVista (1995), Excite (1995), Lycos (1994), and WebCrawler (1994) also implemented crawlers. Some of these services attempted to implement a portal interface, mimicking the popular directory style of Yahoo!. Google's initial rise in popularity has been attributed, at least in part, to its comparatively minimalist interface. When it first launched, Google presented its search service as a simple tool: a textbox beneath a static, unchanging logo that was not linked to a larger service suite. It loaded quickly and made simple, fast queries possible on a variety of bandwidths.

Structurally, search engines like Google have become a focal point for our web reading practices. In place of the networked experience of traversing hyperlinks from site to site in search of information, search engines have effectively flattened the web. Only web crawlers continue to experience web reading as a practice of networked discovery. Instead, most users expect search engines to bring content to them. In this respect, Google's popularity in the long term has been sustained by its PageRank algorithm, which attempts to match search queries against the full text of documents in its web crawler database, ranking them through a series of relevancy scores that weight, among other things, the number of times other users have clicked a link among search results for similar queries. Thus, as Safiya Noble argues, search engines like Google function not merely as tools but as knowledge institutions that manage our relationship to information. Because search engines now serve as primary entry points for a significant portion of our web interactions, they are flattening our perception of the web and centralizing our experience of it around the databases constructed by their web crawlers. Officially, search engine services like Google present their technologies in objective terms, pointing to a series of mathematical and natural language processing theories to justify the structure of their ranking and relevance algorithms. But as Noble's work shows, these algorithms inevitably reproduce the social and cultural biases of their designers – just as any other institution would reproduce the biases of the people working within it, intentionally or otherwise (Noble 2018). Google is not the only service that flattens the web. Social media services like Facebook, Reddit, and Twitter also similarly draw in links to information sources from around the web. Each of these services tries to effect its own kind of web cultural cachet but few, apart from Facebook, have aspired to attain the same level of authority over the web itself as Google.

Together, the automated content management and search systems that now support so much of our experience as authors and readers of the web have fundamentally changed our experience of it. Because URLs now serve as commands for document generation rather as links to static documents, the topology of the web has taken on an M.C. Escher-like quality. When working in concert, the automated writing and reading systems of the modern web erase the boundaries between documents, sites, and services, leading us back to the same locations even as we try to follow links in new directions. Google is not just a website we use to search, as Siva Vaidhyanathan explains, it is a center of power that exerts a kind of "infrastructural imperialism" by weaving itself into other automated web systems (Vaidhyanathan 2012). Consider, for example, Google's AdSense marketing service which links a history of search queries to a person's account – or, failing that, to their browser through identification codes stored as "cookies." AdSense can be woven into a content management system to populate web sites with advertisements micro-targeted towards a reader's presumed interests. When moving from page to page within a web site, or even across sites, AdSense continually presents its content to readers, meaning that the same or similar content will now follow readers across the web. Advertising and recommendation systems shape the web's landscape for us, continually directing us towards content they determine will be of interest to us.

Our sense of the web as a network can also be disrupted through the increasing prevalence of embedded objects. The content management systems developed for use by large media services, like YouTube, now allow for authors to embed objects from those services into their own websites. During earlier web eras, the practice of presenting content from another website as part of one's own was referred to variously as "hotlinking," "direct linking," or "inline linking." The practice was considered a form of theft. Regardless of whether someone managed a website on their own server or used a hosting service, the total amount of information that could transferred during a given period was limited. Thus, each time a visitor to a website viewed an image hotlinked from another website, the source's web server would provide the bandwidth for it. Now, however, larger online services like YouTube, Instagram, SoundCloud, and Twitter encourage this practice by providing source code that can be embedded into a page. Content no longer feels located in any one specific place on the web. Readers no longer need to travel across links because content can places they've never visited on the web can be brought to them, spliced into sites they already know so that they never have to travel away to find something.

In thinking about how to study the web as a media ecology, it is important to recognize that many of the grand narratives surrounding its origins are of limited value in describing our experience of it or the politics of its systems. The web and other hypertext systems that inspired it were intended to be technologies that would democratize communication or empower certain communication styles that their creators valued, but these visions of the web were ultimately realized through specific actors – both human and technological – which have shaped them towards other ends. Decentralization is a powerful myth in narratives of technological development, but decentralized technologies rely on embedded standards or infrastructural systems that themselves operate according to their own agendas. In thinking about how to study the history and future of the web, it is important that we pay attention to how the various technologies that comprise it structure our agency within it and our experiences of it.

References

Ankerson, M. 2018. *Dot-Com Design: The Rise of a Usable, Social, Commercial Web.* New York: New York University Press.

Bardini, T. 2000. *Bootstrapping: Douglas Engelbart, Coevolution, and the Origins of Personal Computing.* Stanford, CA: Stanford University Press.

Barnet, B. 2012. Machine enhanced (re)minding: The development of Storyspace. *Digital Humanities Quarterly* 6(2). www.digitalhumanities.org/dhq/vol/6/2/000128/000128.html.

Berners-Lee, T., and Fischetti, M. 1999. *Weaving the Web: The Original Design and Ultimate Destiny of the World Wide Web By Its Inventor.* San Francisco: Harper.

Brügger, N. 2018. *The Archived Web: Doing History in the Digital Age.* Cambridge, MA: MIT Press.

Bush, V. 1945. As we may think. *Atlantic Monthly* 176(1):101–108.

Cusumano, M. and Yoffie, D. 1998. *Competing on Internet Time: Lessons from Netscape and Its Battle with Microsoft.* New York: Simon and Schuster.

Engelbart, D. 1962. Augmenting Human Intellect: A Conceptual Framework. SRI Summary Report AFOSR-3223. http://dougengelbart.org/content/view/138/000/.

Kittler, F. 1999. *Gramophone, Film Typewriter.* Stanford: Stanford University Press.

Landow, G. 2006. *Hypertext 3.0: Critical Theory and New Media in an Era of Globalization.* Baltimore: Johns Hopkins University Press.

McLuhan, M. 1964. *Understanding Media: The Extensions of Man.* New York: McGraw-Hill.

Nelson, T. 1987. *Computer Lib / Dream Machines.* Redmond, WA: Microsoft Press.

Nelson, Theodore H. 2000. "Xanalogical Structure, Needed Now More than Ever: Parallel Documents, Deep Links to Content, Deep Versioning and Deep Re-Use." www.xanadu.com.au/ted/XUsurvey/xuDation.html.

No Maps for These Territories. 2000. Film. Mark Neale.

Noble, S. 2018. *Algorithms of Oppression: How Search Engines Reinforce Racism.* New York: New York University Press.

O'Reilly, T. 2009. *What Is Web 2.0?: Design Patterns and Business Models for the Next Generation of Software.* Sebastopol, CA: O'Reilly Media.

Sterling, B. 1995. The life and death of media. *Proceedings of the 6th International Symposium on Electronic Art*: 272–282.

Vaidhyanathan, S. 2012. *The Googlization of Everything (And Why We Should Worry).* Berkeley: University of California Press.

Wolf, G. 1995. The Curse of Xanadu. *WIRED* 3(6):137.

4 Robustness and the internet

A geographic fiber-optic infrastructure perspective

Ramakrishnan Durairajan

While it can be easy to conceive of cyberspace as a borderless world consisting of virtual resources, users, and content connected by logical links, the internet strongly depends on a geographically-anchored physical infrastructure: nodes (e.g., datacenters, colocation facilities) and links (e.g., fiber-optic and submarine cables) interconnecting them. Indeed, naïve abstractions can obscure the true robustness (or fragility) of the network. For example, a single fiber optic cable may be multiplexed to provide the capacity to logically distinct service providers and may be routed in a conduit shared with cables from other providers – seemingly disjoint paths that share fate in the physical world. Without understanding the physical manifestation – in particular, the fiber-optic portion – of the network, it can be difficult to correctly predict (and mitigate) the impact of natural disasters (Miller 2006; Cho et al. 2011; Eriksson, Durairajan, and Barford 2013), accidents (McGrattan and Hamins 2006a; Zmijewski 2008a), censorship (Dainotti et al. 2011), or targeted malicious attacks (Hunter 2008a) on the internet's availability.

Recent works on mapping the internet's physical topology have thus been motivated by a number of compelling applications: enhancing robustness, improving security and increasing performance (Durairajan et al. 2013; Eriksson, Durairajan, and Barford 2013; Durairajan, Sommers, and Barford 2014; Durairajan et al. 2015). However, obtaining accurate physical maps remains challenging due to the autonomous and fractured nature of infrastructure ownership and municipal governance, the organic way in which infrastructure has been deployed, variable deployment and management practices, and incomplete and inconsistent right-of-way records. Consider, for example, two logically connected IP routers in different cities. The physical connection between the routers may involve multiple fiber providers, lessors, states, municipalities, and regulators, and may use public and private right-of-ways. Each of these entities may maintain separate and incompatible databases of their portion of the path, if any records are available at all. This is the focus of part 2. At the same time, it is either taken for granted or implicitly assumed that the physical infrastructure of tomorrow's internet will have the capacity, performance, and resilience required to develop and support future services and applications. This opens up a number of issues and implications which are in parts 3 and 4 respectively.

Origins, acquisitions, and current trends

Understanding the true origins of the internet's fiber-optic infrastructure and its evolution over the years is fraught with challenges. Some of the challenges include the autonomous and fractured nature of infrastructure ownership and municipal governance, variable deployment and management practices, and incomplete and inconsistent right-of-way records. Based on the information gathered from bells and whistles (e.g., ATTstory1; ATTstory2; PFNetAndTouchAmerica1; PFNetAndTouchAmerica2; PFNetstory; Pipestory), one can posit that major fiber-optic infrastructure rollouts in the U.S. started with three big infrastructure entities: CapRock communications, PF.Net, and Touch America. Many of tier-1 internet service providers of today relied on these three infrastructure providers for connecting their networks.

Over the years, tier-1 internet service providers made deals with the three fiber infrastructure providers, some of the infrastructure providers went bankrupt for a number of reasons (e.g., revenue loss, expanded customer base, a spike in demands, etc.), and tier-1s acquired them, consequently. Several instances of this phenomenon can be found here: VelocitaStory1; VelocitaStory2; FCCapproves. To avoid this situation, the tier-1 internet service providers found a new market opportunity – called *dark fiber* – fueling the commercialization of the internet. Over the past several years, there have been significant changes among tier-1 service and infrastructure providers due to this commercialization of the internet in general and dark fiber-optic infrastructure assets in particular.

Consolidation of dark fiber providers. Recent trends show that dark fiber providers are merging at an unprecedented rate. Examples include CenturyLink's acquisition of Qwest in 2011, resulting in a combined 190k mile fiber network (centurylink 2011), Zayo's acquisition of Abovenet in 2012, resulting in a combined 6.7M fiber mile network connecting some 800 datacenters (zayo 2012), Level 3's acquisition of TW telecom in '14 (acq2014), Lighttower merging with Fibertech in 2015 (acq2015), CenturyLink's acquisition of Level 3 in 2016 (acq2016CL), and Verizon's recent announcement it will acquire XO communications' fiber-optic network business (acq2016). A clear consequence of these mergers is that there are fewer fiber-optic network providers, but the remaining ones have larger fiber footprints.

Evolution in the datacenter market. Similar to the mergers among dark fiber providers, there has been consolidation as well as expansion within the datacenter market. Among the tier-1 datacenter providers (i.e., serving major metro areas and large cities), examples of consolidation include Equinix' acquisition of Telecity Group (EU/UK) (equinix1 2015) and Bit-Isle (JP) (equinix2 2015) in 2015, Digital Reality Trust's acquisition of Telx (digreality 2015) in 2015, AT&T's announcement it will sell datacenter assets (att 2015) in 2015, and Windstream's announcement to sell its datacenter business to TierPoint (windstreamdc 2015) in 2015. At the same time, the growing demand for cloud services has put pressure on the largest cloud providers to have presence in more locations and also closer to their customers, which has led to the emergence of an increasing number of new second-tier datacenter providers (e.g., EdgeConneX (EdgeCoX)) that are focused on medium-sized markets such as Portland, OR and Pittsburgh, PA.

Dark fiber providers acquiring datacenters. There are recent examples of dark fiber providers acquiring datacenters, which presents an opportunity for one provider to supply high-bandwidth connectivity between datacenter co-location endpoints to customers who need it. One example of a provider with this nascent capability is Lightower (lightower), which acquired ColocationZone in 2015 (lightowerdc2 2015) and Datacenter101 (lightowerdc1 2016) in 2016. Similarly, Allied Fiber aimed to be a network-neutral and dark fiber "superstructure" with a footprint across the United States and offered traditional 20-year and non-traditional 12, 24, and 36-month Indefeasible Rights of Use (IRU) options (alliedfibertiers). These developments indicate that there exist business opportunities for companies that offer integrated (network-neutral) colocation/dark fiber services and that could benefit from available flexible connectivity to boost their existing but may be constrained dark fiber infrastructure.

Issues and debates

In this section, we discuss three key issues that result from a lack of understanding of the fiber-optic internet infrastructure.

Issue 1: Lack of physical infrastructure awareness

While the larger internet is robust in general, small-scale events can have significant effects on the day-to-day operations of the internet including the loss of connectivity for large sections of users for extended periods of time. Events include natural or technological disasters (e.g., earthquakes), accidents (e.g., the Baltimore Howard Street tunnel fire (McGrattan and Hamins 2006b) or Mediterranean cable cuts (Zmijewski 2008b)), misconfigurations (e.g., Pakistani YouTube routing (Hunter 2008)), a potential Electromagnetic Pulse Attack (empAttack), or censorship (e.g., the Egyptian government's response to the 2011 uprising (dainotti11 2011)). These effects indeed suggest that small-scale events can have localized effects on sections of the internet and the underlying mechanisms to ensure internet's robustness are often insufficient.

Thus, the motivating questions are "why is today's Internet far from being robust with numerous loss-of-connectivity episodes and how do we transcend the robustness gap to build a more robust Internet?" The main observation here is that while many other dynamic aspects (like routing) of the internet have been examined in prior work (Andersen et al. 2001; Hansen et al. 2005; Wang et al.

2007; Zhu et al. 2008), the underlying physical infrastructure – specifically, the geographic locations of links (e.g., fiber strands housed in physical conduits) – that make up the internet are, by definition, static and are completely ignored, which leads to a robustness gap (e.g., unpredictability in routing convergence, cascading failures, etc.). Bridging this robustness gap is an open problem.

The aforementioned questions and the observation lead us to hypothesize that the (i) ignored physical infrastructural complexity (i.e., lack of understanding of fiber-optic links and connectivity) and inherent risks on topology are what makes the internet failure-prone and (ii) creating methods and frameworks to unravel the complexity and risks in the internet is the first logical step towards a truly robust internet. We describe these hypotheses further below.

Unravelling structural complexity

Among the factors impeding the comprehensive unraveling of complexity, one stands out: fiber-optic topology. Why? Studies that aim to map the internet's topological structure have been motivated for many years by a number of compelling applications including the possibilities of improving performance, security, and robustness. While these motivations remain as compelling as ever, the ability to accurately and comprehensively map the internet has, for the most part, remained beyond our grasp. In fact, despite some 20 years of research efforts that have focused on understanding aspects of the internet's infrastructure such as its router-level topology or the graph structure resulting from its inter-connected Autonomous Systems (AS), very little is known about the complexity of today's fiber internet.

Enormous size, fractured and distributed ownership, and constant flux are the primary challenges that hinder the thorough mapping of the internet. Faced with these challenges, the most widely used approach to mapping the internet at the network layer (i.e., the network-layer map) has been based on leveraging the TTL-limited traceroute probes. Ideally, network-layer maps reflect a timely representation of network topology as well as the dynamic aspects of management and configurations. While great progress has been made in solving some of the specific problems related to using these network-layer maps for understanding aspects of internet topological characteristics, the fact remains that layer 3 data are inherently tied to the management policies and operational objectives of internet service providers, which may be at odds with comprehensive and accurate mapping of the internet.

In addition, analyzing the interconnection structure of the internet has been the subject of a large number of studies over the past decade. The network scope of these studies ranges from the router-level (e.g., CAIDA 2007; Madhyastha et al. 2006), to POP-level (e.g., Spring, Mahajan, and Wetherall 2002; Shavitt and Shir 2005), to the autonomous system-level (e.g., Mao et al. 2003; Zhang et al. 2005). Most of these studies are based on layer 3 measurements from traceroute-like tools at the router level, and BGP announcements at the AS level (e.g., routeviews). The dynamic nature of the data used in these studies presents significant challenges in recovering details of the underlying physical infrastructure.

There has been a great deal of effort made to harness layer 3 TTL-limited probes for network mapping since the introduction of the traceroute tool (Jacobson and Deering 1989). Some efforts (e.g., Spring, Mahajan, and Wetherall 2002; Sherwood, Bender, and Spring 2008) have focused on the goal of developing a comprehensive network-layer view of the internet, that is, unique identification of nodes and links. Other efforts have focused on developing new probing techniques that expand the ability to collect data and thereby improve accuracy and mapping coverage (e.g., Augustin et al. 2006; Sherwood and Spring 2006; Augustin, Krishnamurthy, and Willinger 2009). More recent efforts have focused on analyzing and addressing various inaccuracies inherent in probe-based network mapping (Zhang et al. 2011; Spinelli, Crovella, and Eriksson 2013). For example, Roughan et al. and Eriksson et al. develop inferential techniques to quantify the nodes and links that are missed through network-layer mapping (Roughan, Tuke, and Maennel 2008; Eriksson et al. 2011). Other researchers have looked closely at the rise of Internet Exchange Points (IXPs) and the effects of IXPs on inaccuracies of network-layer mapping (e.g., Augustin, Krishnamurthy, and Willinger 2009; Ager et al. 2012). Concurrent with the rise of IXPs has come a "flattening" of the internet's peering structure (Gill et al. 2008; Labovitz et al. 2010; Dhamdhere and Dovrolis 2010), which affects the very nature of end-to-end paths through the internet. Still, other researchers have observed that increased use of network virtualization techniques such as MPLS has led to additional inaccuracies in layer 3 mapping, and which are likely to continue to thwart probe-based mapping efforts (Sherwood, Bender, and Spring 2008; Sommers, Barford, and Eriksson 2011; Donnet et al. 2012). We posit that layer 3 mapping efforts will continue to be important sources of internet topology information and that complementary efforts to build repositories of physical internet maps (e.g., Knight et al. 2011; Durairajan et al. 2013) will result in representations of internet topology that are more accurate and applicable to problems of interest than either representation in isolation.

The Internet Atlas (Durairajan et al. 2013) is the first academic effort to establish a GIS-based web portal that includes diverse measurement data layers on top of a map of the physical internet. Internet Atlas is the starting point towards understanding the complexity of the internet is to understand the fiber-optic infrastructure topology. In particular, the Internet Atlas is concerned with the physical aspects of the wired internet, ignoring entirely the wireless access portion as well as satellite or any other form of wireless communication. Moreover, Internet Atlas is exclusively interested in the long-haul fiber-optic portion of the wired internet in the U.S. The detailed metro-level fiber maps (with corresponding colocation and data center facilities) and international undersea cable maps (with corresponding landing stations) are only accounted for to the extent necessary. In contrast to short-haul fiber routes that are specifically built for short distance use and purpose (e.g., to add or drop off network services in many different places within metro-sized areas), long-haul fiber routes (including ultra-long-haul routes) typically run between major city pairs and allow for minimal use of repeaters. Gorman used a similar approach in his Ph.D. dissertation (Gorman 2005) although no maps from that work are available.

The Internet Atlas is predicated on the assumption that detailed information on network infrastructure can be found on publicly available webpages. This implies that the internet search can be used as the primary data gathering tool. In addition to the major search engines, we used search aggregators (e.g., Soovle (soovle) and SidePad (sidePad)) to enhance the ability to find network maps.

The U.S. long-haul fiber map

Durairajan et al. (2015) mapped the long-haul fiber-optic portion of the physical internet using the search-based amassing process described above. The map constructed through the search process is shown in Figure 4.1, and contains 273 nodes/cities, 2411 links, and 542 conduits (with multiple tenants). Prominent features of the map include (i) dense deployments (e.g., the northeast and coastal areas), (ii) long-haul hubs (e.g., Denver and Salt Lake City), (iii) pronounced absence of infrastructure (e.g., the upper plains and four corners regions), (iv) parallel deployments (e.g., Kansas City to Denver), and (v) spurs (e.g., along northern routes).

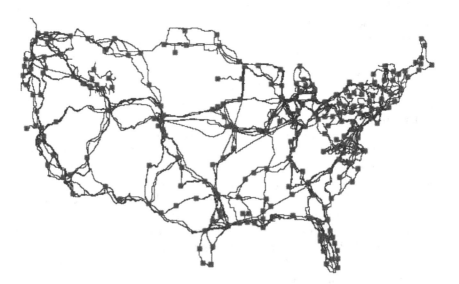

Figure 4.1 Location of physical conduits for networks considered in the continental United States. Image courtesy of (Durairajan et al. 2015).

While the mapping effort described by Durairajan et al. invariably raise the question of completeness (e.g., quality of the map), it is *incomplete* (Durairajan et al. 2015). However, the effort takes a significant first step in creating a first-of-its-kind blueprint of the internet and the constructed map is of sufficient quality for studying issues that do not require local details typically found in metro-level fiber maps. Moreover, as with other internet-related mapping efforts (e.g., AS-level maps), we hope this work will spark a community effort aimed at gradually improving the overall fidelity of our basic map by contributing to a growing database of information about geocoded conduits and their tenants.

In addition, the methodological blueprint outlined in Durairajan et al. shows that constructing such a detailed map of the U.S.'s long-haul fiber infrastructure is feasible, and since all data sources we use are publicly available, the effort is reproducible. The fact that the work can be replicated is not only important from a scientific perspective, but it also suggests that the same effort can be applied more broadly to construct similar maps of the long-haul fiber infrastructure in other countries and on other continents. For instance, (lux) reveals several ISPs like AT&T, Cogent, Intelliquent, Level 3, TeliaSonera, and Verizon sharing infrastructure in Luxembourg. Other examples include documents from (i) locations like Slovakia (slo) and Italy (ita), (ii) ISPs like (gc) at several places and Tata (tataafrica) in Africa, and (iii) initiatives like Open Access (openaccess).

Interestingly, recommendation 6.4 made by the FCC in chapter 6 of the National Broadband Plan (bbplan) states that "the FCC should improve the collection and availability regarding the location and availability of poles, ducts, conduits, and rights-of-way." It also mentions the example of Germany, where such information is being systematically mapped. Clearly, such data would obviate the need to expend significant effort to search for and identify the relevant public records and other documents.

Lastly, it is also important to note that there are commercial (fee-based) services that supply location information for long-haul and metro fiber segments (e.g., fiberlocator). These services typically offer maps of some small number (5–7) of national ISPs, and that, similar to the map created by Durairajan et al. (see map in (geoteltr)), many of these ISPs have substantial overlap in their locations of fiber deployments. Unfortunately, it is not clear how these services obtain their source information and/or how reliable these data are. Although it is not possible to confirm, in the best case these services offer much of the same information that is available from publicly available records, albeit in a convenient but non-free form.

Geography of fiber deployments

While the conduits through which the long-haul fiber-optic links that form the physical infrastructure of the internet are widely assumed to follow a combination of transportation infrastructure locations (i.e., railways and roadways) along with public/private right-of-ways, very few prior studies have attempted to confirm or quantify this assumption (GMUmappingproject). Understanding the relationship between the physical links that make up the internet and the physical pathways that form transportation corridors helps to elucidate the prevalence of conduit sharing by multiple service providers and informs decisions on where future conduits might be deployed.

In Durairajan et al. (2015), an analysis is performed by comparing the physical link locations identified in their constructed map to geocoded information for both roadways and railways from the United States National Atlas website (nationalAtlas). The geographic layout of roadway and railway data sets can be seen in Figure 4.2, respectively. In comparison, the physical link geographic information for the networks under consideration can be seen in the Figure 4.1.

Specifically, Durairajan et al. (2015) use the polygon overlap analysis capability in the ArcGIS (arcgis) to quantify the correspondence between physical links and transportation infrastructure. These results show that a significant fraction of all the physical links are co-located with roadway infrastructure. The plots also show that it is more common for fiber conduits to run alongside roadways than railways, and an even higher percentage are co-located with some combination of roadways and railway infrastructure. Furthermore, for a vast majority of the paths, we find that physical link paths more often follow roadway infrastructure compared with rail infrastructure.

Despite the results reported in Durairajan et al. (2015), there remain conduits in their infrastructure map that are not co-located with transportation ROWs. For example, Level3's network (level3networkmap), where the map shows the existence of a link from (1) Anaheim, CA to Las Vegas, NV, and (2) Houston, TX to Atlanta, GA, but no known transportation infrastructure is co-located. By considering other types of rights-of-way (pipeAll), many of these situations could be explained. Visually, we can verify that the link from Anaheim, CA to Las Vegas, NV is co-located with the refined-products pipeline. Similarly, the link from Houston, TX to Atlanta, GA is deployed along with NGL pipelines.

Issue 2: Lack of understanding of the topological risks

Creating a blueprint of the fiber-optic infrastructure of the internet such as the one depicted in Figure 4.2 opens up an opportunity to pursue qualitative assessment of the inherent risk and its corresponding root causes. Specifically, a striking characteristic of the constructed maps is a significant amount of observed infrastructure sharing (Durairajan et al. 2015) which leads to the problem called Shared Risk: physical conduits shared by many service providers are at an inherently risky situation since damage to those conduits will affect many several providers. Such infrastructure sharing is the result of a common practice among many of the existing ISPs to deploy their fiber

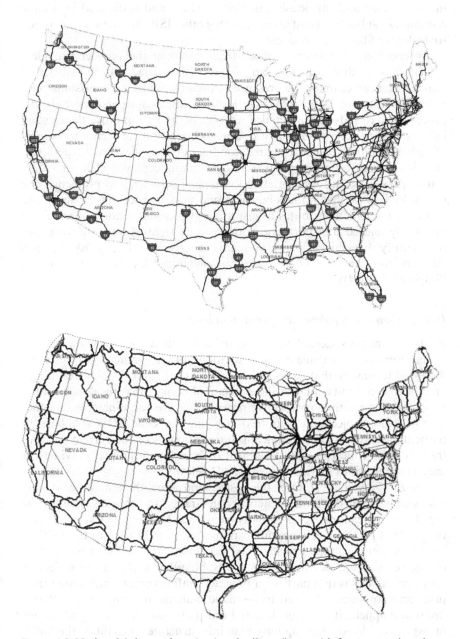

Figure 4.2 National Atlas roadway (top) and railway (bottom) infrastructure locations. Image courtesy of Durairajan et al. (2015).

in jointly used and previously installed conduits and is dictated by simple economics – substantial cost savings, among other ISP objectives, as compared to deploying fiber in newly constructed conduits.

By considering different metrics (e.g., counting the number providers present in a conduit) for measuring the risks associated with infrastructure sharing, one can examine the presence of high-risk links in the existing long-haul infrastructure, both from a connectivity and usage perspective. In the process, one can also do a detailed analysis of how to improve the existing long-haul fiber-optic infrastructure so as to increase its resilience to failures of individual links or entire shared conduits or to achieve better perform-ance in terms of reduced propagation delay along deployed fiber routes. By framing the issues as appropriately formulated optimization problems, one can indeed show that both robustness and performance can be improved by deploying new fiber routes in just a few strategically chosen areas along previously unused transportation corridors and right-of-ways (ROWs), and we quantify the achievable improvements in terms of reduced risk (i.e., less infrastructure sharing) and decreased propagation delay (i.e., faster internet (Singla et al. 2014)).

Issue 3: Conflicting policies and legislation issues

While the metrics described above to understand topological risks are straightforward to calculate, such solutions often conflict with currently discussed legislation that favors policies such as "dig once," "joint trenching," or "shadow conduits" due to the substantial savings that result when fiber builds involve multiple prospective providers or are coordinated with other infrastructure projects (i.e., utilities) targeting the same ROW (bbplan). In particular, we discuss our technical solutions in view of the current net neu-trality debate concerning the treatment of broadband internet providers as telecommunications services under Title II below.

The FCC and Title II. Over the past several years, there have been many discussions about the topic of network neutrality. In those discussions, the U.S. Communications Act of 1934 is mentioned frequently. A key reason is that Title II of the U.S. Communications Act of 1934 enables the FCC to specify communications providers as "common carriers." One implication of the recent FCC decision to reclassify broadband internet providers as common carriers is that parts of a provider's infrastructure, including utility poles and conduits, will need to be made available to third parties. If this decision is upheld, it will likely lead to third-party providers taking advantage of expensive already-existing long-haul infrastructure to facilitate the build-out of their own infrastructure at a considerably lower cost. Indeed, this is exactly the issue that has been raised by Google in their current fiber deploy-ment efforts (googleTitle). Furthermore, an important consequence of the additional sharing of long-haul infrastructure that will likely take place if the Title II classification is upheld is a significant *increase* in shared risk. We argue that this trade-off between broader metro-area fiber deployments (e.g., Google) and the increased risks in shared long-haul infrastructure requires more careful consideration in the broader Title II debate.

We argue that the current debate would benefit from a quantitative assessment of the unavoidable trade-offs that have to be made between the substantial cost savings enjoyed by future Title II regulated service providers (due to their ensuing rights to gain access to existing essential infrastructure owned primarily by utilities) and an increasingly vulnerable national long-haul fiber-optic infrastructure (due to legislation that implicitly reduced overall resilience by explicitly enabling increased infrastructure sharing).

Geographical and technological implications

Implications for service providers

The fiber-optic infrastructure map depicted in Figure 4.2 has several important implications for service providers on the internet. First, the long-haul fiber infrastructure map highlights the fact that fiber conduits are used to transmit data between large population centers. While infrastructure such as content delivery networks and data centers complicate the details of data flows, this map can support and inform decisions by service providers on provisioning and management of their infrastructures. Second, beyond performance and robustness analysis, the map can inform decisions on local and/or regional broadband deployment, peering, and route selection, as well as provide competitive insights. Third, the fact that there is widespread and sometimes significant conduit sharing complicates the task of identifying and configuring backup paths since these critical details are often opaque to higher layers of the ISO-OSI stack. Fourth, enrichment of this map through the addition of long-haul links in other regions around the world, undersea cable maps for inter-continental connectivity, and metro-level fiber maps will improve our global view of the physical internet and will provide valuable insights for all involved players (e.g., regional, national, or global-scale providers). Finally, the map also informs regulatory and oversight activities that focus on ensuring a safe and accessible physical communications infrastructure for service providers.

While much prior work on aspects of (logical) internet connectivity at layer 3 and above points to the dynamic nature of the corresponding graph structures as an invariant, it is important to recognize that the (physical) long-haul infrastructure is comparably static by definition (i.e., deploying new fiber takes time). In that sense, the links reflected in our map can also be considered an internet invariant, and it is instructive to compare the basic structure of our map to the NSFNET backbone circa 1995 (nsfnet).

Enriching the long-haul fiber-optic infrastructure

On the one hand, the study by Durairajan et al. (2015) shows that the addition of a small number of conduits can lead to significant reductions in shared risk and propagation delays. At the same time, their examination of public records also shows that new conduit infrastructure is being deployed at a steady rate. Based on the examination, assuming that the locations for these actual deployments are based on a combination of business-related factors and are not necessarily aligned with the links that our techniques identify, the question that arises is how the conduits identified in their analysis might actually be deployed? Since IXPs largely grew out of efforts by consortia of service providers as a means for keeping local traffic local (Chatzis et al. 2013), perhaps a version of the internet exchange point (IXP) model could be adapted for conduits? A "link exchange" for fiber-optic infrastructure?

Indeed, the deployment of key long-haul links such as those identified by Durairajan et al. (2015) would be compelling for a potentially large number of service providers, especially if the cost for participating providers would be competitive. At the same time, given the implications for shared risk and the critical nature of communications infrastructure, government support may be warranted. In fact, the involvement of some states' DOTs in the build-out and leasing of new conduits can be viewed as an early form of the proposed "link exchange" model (cdot1).

References

acq2014. Level3 Completes Acquisition of tw Telecom. http://investors.level3.com/ investor-relations/press-releases/press-release-details/2014/Level-3-Completes-Acquisition-of-tw-telecom/default.aspx.

acq2015. Lightower Fiber Networks to Merge with Fibertech Networks. www.lightower.com/company/news/press-releases/lightower-fiber-networks-to-merge-with-fibertech-networks/#.V92nIz4rKL0.

acq2016. Verizon to Buy XO's Fiber Unit from Carl Icahn for $1.8 Billion. www.bloomberg.com/news/articles/2016-02-22/verizon-to-acquire- xo-communications-fiber-unit-for-1-8-billion.

acq2016CL. CenturyLink to Buy Level 3 for $34 Billion in Cash, Stock. www.bloomberg.com/news/articles/2016-10-31/centurylink-agrees-to-buy-level-3-for-34-billion-in-cash-stock.

Ager, B., N. Chatzis, A. Feldmann, N. Sarrar, S. Uhlig, and W. Willinger. 2012. *Anatomy of a Large European IXP*. ACM SIGCOMM.

alliedfibertiers. Allied Fiber: Long Haul Dark Fiber. www.alliedfiber.com/products/long-haul-dark-fiber/.

Andersen, D., H. Balakrishnan, F. Kaashoek, and R. Morris. 2001. *Re- silient Overlay Networks*. ACM SOSP.

arcgis. ESRI ArcGIS. www.arcgis.com/features/.

AT&T. 2015. Report: AT&T to Sell $2B Worth of Data Center Assets. www.datacenterknowledge.com/archives/2015/02/03/report-2b-worth-of-att-data-centers-may-be-up-for-sale/.

ATTstory1. ://www.fiberopticsonline.com/doc/att-undertakes-network-expansion-0001.

ATTstory2. www.channelpartnersonline.com/articles/2000/02/at-amp-t-to-build-metro-network-with-help-from-so.aspx.

Augustin, B., C. Xavier, B. Orgogozo, F. Viger, T. Friedman, M. Latapy, C. Magnien, and R. Teixeira. 2006. *Avoiding Traceroute Anomalies with Paris Traceroute.* ACM SIGCOMM IMC.

Augustin, B., B. Krishnamurthy, and W. Willinger. 2009. *IXPs: Mapped?* ACM SIGCOMM IMC.

bbplan. Broadband plan. www.broadband.gov/plan/6-infrastructure/.

CAIDA. 2007. The Skitter Project. www.caida.org/ tools/measurement/skitter/.

cdot1. Colorado Department of Transportation document. www.wmxsystems.com/EndUserFiles/44489.pdf.

centurylink. 2011. CenturyLink and Qwest Complete Merger. http://news.centurylink.com/news/centurylink-and-qwest-complete-merger.

Chatzis, N., G. Smaragdakis, A. Feldmann, and W. Willinger. 2013. There Is More to IXPs than Meets the Eye. SIGCOMM CCR.

Cho, K., C. Pelsser, R. Bush, and Y. Won. 2011. The Japan earthquake: The impact on traffic and routing observed by a local ISP. Proceedings of the Workshop on Internet and Disasters.

Communications Act of 1934. http://transition.fcc.gov/ Reports/1934new.pdf.

Dainotti, A., C. Squarcella, E. Aben, K. C. Claffy, M. Chiesa, M. Russo, and A. Pescapé. 2011. Analysis of country-wide internet outages caused by censorship. Proceedings of ACM IMC.

dainotti11. 2011. *Analysis of Country-wide Internet Outages Caused by Censorship.* Proceedings of ACM IMC.

Dhamdhere, A. and C. Dovrolis. 2010. The Internet Is Flat: Modeling the Transition from a Transit Hierarchy to a Peering Mesh. ACM CoNEXT.

digreality.2015.DigitalRealtyCloses$1.9BTelxAcquisition.www.datacenterknowledge.com/archives/2015/10/12/ digital-realty-closes-1-9b-telx-acquisition/.

Donnet, B., M. Luckie, P. Mérindol, and J.-J. Pansiot. 2012. Revealing MPLS Tunnels Obscured from Traceroute. ACM SIGCOMM CCR.

Durairajan, R., P. Barford, J. Sommers, and W. Willinger. 2015. InterTubes: A Study of the US Long-haul Fiber-optic Infrastructure. ACM SIGCOMM.

Durairajan, R., S. Ghosh, X. Tang, P. Barford, and B. Eriksson. 2013. *Internet Atlas: A Geographic Database of the Internet.* ACM SIGCOMM HotPlanet.

Durairajan, R., J. Sommers, and P. Barford. 2014. Layer 1-Informed Internet Topology Measurement. ACM SIGCOMM IMC.

EdgeCoX. EdgeConneX. www.edgeconnex.com/.

equinix1. 2015. Equinix closes its blockbuster $3.8B acquisition TelecityGroup acquisition. www.datacenterknowledge.com/archives/2016/01/15/equinix-closes-blockbuster-3-8b-telecitygroup-acquisition/.

equinix2. 2015. Equinix Closes Bit-isle Deal, Expands Japan Data Center Footprint. www.datacenterknowledge.com/archives/2015/11/04/equinix-closes-bit-isle-deal-expands-japan-data-center-footprint/.

Eriksson, B., P. Barford, J. Sommers, and R. Nowak. 2011. Inferring Unseen Components of the Internet Core. IEEE JSAC.

Eriksson, B., R. Durairajan, and P. Barford. 2013. *Riskroute: A Framework for Mitigating Network Outage Threats.* ACM CoNEXT.

FCCapproves. www.prnewswire.com/news-releases/fcc-approves-touch-americas-acquisition-of-qwest-communications-14-state-long-distanc html.

fiberlocator. FiberLocator Online. www.fiberlocator.com/product/fiberlocator-online.

gc. Example in Global Crossing. http://media.frnog.org/FRnOG_8/FRnOG_ 8-3.pdf.

geoteltr. Image from Geo-tel. www.techrepublic.com/article/ the-google-fiber-lottery/.

Gill, P., M. Arlitt, Zongpeng Li, and A. Mahanti. 2008. The Flattening Internet Topology: Natural Evolution, Unsightly Barnacles or Contrived Collapse? In PAM.

GMUmappingproject. GMU Mapping Project. http://gembinski.com/ interactive/ GMU/research.html.

googleMaps. Google Maps. http://maps.google.com/.

googleTitle. Google to FCC. www.engadget.com/2015/01/01/google-letter-fcc-title-ii/.

Gorman, S. P. 2005. *Networks, Security and Complexity: The Role of Public Policy in Critical Infrastructure Protection*. London: Edward Elgar.

Hansen, A. Fosselie, A. Kvalbein, T. Cicic, and S. Gjessing. 2005. *Resilient Routing Layers for Network Disaster Planning*. ICN.

Hunter, P. 2008. Pakistan YouTube block exposes fundamental internet security weakness. *Computer Fraud & Security* 4:10–11.

ita. Example in Italy. http://leos.unipv.it/slides/lecture/Valente-LEOS-29-apr-2009-Slide_GARR.pdf.

Jacobson, V. and S. Deering. 1989. Traceroute.

Knight, S., H. X. Nguyen, N. Falkner, R. A. Bowden, and M. Roughan. 2011. *The Internet Topology Zoo*. IEEE JSAC.

Labovitz, C., S. Iekel-Johnson, D. McPherson, J. Oberheide, and F. Jahanian. 2010. Internet Inter-domain Traffic. ACM SIGCOMM.

level3networkmap. Level3 Network Map. http://maps.level3.com/default/.

lightower. Control and Flexibility with Unlimited Scale Through Lightower Dark Fiber Network. www.lightower.com/network-services/dark-fiber- service/#overview.

lightowerdc2. 2015. Lightower Acquires ColocationZone, Enterprise-Class Data Center in Chicago. www.lightower.com/company/news/press-releases/lightower-acquires-colocationzone-enterprise-class-data-center-in- chicago/#.V9yYazsgkdc.

lightowerdc1. 2016. Lightower Fiber Networks Acquires Datacenter101, Leading Data Center in Columbus, Ohio. www.lightower.com/company/news/press-releases/lightower-fiber-networks-acquires-datacenter101-leading-data-center-in-columbus-ohio/#.V9yX5Tsgkdc.

lux. Example in Luxembourg. http://ict.investinluxembourg.lu/ict/sites/ict.investin luxembourg.lu/files/Luxembourg_and_ICT_-_a_ snapshot.pdf.

Madhyastha, H. V., T. Isdal, M. Piatek, C. Dixon, T. Anderson, A. Krishnamurthy, and A. Venkataramani. 2006. *iPlane: An Information Plan for Distributed Services*. USENIX OSDI.

Mao, Z. M., J. Rexford, J. Wang, and R. H. Katz. 2003. *Towards an Accurate AS-Level Traceroute Tool*. ACM SIGCOMM.

McGrattan, K. and A. Hamins. 2006a. Numerical simulation of the Howard Street tunnel fire. *Fire Technology* 42(4): 273–281.

Miller, R. 2006. Hurricane Katrina: Communications and infrastructure impacts. *Technical Report, National Defense University, Fort McNair, DC*.

nationalAtlas. The National Atlas. http://nationalatlas.gov/.

nsfnet. NSFNet. http://en.wikipedia.org/wiki/National_Science_Foundation_Network.

openaccess. Open Access. www.itu.int/wsis/c2/docs/ 2008-May-19/mdocs/Jagun%20 et\%20al\%20_Telecom\%20Africa\%202008_final\%20_2_.pdf.

PFNetAndTouchAmerica1. www.prnewswire.com/news-releases/pfnet-and-touch-america-exchange-fiber-and-conduit-in-a-major-agreement-to-expand-the html.

PFNetAndTouchAmerica2. www.fiberopticsonline.com/doc/el-paso-energy-touch-america-pfnet-plan-1500-0001.

PFNetstory. www.fiberopticsonline.com/doc/pfnet-to-build-cross-country-fiber-optic-netw-0001.

pipeAll. Pipeline 101. www.pipeline101.org/ where-are-pipelines-located.

Pipestory. www.naturalgasintel.com/articles/ 50616-pipe-companies-running-telecom-cable.

Roughan, M., S. J. Tuke, and O. Maennel. 2008. Bigfoot, sasquatch, the yeti and other missing links: what we don't know about the AS graph. Proceedings of ACM Internet Measurement Conference.

routeviews. Route Views Project. www.routeviews.org/.

Shavitt, Y. and E. Shir. 2005. DIMES: Let the Internet Measure Itself. ACM SIGCOMM CCR.

Sherwood, R. and N. Spring. 2006. Touring the Internet in a TCP Sidecar. ACM SIGCOMM IMC.

Sherwood, R., A. Bender, and N. Spring. 2008. *Discarte: A Disjunctive Internet Cartographer*. ACM SIGCOMM.

sidePad. SidePad. www.sidepad.gov/.

Singla, A., B. Chandrasekaran, P. B. Godfrey, and B. Maggs. 2014. *The Internet at the Speed of Light*. ACM Hotnets.

slo. Example in Slovakia. http://sitel.cz/public/upload/other/SITEL_presentation__ 130913en.pdf.

Sommers, J., P. Barford, and B. Eriksson. 2011. *On the Prevalence and Characteristics of MPLS Deployments in the Open Internet*. ACM SIGCOMM IMC.

soovle. Soovle. www.soovle.gov/.

Spinelli, L., M. Crovella, and B. Eriksson. 2013. *AliasCluster: A Lightweight Approach to Interface Disambiguation*. Global Internet Symposium.

Spring, N., R. Mahajan, and D. Wetherall. 2002. Measuring ISP topologies with Rocketfuel. ACM SIGCOMM.

tataafrica. Tata – Africa. www.internet2.edu/presentations/fall09/20091005-southasia-poppe.pdf.

empAttack. 2004. Report of the Commission to Assess the Threat to the United States from Electromagnetic Pulse (EMP) Attack: Critical National Infrastructures. Critical National Infrastructures Report.

VelocitaStory1. www.billingworld.com/articles/2001/06/rights-of-way-drive-velocita-from-start-up-to-non.aspx.

VelocitaStory2. www.nprg.com/Company/ 1602-velocita-communications-inc.

Wang, H., R, Yang, P. H. Liu, J. Wang, A. Gerber, and A. Greenberg. 2007. Reliability as an Interdomain Service. ACM SIGCOMM.

windstreamdc. 2015. *Windstream to Sell Data Center Business for $575M*. www.datacenterknowledge.com/archives/2015/10/19/windstream-to-sell-data-center-business-for-575m/.

zayo. 2012. *Zayo group completes acquisition of Abovenet*. www.zayo.com/news/zayo-group-completes-acquisition-of-abovenet-3/.

Zhang, B., R. Liu, D. Massey, and L. Zhang. 2005. Collecting the Internet AS-Level Topology. ACM SIGCOMM CCR.

Zhang, Y., R. Oliveira, Y. Wang, S. Su, B. Zhang, J. Bi, H. Zhang, and L. Zhang. 2011. A Framework to Quantify the Pitfalls of Using Traceroute in AS-level Topology Measurement. IEEE JSAC.

Zhu, Y., A. Bavier, N. Feamster, S. Rangarajan, and J. Rexford. 2008. UFO: A Resilient Layered Routing Architecture. ACM SIGCOMM CCR.

Zmijewski, E. 2008. *Mediterranean Cable Break*. Renesys Blog, January.

5   The history of broadband

Elizabeth Mack

The internet was born in the Cold War era as part of a government project to develop distributed communications systems for the military that could survive a nuclear bomb (Abbate 1999). The ancestor of today's internet was born on April 30, 1995, when the government ceased operations of the old NSFNET backbone and handed over this infrastructure to private companies (Abbate 1999). In the first days of the internet, people used modems to connect to the internet over telephone lines on a per-use basis at speeds between 33.6 and 56 kbps (Gorsche et al. 2014). To put these speeds in perspective, it would take 1 day, 18 hours and 36 minutes to download a 1 gigabyte file (DTC 2018).

In the early 2000s broadband started to become available to the public. Broadband internet connections are always on, which means users do not have to dial into the network each time they wish to get online (Savage and Waldman 2005). Broadband connections are also much faster than dial-up, which has expanded the range of activities that can be conducted online. This expansion of possibilities for online activities and devices means that broadband has had tremendous impacts on the economy and society. Given these impacts, the purpose of this chapter is to provide a brief review of broadband technologies and the body of research that has examined the effects of these transformative technologies to date.

Broadband platforms

Broadband service is delivered to customers over a variety of platforms, whose popularity varies geographically since the rollout of broadband is an inherently local phenomenon, that depends upon providers' use of existing infrastructure and the policy context for deploying these technologies (Grubesic and Mack 2015). Aside from these overarching factors, there are important differences in these technologies in terms of their reliability, physical security, and speeds (Gorsche et al. 2014). For an extended technical description of these platforms that covers issues of reliability, physical security, and speeds, readers are referred to Gorshe et al. (2014) and Grubesic and Mack (2015).

In the early years, broadband provided over digital subscriber lines (DSL) was popular, which is broadband service delivered over telephone lines

(Gorshe et al. 2014). There are many varieties of DSL but it can be divided according to the asymmetry or symmetry of download and upload speeds (FCC 2014). Asymmetric digital subscriber lines (ADSL) were rolled out in 1995 and are the most popular form of broadband globally (Gorshe et al. 2014). This type of broadband is branded as asymmetric because the down-stream or download rates are much faster than the upload rates (Gorsche et al. 2014). The "s" in SDSL refers to the symmetry in upload and download speeds. This means a user can download and upload (send) data at the same speeds. These connections are important for users that not only download data from the internet but also send a lot of data; videoconferencing is one internet use that requires symmetrical connections (FCC 2014).

A promising alternative to DSL is broadband provided by cable television providers over hybrid fiber coaxial cable (HFC), which provides faster speeds and service at greater distances from a distribution hub (Grubesic and Mack 2015). Another, albeit more expensive alternative to DSL is fiber broadband. This platform is an attractive alternative because of its "unlimited bandwith capabilities" (Grosche et al. 2014), which means the speeds offered by fiber optic technology far exceed those provided by either DSL or cable (FCC 2014). This is because electronic signals are converted to light and transmitted via fine, hair like glass fibers (FCC 2014).

Broadband via satellite is a means of providing service to remote locations (FCC 2014) and is used to provide service in underserved or unserved areas. A downside to this platform is that download and upload speeds are slower than DSL and cable (FCC 2014). Other issues with this platform include variations in service quality related to a customer's line of sight with the sat-ellite and weather-related service interruptions (FCC 2014). Developments in next generation technology such as high-throughput and non-geostationary orbit satellites could ameliorate these issues (BCSD 2017).

The term "wireless" broadband covers a wide range of technologies. This type of connection can be categorized in several ways: short vs. long-range technologies and fixed and mobile technologies (Grubesic and Mack 2015). Fixed wireless connections transmit data from a fixed location and "often require a direct line of sight between transmitter and receiver" (FCC 2014). The extent that fixed connections provide short or long-range service depends on the type of technology used; Wi-Fi networks use short-range technologies to provide wireless service while technologies like WiMAX are longer range technologies to provide community-wide service (Grubesic and Mack 2015).

The ability to communicate and navigate the web on mobile phones is made possible by cellular technologies often referred to as "mobile broad-band." The growth in mobile telephony, particularly in the developing world, is a means of extending broadband-like services to people in the palm of their hand. Over the years, cellular service has undergone tremendous changes. Cellular networks are arranged according to a honeycomb where the size of the coverage area or comb is dictated by "the transmission power of radio transmitters and receivers" (Grubesic and Mack 2015, 53). In 1998,

3G networks were introduced and was the generation of cellular service to which the phrase "mobile broadband" was applied; this generation of technology made it possible to use cell phones for video calls and web-browsing (Fendelman 2018). 2008 marked the arrival of 4G networks, which expanded the functionality of mobile phones to include "gaming services, HD mobile TV, video conferencing, 3D TV" among other uses (Fendelman 2018). 5G is expected to arrive in 2020 and is anticipated to be 1000 times faster than 4G mobile speeds (Dean 2014). The reason 5G will be much faster is because it uses new receiver technology and parts of the spectrum not in use presently (Reynolds 2018). Predictions suggest speeds will be on par with desktop broadband, which has prompted speculation that 5G may offer a less expensive solution to fixed broadband services in the next ten years (Dean 2014).

Broadband-related innovations

Broadband is a general-purpose technology (GPT) because of its enabling capacity, which makes "innovational complementarities" possible in downstream sectors (Bresnahan and Tratjenberg 1995). Innovations in downstream sectors explain the widespread economic and societal impacts of broadband. While the innovations made possible by high-speed broadband connections are numerous, arguably, three of the more important innovations that have helped popularize the internet are web browsers, Web 2.0, and smartphones.

Until the development of hypertext markup language (HTML) in 1990 by Tim Berners-Lee (Zimmerman and Empsak 2017) and subsequent development of web browsers, only the most technically competent people could use the internet. Web browsers helped organize information on the web to make it accessible to the public. In the early 1990s, research attention in the computer science community focused on the development of a series of browsers. In 1993, the first web browser, Mosaic, was developed (Lasar 2011) and was followed by Netscape in 1994 (Heisler 2014) and Internet Explorer in 1995 (Daly and Winkler 2013). In 1996, Google released the first version of its search engine (Kollewe 2008). Over a decade later in 2008, Google Chrome launched (Kollewe 2008).

While browsers made information on the web accessible to people, in the early 2000s, the arrival of the Web 2.0, also referred to as the "participatory web" (Blanck and Reisdorf 2012) allowed users to contribute to online content and use the internet to make social connections (O'Reilly 2005; Cogburn and Espinoza Vazquez 2011). This change in webpage development and web content, coupled with the development of the smartphone, changed the digital landscape forever. While personal assistants and antecedents to the smartphone such as the Blackberry had been available for quite some time, the era of the smartphone could be said to have started in 2007 when Apple sold its first iPhone (Heathman 2017).

An invention around this time that enhanced the popularity of smartphones was the iTunes store which launched in 2003 (Chen 2003). In 2008, the Apple app store launched, enabling users to personalize their phones (Apple 2008). The launch was a fantastic success; in its first three days of operation, customers downloaded over 10 million applications (Apple 2008). Today people spend several hours on their phone each day. In the U.K., recent statistics show that young adults spend 4 hours each day online and adults spend 2 hours and 49 minutes on their phones (Hymas 2018). In the United States, a Nielsen (2018) survey finds Americans spend a significant portion of their day online, particularly on smartphones; the average for all adults is about 21% of the day.

Economic benefits

Broadband's status as a general-purpose technology (GPT) means that adopting sectors and regions have been endowed with a variety of economic benefits (Gruber, Hätonen, and Koutroumpis 2014). To understand these impacts, research has examined several benefits of the availability and adoption of broadband using several indicators including economic growth, employment, productivity, and business presence.

Economic growth

Studies of broadband and economic growth across different countries and time periods have produced varied results. A recent study of U.S. counties found no link between broadband and economic growth, as specified by employment, personal income, or earnings (Ford 2018). Many more studies have found positive impacts however (Ford and Koutsky 2005; Koutroumpis 2009; Czernich 2011; Kongaut and Bohlin 2017). For example, an OECD study over an 11-year time period (1996–2007) found broadband positively contributed to per capita income growth (Czernich et al. 2011).

In terms of the exact contribution that broadband makes to gross domestic product (GDP), an older study of the United States using data between 1999 and 2006 estimated that broadband accounted for 72% of internet-generated GDP (Greenstein and McDevitt 2011). They also found that broadband contributed 40–50% of total GDP. A study of mobile 3G/4G broadband between 2001 and 2014 also uncovered a positive linkage with economic growth, but found that growth was achieved through mechanisms such as prices, entrepreneurship, and R&D activity (Ghosh 2017). In the EU, studies suggest that the positive economic impacts of broadband justify additional spending on infrastructure expansion (Gruber, Hätonen, and Koutroumpis 2014). Combined, these studies provide economic support for the expansion and improvement of broadband infrastructure.

That said, there are important nuances to this relationship that are important to recognize. For example, in a study of OECD countries, Kongaut and Bohlin (2017) found that the association between broadband speed and GDP is stronger for low-income countries. A study by Whitacre, Gallardo, and Strover (2014) highlighted the importance of broadband adoption over mere availability in producing positive economic outcomes. Specifically, broadband adoption produced positive economic outcomes in terms of reduced unemployment and increased income growth.

Employment benefits

Aside from broadband's links with economic growth, studies have uncovered several other economic benefits of the availability of broadband internet connections. The Whitacre, Gallardo, and Strover (2014) discussed above is one of several studies investigating broadband's impacts on employment (Crandall, Lehr, and Litan 2007; Jayakar and Park 2013). This is a harder question to answer because broadband may destroy old jobs but also create new ones. Consequently, studies find that broadband does not uniformly increase employment levels, but contributes to employment gains in particular industries (Shideler, Badasyan, and Taylor 2007). Specifically, broadband was weakly and positively associated with growth in services and indirectly impacted employment growth in supporting industries such as a health care, retail, and professional, scientific, and technical services. This same study also suggested broadband was likely to reduce employment in accommodation and food services because broadband was a substitute for work functions in that industry. Similarly, Kolko (2012) suggested that technology-oriented industries experienced more employment gains from broadband availability than other industries.

Alternatively, a study of German municipalities in the early 2000s found no impact of broadband on municipal unemployment rates (Czernich 2014). The author goes on to state that there are potential labor market impacts of broadband not measured in his assessment which focused on unemployment. Possible labor market impacts include increased efficiencies in matching employed persons with jobs and increases in the local labor supply from outside the region to take advantage of alternative work setups (telecommuting or freelance work) that are enabled by broadband (Czernich 2014). This study speaks to earlier work that suggested the internet can promote labor market mobility into better paying jobs for less educated people (Mossberger, Tolbert, and McNeal 2008).

Productivity impacts

Aside from employment impacts, another important and related consideration is the impact of broadband on the workforce in terms of productivity gains. Despite the large body of work about ICTs and productivity, the articles dedicated to broadband are surprisingly sparse. Yet, there are some articles that evaluate broadband specifically (Majumdar, Carare, and Chang 2009; Mack and Faggian 2013; Colombo, Croce, and Grilli 2013). At the firm level, Majumdar et al. (2009) found that broadband deployment positively impacts firm productivity, but that these benefits are not necessarily universal.

An examination of small and medium sized enterprises (SMEs) finds that positive benefits are contingent upon several factors including the industry of the firm, relevance of adopted applications to business processes, and whether or not adoption is accompanied by organizational changes (Colombo, Croce, and Grilli 2013). At the regional level, the presence of productivity impacts is dependent upon levels of human capital. Specifically, positive productivity impacts are only possible when workers with high levels of knowledge and skills are present (Mack and Faggian 2013). Based on this sample of studies, the mere presence of broadband does not produce positive productivity impacts; complementary factors are necessary.

Broadband and business presence

In addition to several of the economic benefits described above, researchers have also examined the extent that broadband availability impacts the presence of businesses generally and also businesses in specific sectors. Early work on this topic found that the urban bias in broadband provision may impact the ability of firms to relocate to lower cost non-urban locations (Mack and Grubesic 2009). Later work found evidence of firm specific impacts (Kandilov and Renkow 2010; Mack 2015). For example, the broadband loan program (BLP) from the United States Department of Agriculture (USDA) positively impacted employment and payroll in some sectors such as transportation and warehousing, but negatively impacted others such as the information sector via a possible substitution effect (Kandilov and Renkow 2010). In an evaluation of more nuanced definitions of the urban hierarchy in addition to industry impacts, Mack (2015) found positive linkages with broadband availability and business presence for several industries (agriculture, manufacturing, and knowledge). The same study found these industry linkages were strongest in intermediate portions of the urban hierarchy such as adjacent counties to metropolitan areas and counties within micropolitan areas. More recently, research has uncovered long-lasting impacts of broadband availability on business presence. A county-level study of this relationship found that locales with broadband in the early years had higher levels of business growth; business growth was also evident from spatial spillovers in broadband availability in core areas to nearby locations (Mack and Wentz 2017).

Social benefits

Aside from economic benefits, broadband is also recognized as important to social transformation. Research in the MENA region found that broadband enabled social networking tools promote social development in the form of enhanced civic engagement and social inclusion (Gelvanovska, Rogy, and Rossotto 2014). Of the social benefits associated with broadband, political participation, and civic engagement have received considerable attention. In

2008, Mossberger, Tolbert, and McNeal coined the phrase "digital citizenship" to refer to online participation in society via internet connections. In their examination of digital citizenship, they find that internet use increases civic engagement and the probability of voting. It can also help underrepresented minorities, particularly African Americans and Latinos, enhance household income. More recently, political participation via Web 2.0 platforms has received a lot of research attention (Jackson and Lilleker 2009).

Specifically, studies of online political participation have found that online engagement translates to offline engagement (Conroy, Feezell, and Guerrero, 2012). For example, a European study of political participation via Facebook (Vesnic-Alujevic 2012) found positive correlations between online and off-line participation. Similarly, a study of "cyber participation" or political par-ticipation via social networking outlets found that "cyber-participation" is positively associated with voter turnout (Steinberg 2015). Visible examples of the power of Web 2.0 technologies to mobilize people include the use of social media platforms such as Facebook, Twitter, and YouTube in the his-toric election of Barack Obama to the U.S. presidency in 2008 (Cogburn and Espinoza-Vazquez 2011), and social media facilitated protests associated with the Arab Spring (Ghosh 2017).

A social benefit of broadband that can also translate into economic outcomes is the ability to leverage the space-time flexibility of internet connections to educate and train people (BCSD 2017). Since the early 2000s, studies have analyzed various aspects of distance education (Grubesic 2003; Bolliger and Halupa 2018). One notable example of online learning enabled by broadband technology are Massive Online Open Courses or MOOCs. Stephen Downes and George Siemens offered the first MOOC in 2008 called "Connectivism and Connective Knowledge," at the University of Manitoba. The course utilized several Web 2.0 tools including Facebook groups, Wiki pages, and blogs to communicate and engage with students (Marques 2013). Since this first course offering, some sources estimate global enrollment in MOOCS to be about 78 million at over 800 institutions and 9,400 courses (Lederman 2018).

Recent studies of distance education highlight that it serves as a com-plement to in-person education. This research highlights that over 50% of students enrolled in an online course (52.8%) also took an on-campus course at the same institution (Seaman, Allen, and Seaman 2018). This same study also found high levels of concentration in enrollments among distance edu-cation students; 5% of all institutions account for 50% of distance education enrollment (Seaman, Allen, and Seaman 2018).

Digital divide

While broadband has undoubtedly benefitted millions of people around the globe, the benefits derived from high-speed internet connections are unequally distributed and these inequities are manifested geographically, socially, and economically. From a global perspective, over 50% of people around the world do not have broadband internet service (BCSD 2017). Many of these people are located in the developing world. For example, a study of Middle Eastern and North African countries noted that many of them (10 out of 19) are in the initial stages of both fixed and mobile broadband deployment (Gelvanoska et al. 2014). In terms of mobile access, large segments of the global community do not have access to higher speed connections; only 76% and 43% of people have access to 3G and 4G mobile connections respectively (BCSD 2017).

The phrase "digital divide" refers to uneven access to the most recent group of popularly used information and communications technologies (ICTs). Historical uses of this term have been used with respect to rates of computer ownership (NTIA 1995) and dial-up internet access (NTIA 2000). Over the years, however, use of this term became common in reference to broadband (Prieger 2003; Prieger and Hu 2008), and this phrase now encompasses a complex range of access and use issues. That said, since the early days of broadband availability, the research community has devoted a lot of attention to understanding who, what, where, and when people obtained access to the internet.

Early research on broadband access analyzed a variety of factors, including socio-economic (Prieger 2003; Prieger and Hu 2008), demographic (Grubesic 2004; Prieger and Hu 2008), and geographic dimensions of the digital divide (Strover 2001; Grubesic 2004). For example, Prieger and Hu (2008) find racial differences in DSL adoption even after controlling for income and education, and attribute these differences to a lack of online content for these populations and a lack of time to participate in online activities. A look at the geodemographic correlates of broadband access highlighted groups of people with specific blends of socio-economic and demographic characteristics that had improved levels of broadband access over other groups (Grubesic 2004). A large body of research also highlights geographic unevenness to broadband availability, with a large focus on urban/rural disparities (Strover 2001; LaRose et al. 2007; Whitacre et al. 2014). On a global scale, there are also differences in broadband adoption between developed and developing countries (Andres et al. 2010).

While global and urban/rural comparisons of broadband access highlight persistent geographic dimensions, more recently, this divide has become much more nuanced and multifaceted. Studies have suggested that the divide is no longer a case of the haves and have nots, but is multi-dimensional and involves access to particular platforms, speed of access, provider choice, and quality of service (Grubesic 2015). Recently, studies have focused on the speed of broadband connections (Riddlesden and Singleton 2014; Mack 2014; BCSD 2017) as well as questions of use (Savage and Waldman 2005; Lenhart et al. 2010; Mack et al. 2017). Differences in the use of broadband are an important but understudied dimension of the digital divide. This dimension is critical to unravel if we are to understand the diffusion or lack thereof of the economic benefits associated with internet use. An international study of internet connectivity highlighted that lack of skills was one of four reasons for lack of internet connectivity (BCSD 2017).

One of the other "dark sides" of broadband research is the lack of data available to the research and policy community. A number of studies have noted several inadequacies with public sources of data (Lehr et al. 2006; Greenstein 2007; Xiao and Orazem 2011). Substituting private data sources can be an expensive and not necessarily better alternative to overcoming issues with public data given a lack of transparency about data collection and processing procedures (Malone and Nguyen 2018). For reviews of these specific data limitations readers are directed to Ford (2011), Grubesic (2012), and Grubesic and Mack (2015).

One of the notable limitations of broadband data is a lack of public data about use. It is with respect to broadband use that more work is needed given the downsides to internet use. Studies have noted problems with internet addiction (Young 1998; Young and Rogers 1998) and linkages between depression and internet use (LaRose et al. 2001; Selfhout et al. 2009). More recent studies of negative aspects of internet use are focusing on the idea of technostress, which stems from the inability to cope with computer related technologies (Nimrod 2018).

Aside from the negative personal impacts of technology overuse, there are larger societal issues associated with internet use including surveillance by the government and employers (Cammaerts 2008) and privacy issues associated with online activities (Mineo 2017). The sharing of data about Facebook users is a recent example of these concerns (Singer 2018). Another major concern is the use of broadband connections by terrorist organizations to promote their activities with online propaganda (Kohlmann 2006) and to commit acts of cyber terrorism (Furnell and Warren 1999). Related problematic aspects of online activities enabled by broadband internet connections are cybersecurity and cybercrime (Furnell 2002; Wall 2007).

Future of broadband

Despite these problems, broadband internet connections have become part of our everyday lives. Due to these tremendous impacts, the goal of this chapter was to provide a brief history of various broadband technologies and highlight how, as a general-purpose technology, broadband has transformed the economy and society. These transformations have a dual nature however and endow both advantages and disadvantages. As innovation continues at an increasingly rapid pace, it is important to keep this duality in mind, as new applications for this transformative set of technologies are developed.

References

Abbate, J. 1999. *Inventing the Internet*. Cambridge, MA: MIT Press.

Adams, P., Farrell, M., Dalgarno, B., and Oczkowski, E. 2017. Household adoption of technology: The case of high-speed broadband adoption in Australia. *Technology in Society 49*:37–47.

Aker, J.C., and I.M. Mbiti. 2010. Mobile phones and economic development in Africa. *Journal of Economic Perspectives 24*(3):207–232.

Andrés, L., Cuberes, D., Diouf, M., and Serebrisky, T. 2010. The diffusion of the internet: A cross-country analysis. *Telecommunications Policy 34*(5–6):323–340.

Apple. 2008. iPhone App Store Downloads Top 10 Million in First Weekend. Press Release July 14, 2008. www.apple.com/newsroom/2008/07/14iPhone-App-Store-Downloads-Top-10-Million-in-First-Weekend/.

BCSD. 2017. The State of Broadband: Broadband catalyzing sustainable development. Broadband Commission for Sustainable Development. www.itu.int/dms_pub/itu-s/opb/pol/S-POL-BROADBAND.18-2017-PDF-E.pdf.

Blank, G., and Reisdorf, B. 2012. The participatory web: A user perspective on Web 2.0. *Information, Communication and Society 15*(4):537–554.

Bolliger, D., and Halupa, C. 2018. Online student perceptions of engagement, transactional distance, and outcomes. *Distance Education 39*(3): 299–316.

Bresnahan, T., and Trajtenberg, M. 1995. General purpose technologies: 'Engines of growth'? *Journal of Econometrics 65*(1):83–108.

Cammaerts, B. 2008. Critiques on the participatory potentials of Web 2.0. *Communication, Culture & Critique 1*(4):358–377.

Chen, B.X. 2003. Apple opens iTunes store. *Wired* (April 28). www.wired.com/2010/04/0428itunes-music-store-opens/.

Cogburn, D., and Espinoza-Vasquez, F. 2011. From networked nominee to networked nation: Examining the impact of Web 2.0 and social media on political participation and civic engagement in the 2008 Obama campaign. *Journal of Political Marketing 10*(1–2):189–213.

Colombo, M., Croce, A., and Grilli, L. 2013. ICT services and small businesses' productivity gains: An analysis of the adoption of broadband internet technology. *Information Economics and Policy 25*(3):171–189.

Conroy, M., Feezell, J., and Guerrero, M. 2012. Facebook and political engagement: A study of online political group membership and offline political engagement. *Computers in Human Behavior 28*(5):1535–1546.

Crandall, R., Lehr, W., and Litan, R. 2007. The effects of broadband deployment on output and employment: A cross-sectional analysis of US data. http://citeseerx.ist.psu.edu/viewdoc/download?doi=10.1.1.622.9097&rep=rep1&type=pdf.

Czernich, N. 2014. Does broadband internet reduce the unemployment rate? Evidence for Germany. *Information Economics and Policy 29*:32–45.

Czernich, N., Falck, O., Kretschmer, T., and Woessmann, L. 2011. Broadband infrastructure and economic growth. *Economic Journal 121*(552):505–532.

Daly, J. and C. Winkler. 2013. A visual history of Internet Explorer. *State Tech* (Aug. 16). https://statetechmagazine.com/article/2013/08/visual-history-Internet-explorer.

Dean, J. 2014. 4G vs 5G Mobile Technology. *Raconteur* (Dec. 7). www.raconteur.net/technology/4g-vs-5g-mobile-technology.

DTC. 2018. Download Time Calculator. https://downloadtimecalculator.com/.

FCC. 2014. Types of Broadband Connections. Federal Communications Commission. www.fcc.gov/general/types-broadband-connections.

Fendelman, A. 2018. 1G, 2G, 3G, 4G, & 5G Explained. *Lifewire* (Oct. 2). www.lifewire.com/1g-vs-2g-vs-2-5g-vs-3g-vs-4g-578681.

Ford, G. 2018. Is faster better? Quantifying the relationship between broadband speed and economic growth. *Telecommunications Policy 42*(9):766–777.

Ford, G.. 2011. Challenges in using the National Broadband Map's data. www.researchgate.net/publication/228199292_Challenges_in_Using_the_National_Broadband_Map's_Data.

Ford, G., and Koutsky, T. 2005. Broadband and economic development: A municipal case study from Florida. *Review of Urban & Regional Development Studies 17*(3):216–229. Melbourne, Australia: Blackwell Publishing Asia.

Furnell, S. 2002. *Cybercrime: Vandalizing the Information Society.* Boston, MA: Addison-Wesley.

Furnell, S. M., and Warren, M. 1999. Computer hacking and cyber terrorism: The real threats in the new millennium? *Computers & Security 18*(1):28–34.

GAO. 2018. Wireless Broadband Spectrum Management. U.S. Government Accountability Office. www.gao.gov/key_issues/wireless_broadband_spectrum_management/issue_summary#t=0.

Gelvanovska, N., Rogy, M., and Rossotto, C. 2014. *Broadband Networks in the Middle East and North Africa: Accelerating High-speed Internet Access.* Washington, DC: World Bank.

Ghosh, S. 2017. Broadband penetration and economic growth: Do policies matter? *Telematics and Informatics 34*(5):676–693.

Gorsche, S., Raghavan, A.R., Starr, T., and Galli, S. 2014. *Broadband Access: Wireline and Wireless-Alternatives for Internet Services.* New York: Wiley.

Greenstein, S. 2007 Data constraints and the internet economy: Impressions and imprecision. *NSF/OECD Meeting on Factors Shaping the Future of the Internet.* www.oecd.org/dataoecd/5/54/38151520.pdf.

Greenstein, S., and McDevitt, R. 2011. The broadband bonus: Estimating broadband internet's economic value. *Telecommunications Policy 35*(7):617–632.

Grubesic, T. 2003. Inequities in the broadband revolution. *Annals of Regional Science, 37*(2):263–289.

Grubesic, T. 2004. The geodemographic correlates of broadband access and availability in the United States. *Telematics and Informatics 21*(4):335–358.

Grubesic, T. 2012. The U.S. national broadband map: Data limitations and implications. *Telecommunications Policy 36*(2):113–126.

Grubesic, T. 2015. The broadband provision tensor. *Growth and Change 46*(1):58–80.

Grubesic, T., and Horner, M. 2006. Deconstructing the divide: extending broadband xDSL services to the periphery. *Environment and Planning B: Planning and Design 33*(5):685–704.

Grubesic, T., and Mack, E. 2015. *Broadband Telecommunications and Regional Development.* London: Routledge.

Grubesic, T., and Murray, A. 2002. Constructing the divide: Spatial disparities in broadband access. *Papers in Regional Science 81*(2):197–221.

Gruber, H., Hätönen, J., and Koutroumpis, P. 2014. Broadband access in the EU: An assessment of future economic benefits. *Telecommunications Policy 38*(11):1046–1058.

Heathman, A. 2017. The Smartphone turns 25. www.verdict.co.uk/smartphone-invented-25-years/.

Heisler, Y. 2014. A visual history of Netscape Navigator. *Network World* (Oct. 14). www.networkworld.com/article/2833526/software/a-visual-history-of-netscape-navigator.html#slide10.

Hymas, C. 2018. A decade of smartphones: We now spend an entire day every week online. *The Telegraph* (Aug. 2). www.telegraph.co.uk/news/2018/08/01/decade-smartphones-now-spend-entire-day-every-week-online/.

Jackson, N. A., and Lilleker, D. 2009. Building an architecture of participation? Political parties and Web 2.0 in Britain. *Journal of Information Technology & Politics* 6(3–4):232–250.

Jayakar, K., and Park, E. 2013. Broadband availability and employment: An analysis of county-level data from the National Broadband Map. *Journal of Information Policy* 3:181–200.

Kandilov, I., and Renkow, M. 2010. Infrastructure investment and rural economic development: an evaluation of USDA's broadband loan program. *Growth and Change* 41(2):165–191.

Kohlmann, E. 2006. The real online terrorist threat. *Foreign Affairs*:115–124.

Kolko, J. 2012. Broadband and local growth. *Journal of Urban Economics* 71(1):100–113.

Kollewe, J. 2008. Google timeline: A 10 year anniversary. *The Guardian* (Sept. 5). www.theguardian.com/business/2008/sep/05/google.google.

Kongaut, C., and Bohlin, E. 2017. Impact of broadband speed on economic outputs: An empirical study of OECD countries. *Economics and Business Review* 3(2):12–32.

Koutroumpis, P. 2009. The economic impact of broadband on growth: A simultaneous approach. *Telecommunications Policy* 33(9):471–485.

LaRose, R., Eastin, M., and Gregg, J. 2001. Reformulating the internet paradox: Social cognitive explanations of internet use and depression. *Journal of Online Behavior* 1(2):1092–4790.

LaRose, R., Gregg, J., Strover, S., Straubhaar, J., and Carpenter, S. 2007. Closing the rural broadband gap: Promoting adoption of the internet in rural America. *Telecommunications Policy* 31(6–7):359–373.

Lasar, M. 2011. Before Netscape: The forgotten Web browsers of the early 1990s. *Ars Technica* (Oct. 11). https://arstechnica.com/information-technology/2011/10/before-netscape-forgotten-web-browsers-of-the-early-1990s/.

Lederman, D. 2018. MOOCs: Fewer new students, but more are paying. *Inside Higher Ed* (Feb. 14). www.insidehighered.com/digital-learning/article/2018/02/14/moocs-are-enrolling-fewer-new-students-more-are-paying-courses.

Lehr, W. H., Osorio, C., Gillett, S., and Sirbu, M. 2006. Measuring broadband's economic impact. MIT Engineering Systems Division Working Paper ESD-WP-2006-02. https://dspace.mit.edu/bitstream/handle/1721.1/102779/esd-wp-2006-02.pdf?sequence=1.

Lenhart, A., Purcell, K., Smith, A., and Zickuhr, K. 2010. *Social Media & Mobile Internet Use among Teens and Young Adults*. Millennials. Pew Internet & American Life Project.

Mack, E. 2014. Broadband and knowledge intensive firm clusters: Essential link or auxiliary connection? *Papers in Regional Science* 93(1):3–29.

Mack, E. 2015. Variations in the broadband-business connection across the urban hierarchy. *Growth and Change 46*(3):400–423.

Mack, E., and Faggian, A. 2013. Productivity and broadband: The human factor. *International Regional Science Review 36*(3):392–423.

Mack, E., and Grubesic, T. 2009. Broadband provision and firm location in Ohio: An exploratory spatial analysis. *Tijdschrift voor Economische en Sociale Geografie 100*(3):298–315.

Mack, E., and Wentz, E. 2017. Industry variations in the broadband business nexus. *Socio-Economic Planning Sciences 58*:51–62.

Mack, E. A., Marie-Pierre, L., and Redican, K. 2017. Entrepreneurs' use of internet and social media applications. *Telecommunications Policy 41*(2):120–139.

Majumdar, S., Carare, O., and Chang, H. 2009. Broadband adoption and firm productivity: Evaluating the benefits of general purpose technology. *Industrial and Corporate Change 19*(3):641–674.

Malone, C., and Nguyen, M. 2018. We used broadband data we shouldn't have – Here's what went wrong. *FiveThirtyEight.* https://fivethirtyeight.com/features/we-used-broadband-data-we-shouldnt-have-heres-what-went-wrong/?mc_cid=2680ac13ec&mc_eid=558f45e107.

Marques, J. 2013. A short history of MOOCs and distance learning. *MOOC News & Reviews.* http://moocnewsandreviews.com/a-short-history-of-moocs-and-distance-learning/.

Mineo, L. 2017. On internet privacy, be very afraid. *Harvard Law Today* (Aug. 25). https://today.law.harvard.edu/internet-privacy-afraid/.

Mossberger, K., Tolbert, C., and McNeal, R. 2008. *Digital Citizenship. The Internet, Society, and Participation.* Cambridge, MA: MIT Press.

Nielsen. 2018. Time Flies: U.S. Adults Now Spend Nearly Half a Day Interacting with Media. July 31, 2018. www.nielsen.com/us/en/insights/news/2018/time-flies-us-adults-now-spend-nearly-half-a-day-interacting-with-media.print.html.

Nimrod, G. 2018. Technostress: Measuring a new threat to well-being in later life. *Aging & Mental Health 22*(8):1080–1087.

NTIA. 1995. Falling Through the Net: A Survey of the "Have Nots" in Rural and Urban America. National Telecommunications and Information Administration (NTIA). www.ntia.doc.gov/ntiahome/fallingthru.html.

NTIA. 2000. Fall Through the Net: Towards Digital Inclusion. National Telecommunications and Information Administration. www.ntia.doc.gov/files/ntia/publications/fttn00.pdf.

O'Reilly, T. 2005. What is Web 2.0: Design patterns and business models for the next generation of software. http://oreilly.com/web2/archive/what-is-web-20.html.

Prieger, J. 2003. The supply side of the digital divide: Is there equal availability in the broadband internet access market? *Economic Inquiry 41*(2):346–363.

Prieger, J., and Hu, W.M. 2008. The broadband digital divide and the nexus of race, competition, and quality. *Information economics and Policy 20*(2):150–167.

Reynolds, M. 2018. What is 5G and when is it coming to the UK? WIRED explains. *WIRED UK* (Sept. 4). www.wired.co.uk/article/what-is-5g-launch-date-mobile-networks-uk.

Riddlesden, D., and Singleton, A. 2014. Broadband speed equity: A new digital divide? *Applied Geography 52*:25–33.

Savage, S., and Waldman, D. 2005. Broadband internet access, awareness, and use: Analysis of United States household data. *Telecommunications Policy 29*(8):615–633.

Seaman, J., Allen, I., and Seaman, J. 2018. Grade Increase: Tracking Distance Education in the United States. Babson Survey Research Group. https://onlinelearningsurvey.com/reports/gradeincrease.pdf.

Selfhout, M., Branje, S., Delsing, M., ter Bogt, T., and Meeus, W. 2009. Different types of internet use, depression, and social anxiety: The role of perceived friendship quality. *Journal of Adolescence 32*(4):819–833.

Shideler, D., Badasyan, N., and Taylor, L. 2007. The Economic Impact of Broadband Deployment in Kentucky. http://citeseerx.ist.psu.edu/viewdoc/download?doi=10.1.1.575.4879&rep=rep1&type=pdf.

Singer, N. 2018. What you don't know about how Facebook uses your data. *New York Times* (April 11). www.nytimes.com/2018/04/11/technology/facebook-privacy-hearings.html.

Steinberg, A. 2015. Exploring Web 2.0 political engagement: Is new technology reducing the biases of political participation? *Electoral Studies 39*:102–116.

Strover, S. 2001. Rural internet connectivity. *Telecommunications Policy 25*(5):331–347.

Talbot, D. 2012. The spectrum crunch that wasn't. *MIT Technology Review* (Nov. 26). www.technologyreview.com/s/507486/the-spectrum-crunch-that-wasnt/.

Vesnic-Alujevic, L. 2012. Political participation and Web 2.0 in Europe: A case study of Facebook. *Public Relations Review 38*(3):466–470.

Wall, D. 2007. *Cybercrime: The Transformation of Crime in the Information Age.* Cambridge: Polity Press.

Whitacre, B., Gallardo, R., and Strover, S. 2014. Broadband's contribution to economic growth in rural areas: Moving towards a causal relationship. *Telecommunications Policy 38*(11):1011–1023.

Worstall, T. 2013. Will the mobile revolution mean we'll run out of spectrum? *Forbes* (Jan. 21) www.forbes.com/sites/timworstall/2013/01/21/will-the-mobile-revolution-mean-well-run-out-of-spectrum/#439e0fc930c4.

Xiao, M., and Orazem, P. 2011. Does the fourth entrant make any difference? Entry and competition in the early US broadband market. *International Journal of Industrial Organization 29*(5):547–561.

Young, K. S. 1998. Internet addiction: The emergence of a new clinical disorder. *Cyberpsychology & Behavior 1*(3):237–244.

Young, K., and Rogers, R. 1998. The relationship between depression and internet addiction. *Cyberpsychology & Behavior 1*(1):25–28.

Zimmermann, K., and J. Emspak. 2017. Internet History Timeline: ARPANET to the World Wide Web Live Science. www.livescience.com/20727-internet-history.html.

6 The mobile internet

Matthew Kelley

A decade after the initial release of the first iPhone (June 29, 2007) and the beta release of the first Android operating system (November 5, 2007) the mobile internet is at once a ubiquitous and increasingly foundational part of everyday life in the developed world (Galloway 2004; Wilson 2012; Kelley 2014b; Kitchin and Perng 2016). On the heels of the introduction of iPhones and the Android OS, a universe of possibility blossomed as the pervasive streaming data and passive internet-connectivity of mobile technology became threaded through the fabric of the infrastructures that compose contemporary society (Figure 6.1). From safer and more accessible transportation via location-based-services (Batty et al. 2012; Almobaideen et al. 2017; Mahtot and Thomas 2018), to "smarter" engagement with household utilities by way of the Internet of Things (Khan, Silva, and Han 2016; Wiltz 2016; Bhati, Hansen, and Chan 2017; Chi-Hua and Kuen-Rong 2018), by the late 2010s there were few aspects of contemporary society not touched or altered in some way by the mobile internet. Yet, for all the prominently placed marketing material and publicly celebrated releases of the next-best iteration of ever-smaller and ever-more-powerful mobile devices, the undercurrents of digital exclusion, privacy issues, and unchecked consolidations of technological power remain ever-present (van Dijk 2010; Leszczynski 2016; Kelley 2018).

The aim of this chapter is to engage with the mobile internet as an emergent technological phenomenon that is changing, and has changed, society in myriad ways – and to do so while considering, from a critical standpoint, the consequences of these changes. Thus, this chapter takes a three-part approach to discussing the mobile internet. First, it serves to chronicle the emergence of the mobile internet from the earliest (late 1990s) successes through early 2000s municipal wifi programs, to contemporary (late 2010s) advancements in high-speed mobile networks. Second, this chapter situates the mobile internet as one of several components of a technological revolution that has had far-reaching impacts on the socio-spatial landscapes of cities, towns and neighborhoods. And third, the mobile internet is situated in a critical and ongoing set of scholarly conversations that address the outcomes and consequences of new technologies in everyday life.

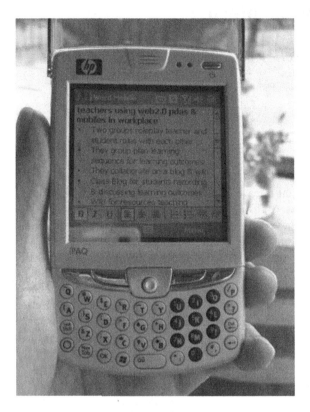

Figure 6.1 An example of internet access via a mobile phone.
Source: Wikicommons.

The emergence of the mobile internet

Web 1.0, Web 2.0, Web 3.0

Without digging into the technological underpinnings of the internet, its evolution can be mapped by drawing attention to and characterizing three somewhat distinct phases. Those phases have been widely referred to as Web 1.0, Web 2.0, and Web 3.0 in popular and scholarly literature and are used to designate paradigmatic shifts in the ways that the internet has functioned over time (O'Reilly 2005; Allen 2013; John 2013; Kelley 2014b; Kitchin and Perng 2016).

The first phase of the internet, Web 1.0, has roots in 1960s efforts by the U.S. Defense Advanced Research Projects Agency (DARPA) to improve communications among spatially disparate nodes in a network (research centers across the U.S. (Perry, Blumenthal, and Hinden 1988). DARPA efforts resulted in the ARPANET (Advanced Research Projects Agency Network), which was the earliest precursor to Tim Berners-Lee's development of the World Wide Web in the early 1980s at the CERN (European Council for Nuclear Research) research laboratory in Switzerland (Perry, Blumenthal, and Hinden 1988; Hillstrom 2005). From the 1960s until the early 1990s, Web 1.0 was characterized by experimental implementation among research universities and government agencies before gradually rolling out to the broader base of consumers in the mid-1990s. Used almost exclusively to consume information, Web 1.0 resembled an asynchronous digital store of content accessible through static html-based web pages and early email services. Users of Web 1.0 tended to consume content on personal computers that were tethered to a geographic location by virtue of the necessity of a physical modem (Hillstrom 2005; Zittrain 2006).

Though the term Web 2.0 was not first uttered until 2004, the internet had begun to take on its defining participative characteristics several years prior (Blank and Reisdorf 2012). By the early 2000s, blogs, wikis, and the roots of social media were supplanting static personal websites and one-way-interaction with web content. Napster and Wikipedia are two culturally significant examples of the origins of Web 2.0. Napster, first launched in 1999, was a peer-to-peer file sharing network that encouraged users to actively participate both in the consumption *and* production of content (primarily music stored as audio files). Wikipedia, launched in 2001, fashioned itself as an alternative to the classical encyclopedia by encouraging the production of user-generated, curated, and maintained content through Web 2.0 technologies. Of course, subsequent innovations such as MySpace and Facebook cemented in the popular imagination the participative nature of the internet and the seemingly irreplaceable role that social media can play in everyday life. It was also during the evolution of Web 2.0 that wireless internet connectivity was made more widely available (commonly referred to as wifi) in homes and, in some cases, across parts of cities (Jassem 2010).

Yet, as with Web 1.0, early Web 2.0 internet activity was still broadly tethered to particular geographic locations (Gorman and Mcintee 2003). While wifi "cut the cord" between the computer and the modem, internet connectivity remained a function of limited wifi range. Mobile computing devices (handhelds, tablets, phones) and mobile (cellular) internet connectivity became ubiquitous in the early 2010s, and the application of the internet in everyday life began also to shift. The mobile turn characteristic of later-stage Web 2.0 (call it Web 2.5) not only decoupled the internet user from location and conventional computing device (laptop/desktop), but also from conventional interaction with web-based content (Blank and Reisdorf 2012; Kelley 2013). As the mobile internet evolved, a host of new technologies (both hardware and software) reimagined how users engaged and interacted with it – shifting the focus from interactions with the *web* to interactions with *data* (Kitchin and Perng 2016; Roche 2016; Zook 2017).

Web 3.0, referred also to as the semantic web, looms large as mobile technologies and the mobile internet become pervasive elements of everyday life (Kelley 2014b). Characterized by passive, anticipatory, and ever-more intelligent software, the mobile internet might best be understood as mutually constituted by code, data, and location (Forlano 2009; Kitchin and Dodge 2011; Kinsley 2012). In as much as the internet of 1995 resembled a virtual library of resources waiting for the user to stumble upon, the internet of the 21st century is designed to augment the user's daily life through an automated awareness and intelligence that is made possible by pervasive connectivity to unimaginably large, dynamic, and constantly evolving stores of data.

Mobile hardware, infrastructure, and regulation

Of course, the mobile turn could not have been possible without public and private investments during the late 1990s to mid-2000s to develop wireless infrastructure and consumer-oriented handheld computing devices (Williams and Dourish 2006; Nam and Pardo 2011; Racherla and Mandviwalla 2013). Leading the way, the first successful mobile internet ventures were launched in Finland (1996) and Japan (1999) on 2G (second generation) cellular phone networks using early Nokia and NTT Docomo mobile devices. The 2G mobile internet connections deployed in Finland and Japan served data at speeds ranging from 50–500 Kbit/s; by way of contrast, contemporary 4G (fourth generation) cellular connections serve data at speeds upwards of 1000 Mbit/s (1,000,000 Kbit/s). So, although the mobile internet in the late 1990s was only a shadow of its contemporary self, these efforts showcased the possibility of serving internet-based information to consumers via wireless connections.

Further development and improvement of private cellular phone networks have continued into the present, with 5G (fifth generation) cellular service on the horizon in the late 2010s. Notably, there has also been considerable investment and experimentation from the public sector in free and/or low-cost public wifi around the world (*The Economist* 2004; Torrens 2008). Countries such as Kenya (Malakata 2015), Ireland (O'Brien 2018), India (Christopher and Shaikh 2018), and the U.S. (e.g., New York City) (Fingas 2018) have initiated public wifi projects that involve considerable investment in infrastructure in order to provide equitable access to broadband internet. And while not analogous to broadband cellular mobile internet, public investment signals a recognition that internet access might be considered as essential as services such as public transportation, water, and electricity.

The expansion of the mobile internet coincided with debates over regulation and the ownership of digital information that first became public during the highly visible censure and dismantling of the Napster file-sharing network in the early 2000s. The Digital Millennium Copyright Act (DMCA), a U.S. copyright law passed in 1998, was intended to prevent copyright infringement via new information technologies. It was, in essence, a regulatory device for the nascent information age (Mableson 2018). Napster (referred to as a peer-to-peer "P2P" network) was only the most high-profile of several similar P2P services that were shuttered following DMCA investigations. Gnutella, Limewire, and Kazaa were other high-profile startups also forced either to adjust their services or to shut down entirely as digital information fell under ever-greater scrutiny.

Inasmuch as the early 2000s were host to efforts to better understand, manage, and regulate the flows of information between and among individual users of the internet, the early 2010s were marked by competing efforts to manage and regulate the digital pipelines that are the backbone of the mobile internet. Coined in 2003, the concept of net neutrality speaks both to debates over the necessity (or not) for information to remain free on the internet as well as to debates over the necessity for all users on the internet to stand on equal footing (Síthigh 2011; Roberts 2019). In the 2010s, debate centered on the legal and ethical dilemmas of permitting internet providers the freedom to "throttle" (slow) connectivity to some web-based content while maintaining higher-speed connections to others. At issue is not simply that some content might be less-accessible than other content, but rather that beneath an apparently free and open internet, information could be promoted or hidden by virtue of a given provider's policy (Roberts 2019).

For a mobile internet verging on web 3.0, debates over net neutrality are essentially debates over who, if anyone, should wield the power to control which information becomes a component of the public imaginary. And evidenced below, the control of information in the era of the mobile internet is akin to holding the controls of an airplane in mid-flight. From anticipatory prompts alongside driving directions, to suggestions for dining reservations or medical clinics, the mobile internet is an ever-present companion in everyday life; and the information that seeds its intelligence is surfaced through channels only loosely regulated by net neutrality law.

The mobile internet and everyday life

Locative technologies

Beginning in the early 2000s, and among the host of technological advancements that decoupled internet usage from personal computers and wired connections, consumer-grade GPS (global positioning system) sensors began finding their way into myriad portable computing devices (Wilson 2012; Kelley 2014b). The rise of location-based technologies was coincident with U.S. efforts in the late 1990s and early 2000s to enhance the accuracy of, and improve public access to, the constellation of U.S.-funded GPS satellites. As GPS sensing technologies such as mobile phones, tablets, and wearables became more accessible and accurate, locative data became a ubiquitous feature of the mobile internet. Thus, alongside the emergence of the mobile internet and portable computing devices, locative technology was embedded in the tacit functions of daily life for much of the developed and developing world (Brighenti 2012; Kitchin and Perng 2016). By simple virtue of the fact that mobile devices (phones, wearables) remain pervasively connected to the internet, data are not only immediately accessible, but also geographically relevant to average internet users (Kelley 2014b). The internet, born in the 1960s out of a desire to share information across space via digital channels, thereby gained new relevance in the late 2000s and 2010s as a mode by which users could augment and enhance their experience of lived space.

Contextual data and the public imaginary

For a technology to have notable, tacit impacts on the user's daily experience of lived space, information had to be made relevant. And so, for mobile internet users context became king in the 2010s. Context, in this regard, referring to the *contextual relevance* of information provided by internet-based services. Again, if Web 1.0 was analogous to a classical library that required the user to construct direct, explicit, and well-formed queries in order to acquire a relevant piece of information, then Web 2.5 is closer in practice to a well-informed assistant prepared to proffer information that is contextually relevant to a given scenario. This evolution is due largely to the combination of big-data from existing social-networking and media services with increasingly accurate locative data – a feat made possible by software algorithms that were designed to harvest and manage user-generated data (Thrift and French 2002; Graham 2005; Kitchin and Dodge 2011; Roche 2016). Algorithms governing the collection, indexing, and filtering of these "big data" grew in sophistication during the same period as developers recognized the possibilities implicit in the socio-cultural normalization of ubiquitous (locative) mobile technologies and pervasive internet connectivity (Kitchin 2014).

The algorithmic shift endemic to the mobile internet of the 2010s also began to situate the user less as active *consumer* of information and more as intermittently active/passive hybrid producer/consumer (*prosumer*) of data (Ritzer and Jurgenson 2010). Through little-to-no direct effort, users who carried mobile technology became active nodes in an amalgam of data consuming and producing activities that happened largely in the background of their daily life. Elaborate terms of service agreements attached to software applications that are installed on mobile devices grant the developers of those applications wide-ranging access to data about the user (including, but not limited to, lists of contacts, email messages, photos, geographic location, musical preferences, calendar events, and search history). Data harvested from individual users are stored in ever-larger and more-complex databases that are subsequently used to generate ever-more-detailed and contextually relevant information in response to user queries (Kelley 2014b; Leszczynski 2016).

One product of the data being harvested from users of the mobile internet is the virtual development of what amounts to living snapshot of the material world. Drawing on millions of individual users, the internet-based data that are now pervasively accessible to mobile devices reflect the most detailed composition of the public imaginary ever assembled (Kelley 2013). Of course not without severe limitations (see the discussion of digital exclusion and privacy in the following section of this chapter), this *digital* imaginary functions as the cortex of nascent efforts to leverage the capacities of the mobile internet and ubiquitous mobile technology to silently augment nearly all facets of daily life.

Artificial intelligence, anticipatory mobile technologies, and smart(er) spaces

If the mobile internet is characterized by the chaos and virtual noise of pervasive connectivity, then silence is its holy grail. Not silence as literal absence of sound, but silence as metaphor for the calm that exists without interaction with technology (Kukka et al. 2014). For the mobile internet to be most effective, it must be embedded in daily life silently; requiring minimal explicit input from the user while offering maximum benefit to the user (Weiser 1991). As Weiser suggested in his early reflection on computing technology in the 21st century, "[t]he most profound technologies ... weave themselves into the fabric of everyday life until they are indistinguishable from it" (1991, 94). At that point, and from a user's perspective, ubiquitous technology and pervasive internet connectivity are no longer a remarkable aspect of life. They simply exist in the same domain as any other functional part of the individual – shoes, jacket, credit card, keys ... internet. But to arrive at the moment Weiser described required decades of technological change and development that is ongoing in the late 2010s.

By connecting users to the vast and living stores of data that compose the digital imaginary, pervasive connections to the internet open possibilities for algorithms that are intimately acquainted with their users to anticipate those users' needs and that are designed to mediate the occurrence of everyday life without drawing the user's attention to the interaction. The artificial intelligences governing these actions, while still in nascent form as of the late 2010s, are expected to recognize and react to dynamic scenarios full of many layers of contextual variables by interacting simultaneously with the user's personal data as well as the broader internet via pervasive data connections. In short, the combination of mobile computing devices and the mobile internet becomes a sort of contingency generator – anticipating outcomes by remaining silently aware of context. And per Weiser's suggestion, these digital interactions between user and internet are becoming indistinguishable elements of everyday life.

Lived "smart" spaces, like individual users in the late 2010s, are also in constant silent conversation with the mobile internet. *Smart cities* and *smart homes* depend on a blanket of connectivity that enables interaction between the devices, services, and processes that govern the lived environment and the data that are stored on the internet (Batty et al. 2012; Hancke, de Carvalho e Silva, and Hancke 2013; Roche 2016). The movement of data between service provider and consumer (such as when monitoring power consumption or managing public transportation) is mediated by web-based technology. And cities around the world (Guardian 2018) have been recognized for their efforts to leverage "smart" (locative and internet-connected) technologies to streamline municipal processes and improve livability and quality of life. Among other things, this has meant that city and town planning can be more responsive to the needs of residents, and that these same residents can enjoy more effective engagement with municipal officials. Systems, such as traffic management, can be made to be more adaptive to real-time situations. And utilities, such as waste-water treatment and public power, can be managed more sustainably through better monitoring (Nam and Pardo 2011; Batty et al. 2012; Zook 2017).

Similarly, nascent smart home technologies such as web-connected thermostats, cameras, light bulbs, networking, and alarm systems came to market during the 2010s. In much the same way that smart city systems opened opportunities for cities to better manage themselves through pervasive data connectivity, smart home spaces provide individuals with the capacity to manage *home systems* through applications on mobile devices (Khan, Silva, and Han 2016; Wiltz 2016). Silence, again, is key to smart systems as they are designed to leverage the pervasive connectivity of the mobile internet to manage everything from electrical consumption to home security by understanding and reacting to contextual variables in dynamic environments. Users are engaged only when the smart systems recognize the need for additional input. Mundane tasks, such as turning off light bulbs or ensuring that doors are locked when residents are not present, are transmitted wirelessly and governed by smart home algorithms (Mechant, All, and De Marez 2018).

Stepping back from the particular technologies that are characteristic of the mobile internet, it is useful to reconsider the trajectory of mobile development as this chapter moves into a consideration of the consequences of the mobile internet. As noted above, mobile technologies in the 2010s are designed to mesh quietly with the daily life of the user. By demanding fewer inputs from the user, the mobile internet becomes a silent partner as she/he moves through the day, offering suggestions, advice, direction based on consideration of contextual variables. Mobile tech is, in other words, designed to *anticipate* and *predict* the needs and wants of the user, and to provide information and guidance relative to the user's current site and situation. For this to be possible, pervasive mobile internet connectivity blends with artificial intelligence software applications that become *smarter* by virtue of their access to the digital public imaginary.

Seen and unseen consequences of the mobile internet

Uneven digital production of space

There is no doubt that technologies become smarter (more contextually aware) the longer they persist on the mobile internet. For lived spaces this led to incremental experimentation with systems that govern commonplace aspects of everyday life (transportation, utilities, communications) – smart cities, and smart homes. But the impact of the mobile internet is not limited to the benign management of municipal and residential utility systems. Rather, there are considerably less benign outcomes tied to the ubiquitous, and increasingly silent, mobile device that connects individuals to the digital imaginary. As individual users become more trusting of the information proffered by their mobile devices, the actions that they take in their everyday lives are augmented (for better or worse) by the collective intelligence generated by the mobile internet and distributed by software applications (Kinsley 2012; Kelley 2014a; Leszczynski 2016).

Lived spaces – homes, neighborhoods, cities and regions – are not just lived-in by residents. They also become what they are as a consequence of the ways in which people use them. Lived spaces are, in other words, a product of the activities, perceptions, experiences, and imaginations of the people who inhabit them. They are *produced by* their inhabitants (Lefebvre 1991). For instance, no matter how much capital is invested in a town square, and no matter how lovely the design of the town square, it does not *become* the center of a town until and unless inhabitants use it in that way. If people decided, for whatever reason, to use the town square as a parking lot and to then walk elsewhere to spend their recreational time then the town square acquires collective meaning as a parking lot.

And interestingly, in an era of pervasive internet connectivity, the town square does not just acquire local meaning as parking lot. The digital imaginary mirrors the perception/experience of those who happen upon the town center by virtue of the reviews, comments, pictures, and other media shared on the mobile internet. Consequently, the collective intelligence drawn on by mobile applications about the town square is self-reinforcing and generative, as the actions of more users are mediated by their devices (e.g., they are directed to park, rather than to spend time, in the town center), their activity becomes additional data points. The town square is re-produced as parking lot in the minds, memories, actions of users, and the imaginary of the mobile internet.

While the reproduction of a town center as parking lot is unfortunate, the consequences of producing entire urban neighborhoods as unsafe based on the digitally shared misgivings or biases of a handful of individuals are far worse (Leszczynski 2016; Kelley 2014a). To draw on an oft-used example, Ghetto Tracker was a mobile application released in the early 2010s with the intent to provide users a way to avoid "ghettos" in their movements through cities. The overtly racist and classist application existed only for a short while before it was taken down following considerable outrage, but it is emblematic of how the mobile internet can become an unintentional culprit in the harmful production of space (Narula 2013). And while Ghetto Tracker was relatively easy to erase from the collective intelligence, it is far more difficult to overcome the generative accumulation of bias and partial truths that are much more common in the digital imaginary – particularly as users become less skeptical of the information shared with them by their mobile devices.

Emergent digital exclusion

A glance at data in the late 2010s about access to mobile technology and internet access suggests that the first order digital divide, the gap between those who have access to digital technology and those who do not, has narrowed (Warschauer 2003; Stevenson 2009; Warf 2013; Kelley 2018). This is not to say that a computer, laptop, and broadband internet connection are present in every home. Rather, that ongoing efforts (beginning in the early 2000s) to close the first order divide combined with the introduction of more affordable mobile computing technologies and mobile internet connectivity have made the presence of these technologies more commonplace. Looking past the binary haves/have-nots divide, subsequent modes of digital exclusion were characterized by uneven levels of digital literacy rather than access to digital technology. After investing time and effort into programs to close the divide, questions such as "how would new users become fluent in digital technologies?" were more common. Technology training programs, community technology centers, and a host of initiatives were implemented to address the new problem of literacy (Kvasny and Keil 2006; Steyaert and Gould 2009; Kelley 2018).

Yet with efforts directed at issues such as access and literacy, there remain open questions about new modes of digital exclusion that do not have straight-forward solutions. In particular, assuming that issues of access and fluency are being engaged by policy then it is necessary to begin conversation about *modes of participation* in the mobile internet (Stewart et al. 2006; Sylvester and McGlynn 2010; Kelley 2018). Because at issue is not simply that participation is made an option for all, but instead *how* and *why* participation tends to occur across the demographic spectrum. Digital exclusion can take many forms, and as the mobile internet grows in complexity and is woven further into the fabric of everyday life, the user-generated data that form the digital imaginary have ever-greater potential to reinforce and reproduce deeply seeded socio-cultural, spatial, and economic biases (Kelley 2014a; Graham, Zook, and Boulton 2013). Supposing that participation in the generative mobile internet is uneven demographically (including race, class, sex, gender, age, geography) then there is work to do to not only understand why certain groups are marginalized while others are not, but also to better understand the consequences (socio-culturally and spatially) of being marginalized by mobile technologies.

Interventions that are designed to overcome new modes of exclusion are less-likely to be the sort of "one-size-fits-all" programs that were characteristic of the first order (access) and second order (literacy) divides. Instead, agencies will find themselves needing to narrow the scale of their activities in order to recognize and valorize the unique socio-cultural, economic, and geographic attributes of sub-groups who have access to and are fluent with, but who remain *marginalized* by mobile technologies. Making strides to increase the diversity of the pool of mobile developers would, for instance, enable professionals from a wider range of backgrounds and experiences to challenge normative assumptions about how, when, where, and why the mobile internet is woven into daily life, and to work toward the development of applications that cater more explicitly to the needs and interests of marginalized groups.

Privacy concerns and data security

The erosion of digital privacy is perhaps the most visible and well-documented consequence of the mobile internet (van Dijk 2010; Andrejevic and Gates 2014; Anthony 2015; Leith 2015; Kitchin 2016; Vincent 2016). Because while participation is all-but-required in order to function in contemporary society, it comes at the cost of carrying a device loaded with software and sensors that necessarily compromise the privacy of the user; and that do so with the permission of the user. In a 21st-century Faustian bargain, search histories, contact lists, entertainment preferences, spending habits, travel patterns, and spatial location are among the many forms of data that are traded by users of the mobile internet in exchange for free, or exceedingly low-cost, access to pervasive connectivity.

Mobile devices in the late 2010s are incredibly sophisticated data-loggers, and mobile applications are designed to mediate the flow of data from user to database as seamlessly and silently as possible. Search histories, for instance, are captured silently alongside normal device usage and in exchange for information. Spatial location and travel patterns are logged silently in exchange for unprecedented access to interactive maps and location-based-services. Contact lists are exchanged for access to communication services that provide instant access to friends and family via all manner of media. By design, the mobile internet can only function if it is in a perpetual state of data collection, and as mobile technology has become more ubiquitous, users have proven amenable to the loss of privacy (Boyles, Smith, and Madden 2012; BBC News 2015). How personal data are accessed, stored, shared and kept secure on the mobile internet remain open questions that demand attention and become more urgent each time a data center is breached.

Concerns over privacy, tracking and security have been articulated by scholars and critics from a range of disciplinary backgrounds. Within disciplines such as sociology (Custers 2013; Berger et al. 2014; Canossa 2014), computer science and communications (Leenes et al. 2017; Francis 2017), geography (Monahan and Mokos 2013; Cinnamon 2017), and law (Lind and Rankin 2015; Lamoureux 2016), scholars have worked for more than a decade to detangle the domains digital privacy, ubiquitous technology, and the mobile internet. And common among them is a sense of uncertainty – not over the risks posed by the loss of privacy and the ability to track individuals through mobile technologies, but uncertainty about how best to reclaim privacy and to re-establish among users the expectation that privacy and security are paramount to participation in the mobile internet. Moving into the next decade, it is inarguable that developers and policy-makers must engage more directly the issues of privacy and security on the mobile internet.

Conclusion

With roots in web 1.0 and a prominent role in the development in smart cities and augmented reality, the mobile internet is a project that remains in progress. And even in a state of perpetual evolution, in a strikingly short period of time the mobile internet became foundational to most every facet of everyday life in the developed world. Given the centrality of data to contemporary society, it is difficult to imagine a path forward that does not include pervasive connectivity to the internet and ubiquitous mobile technology. For this reason, it is also necessary for scholars, developers, policy-makers, and advocates continue to examine the consequences of the diffusion of these technologies and to engage in ongoing conversations about how best to plan for appropriate and equitable future iterations of the mobile internet.

References

Allen, M. 2013. What was Web 2.0? Versions as the dominant mode of internet history. *New Media & Society* 15(2):260–275.

Almobaideen, W., R. Krayshan, M. Allan, and M. Saadeh. 2017. Internet of Things: Geographical routing based on healthcare centers vicinity for mobile smart tourism destination. *Technological Forecasting and Social Change* 123:342.

Andrejevic, M., and K. Gates. 2014. Big data surveillance: Introduction. *Surveillance & Society* 12(2):185–296.

Anthony, K. 2015. Digital privacy: Subverting surveillance. *Nature* 524(7566):413.

Batty, M., K. Axhausen, F. Giannotti, A. Pozdnoukhov, A. Bazzani, M. Wachowicz, G. Ouzounis, and Y. Portugali. 2012. Smart cities of the future. *European Physical Journal. Special Topics* 214(1):481–518.

BBC News. 2015. Warnings over smart device privacy. February 10, 2015. www.bbc. com/news/technology-31360870.

Berger, P., R. Brumme, C. Cap, and D. Otto. 2014. Surveillance in digital space and changes in user behaviour. *Soziale Welt-Zeitschrift Fur Sozialwissenschaftliche Forschung Und Praxis* 65(2):221–245.

Bhati, A., M. Hansen, and C. Chan. 2017. Energy conservation through smart homes in a smart city: A lesson for Singapore households. *Energy Policy* 104(May): 230.

Blank, G., and B. Reisdorf. 2012. The participatory web. *Information, Communication and Society* 15(4):537–554.

Boyles, J., A. Smith, and M. Madden. 2012. Privacy and data management on mobile devices. Pew Research Center: Internet, Science & Tech. September 5. www. pewinternet.org/2012/09/05/privacy-and-data-management-on-mobile-devices/.

Brighenti, A. 2012. New media and urban motilities: A territoriologic point of view. *Urban Studies* 49(2):399–414.

Canossa, A. 2014. Reporting from the snooping trenches: Changes in attitudes and perceptions towards behavior tracking in digital games. *Surveillance & Society* 12(3):433–436.

Chi-Hua, C., and L. Kuen-Rong. 2018. Applications of Internet of Things. *ISPRS International Journal of Geo-Information* 7(9):334.

Christopher, N., and S. Shaikh. 2018. Public WiFi to contribute $20 billion to India's GDP by 2019: Study. *Economic Times*. July 4. https://economictimes.indiatimes. com/tech/internet/public-wifi-to-contribute-20-billion-to-indias-gdp-by-2019-study/articleshow/64856557.cms.

Cinnamon, J. 2017. Social injustice in surveillance capitalism. *Surveillance & Society* 15(5):609–625.

Custers, B. 2013. *Discrimination and Privacy in the Information Society: Data Mining and Profiling in Large Databases*. Studies in Applied Philosophy, Epistemology and Rational Ethics, v. 3. New York: Springer.

Fingas, J. 2018. New York City's WiFi kiosks have over 5 million users. Engadget. September 30. www.engadget.com/2018/09/30/new-york-citys-wifi-kiosks-have-over-5-million-users/.

Forlano, L. 2009. WiFi geographies: When code meets place. *Information Society* 25(5):344–352.

Francis, L. 2017. *Privacy: What Everyone Needs to Know*. New York: Oxford University Press.

Galloway, A. 2004. Intimations of everyday life – Ubiquitous computing and the city. *Cultural Studies of Science Education* 18(2–3):384–408.

Gorman, S., and A. Mcintee. 2003. Tethered connectivity? The spatial distribution of wireless infrastructure. *Environment and Planning A* 35(7):1157–1171.

Graham, M., M. Zook, and A. Boulton. 2013. Augmented reality in urban places: Contested content and the duplicity of code. *Transactions of the Institute of British Geographers* 38(3):464–479.

Graham, S. 2005. Software-sorted geographies. *Progress in Human Geography* 29(5):562–580.

Guardian. 2018. Smart cities. Guardian. www.theguardian.com/cities/smart-cities.

Hancke, G., B. de Carvalho e Silva, and G. Hancke. 2013. The role of advanced sensing in smart cities. *Sensors* 13(1):393–425.

Hillstrom, K. 2005. *The Internet Revolution*. Detroit: Omnigraphics.

Jassem, H. 2010. Municipal WiFi: The coda. *Journal of Urban Technology* 17(2):3–20.

John, N. 2013. Sharing and Web 2.0: The emergence of a keyword. *New Media & Society* 15(2):167–182.

Kelley, M. 2013. The emergent urban imaginaries of geosocial media. *Geojournal* 78(1):181–203.

Kelley, M. 2014a. Urban experience takes an informational turn: Mobile internet usage and the unevenness of geosocial activity. *Geojournal* 79(1):15–29.

Kelley, M. 2014b. The semantic production of space: Pervasive computing and the urban landscape. *Environment and Planning A* 46(4):837–851.

Kelley, M. 2018. Framing digital exclusion in technologically mediated urban spaces. In E. Shears (ed.) *Thinking Big Data in Geography*. pp. 178–193. Lincoln, NE: University of Nebraska Press.

Khan, M., B. Nathali Silva, and K. Han. 2016. Internet of Things based energy aware smart home control system. *Access, IEEE* 4:1–1.

Kinsley, S. 2012. Futures in the making: Practices to anticipate "ubiquitous computing." *Environment and Planning A* 44(7):1554–1569.

Kitchin, R. 2016. *Getting Smarter about Smart Cities: Improving Data Privacy and Data Security*. Data Protection Unit, Department of the Taoiseach: Dublin, Ireland.

Kitchin, R. 2014. The real-time city? Big data and smart urbanism. *Geojournal* 79(1):1–14.

Kitchin, R., and M. Dodge. 2011. *Code/Space: Software and Everyday Life*. Cambridge, MA: MIT Press.

Kitchin, R., and S.-Y. Perng. 2016. *Code and the City*. London; New York: Taylor & Francis.

Kukka, H., A. Luusua, J. Ylipulli, T. Suopajärvi, V. Kostakos, and T. Ojala. 2014. From cyberpunk to calm urban computing: Exploring the role of technology in the future cityscape. *Technological Forecasting and Social Change* 84:29–42.

Kvasny, L., and M. Keil. 2006. The challenges of redressing the digital divide: A tale of two US cities. *Information Systems Journal* 16(1):23–53.

Lamoureux, E. 2016. *Privacy, Surveillance, and the New Media You*. New York: Peter Lang.

Leenes, R., R. van Brakel, S. Gutwirth, and P. de Hert (eds.) 2017. *Data Protection and Privacy: (In)visibilities and Infrastructures*. Issues in Privacy and Data Protection. Cham, Switzerland: Springer.

Lefebvre, H. 1991. *The Production of Space*. Oxford, UK: Blackwell.

Leith, P. 2015. *Privacy in the Information Society*. Farnham, Surrey, UK: Ashgate.

Leszczynski, A. 2016. Speculative futures: Cities, data, and governance beyond smart urbanism. *Environment and Planning A* 48(9):1691–1708.

Lind, N., and E. Rankin. 2015. Privacy in the digital age: 21st-century challenges to the Fourth Amendment. https://bit.ly/2VFOyO7.

Mableson, C. 2018. *DMCA Handbook for Online Service Providers, Websites, and Copyright Owners*. 2nd ed. Chicago: ABA, Section of Intellectual Property Law.

Mahtot, G., and W. Sanchez Thomas. 2018. "Smart" tools for socially sustainable transport: A review of mobility apps. *Urban Science* 2(2):45.

Malakata, M. 2015. Kenya becomes second country in East Africa to enjoy free wi-fi. *Network World*. March 20. www.networkworld.com/article/2899974/kenya-becomes-second-country-in-east-africa-to-enjoy-free-wifi.html.

Mechant, P., A. All, and L. De Marez. 2018. Evaluating user experience in smart home contexts: A methodological framework. In N. Streitz and S. Konomi, *Distributed, Ambient and Pervasive Interactions: Understanding Humans*. Lecture Notes in Computer Science, vol. 10921. pp. 91–102. Springer.

Monahan, T., and J. Mokos. 2013. Crowdsourcing urban surveillance: The development of homeland security markets for environmental sensor networks. *Geoforum* 49(C):279–288.

Nam, T., and T. Pardo. 2011. *Conceptualizing Smart City with Dimensions of Technology, People, and Institutions*. Albany, NY: Center for Technology in Government.

Narula, S. 2013. The real problem with a service called "Ghetto Tracker." *The Atlantic* (Sept. 6). www.theatlantic.com/technology/archive/2013/09/the-real-problem-with-a-service-called-ghetto-tracker/279403/.

O'Brien, C. 2018. New free public WiFi project opens for registrations. *The Irish Times* (Mar. 20). www.irishtimes.com/business/technology/new-free-public-wifi-project-opens-for-registrations-1.3433725.

O'Reilly, T. 2005. What Is Web 2.0: Design Patterns and Business Models for the Next Generation of Software. www.oreilly.com. 2005.

Perry, D., S. Blumenthal, and R. Hinden. 1988. The ARPANET and the DARPA internet. *Library Hi Tech* 6(2):51–62.

Racherla, P., and M. Mandviwalla. 2013. Moving from access to use of the information infrastructure: A multilevel sociotechnical framework. *Information Systems Research* 24(3):709–730.

Ritzer, G., and N. Jurgenson. 2010. Production, consumption, prosumption: The nature of capitalism in the age of the digital "prosumer." *Journal of Consumer Culture* 10(1):13–36.

Roberts, K. 2019. *Net Neutrality*. New York: Greenhaven Publishing.

Roche, S. 2016. Geographic information science III: Spatial thinking, interfaces and algorithmic urban places: Toward smart cities. *Progress in Human Geography* 41(5):657–666.

Síthigh, Daithí. 2011. Regulating the medium: Reactions to network neutrality in the European Union and Canada. *Journal of Internet Law* 14(8):3–14.

Stevenson, S. 2009. Digital divide: A discursive move away from the real inequities. *Information Society* 25(1).

Stewart, C., G. Gil-Egui, Y. Tian, and M. Pilleggi. 2006. Framing the digital divide: A comparison of US and EU policy approaches. *New Media & Society* 8(5):731–751.

Steyaert, J., and N. Gould. 2009. Social work and the changing face of the digital divide. *British Journal of Social Work* 39(4):740–753.

Sylvester, D., and A. McGlynn. 2010. The digital divide, political participation, and place. *Social Science Computer Review* 28(1):64–74.

The Economist. 2004. A brief history of wi-fi. www.economist.com/technology-quarterly/2004/06/10/a-brief-history-of-wi-fi.

Thrift, N., and S. French. 2002. The automatic production of space. *Transactions of the Institute of British Geographers* 27(3):309–335.

Torrens, P. 2008. Wi-fi geographies. *Annals of the Association of American Geographers*. 98(1):59–84.

van Dijk, N. 2010. Property, privacy and personhood in a world of ambient intelligence. *Ethics and Information Technology* 12(1):57–69.

Vincent, D. 2016. *Privacy: A Short History*. Cambridge, UK; Malden, MA: Polity Press.

Warf, B. 2013. Contemporary digital divides in the United States. *Tijdschrift voor Economische en Sociale Geografie* 104(1):1–17.

Warschauer, M. 2003. *Technology and Social Inclusion: Rethinking the Digital Divide*. Cambridge, MA: MIT Press.

Weiser, M. 1991. The computer for the 21st century. *Scientific American* 265(3):94–104.

Wikicommons. https://commons.wikimedia.org/w/index.php?search=%22mobile+internet%22&title=Special:Search&profile=advanced&fulltext=1&advancedSearch-current=%7B%7D&ns0=1&ns6=1&ns12=1&ns14=1&ns100=1&ns106=1#/media/File:Mlearning_Steven_Parker.jpg.

Williams, A., and P. Dourish. 2006. Imagining the city: The cultural dimensions of urban computing. *Computer* 39(9):38–43.

Wilson, M. 2012. Location-based services, conspicuous mobility, and the location-aware future. *Geoforum* 43(6):1266–1275.

Wiltz, C. 2016. Meet Otto, the robot that wants to run your home: Samsung is hoping Otto, its "home companion robot," will introduce consumers to the friendlier side of the Internet of Things. *Design News* 71(6): M18.

Zittrain, J. 2006. The generative internet. *Harvard Law Review* 119(7):1974–2040.

Zook, M. 2017. Crowd-sourcing the smart city: Using big geosocial media metrics in urban governance. *Big Data & Society* 4(1):1–13.

7 Geographies of the internet in rural areas in developing countries

Jeffrey James

There are at least three compelling reasons for singling out rural areas in developing countries, if one is concerned with the global dissemination of the internet. One of them is that these areas, especially in the poorest developing countries, have lagged considerably behind in the adoption of this technology. For example, of the global offline population, equal to some 4 billion people, 64% resides in rural areas as opposed to the 36% who inhabit urban regions (McKinsey 2014). Table 7.1 disaggregates these global data into estimates of the rural–urban divide in the internet, for a selected sample of countries in which the non-using population is in most cases especially highly concentrated in non-urban areas. Note also in this regard that according to research undertaken by the ITU "four-fifths of the offline population are located in Asia-Pacific and in Africa" (ITU 2017a, 4). Moreover, "when grouping countries by their level of development, the greatest connectivity shortfall is exhibited in Least Developed Countries – where 85% of the population is still offline set against only 22% in Developed Countries" (ITU 2017a, 4). A final reason for focusing especially on rural areas has to do with global poverty and in particular the recognition that no less than 80% of the world's poor are located in these areas (ITU 2017a). Achieving the UN's Sustainability Development Goals, therefore, has necessarily to do heavily with what can be done in the rural parts of the developing world, especially in Africa and Asia-Pacific. These are regions with especially high rates of poverty outside urban areas (see below).

Yet, and this is the second reason for singling out rural areas in developing countries for attention, there are reasons to suppose that the internet will confer particular benefits on these, as opposed to urban areas and developed countries in general. For one thing, the sheer remoteness of outlying areas[1] means that they will tend to be exceptionally responsive to the distance-shrinking effect of information technologies such as the internet (see more on this below). More specifically,

> The effects of extending Internet access could be particularly important for rural communities. Constraints on the flow of information have limited these communities access to wider markets and to a variety

of employment opportunities. Access to mobile and Internet-based applications can extend the range of business services that become available to these communities. Internet access is also valuable to rural development-oriented organisations that act as local communication conduits or intermediaries.

(Deloitte 2014, 8)

If there are thus reasons to think that the internet may be especially relevant to rural areas of developing countries, it is also true that the constraints on realizing the potential thus afforded, are particularly severe in these same areas. One has only to think, for example, of low incomes and affordability, and a typically severe shortage of electricity and rural internet networks. The severity of these problems (in relation to the prospective gains) constitutes yet another reason to single out rural areas of developing countries, when considering the geographies of the internet.

The discussion in the rest of the chapter follows and elaborates on the arguments that have just been presented. That is, I begin with the notion that this technology affords special opportunities to the rural areas of most developing countries and to the poorest members of these areas. I progress to consider the difficulties entailed in the realization of these opportunities. The final section, in turn, addresses the main policy issues involved in ameliorating or solving these difficulties (though space considerations limit the discussion of what are very large literatures).[2]

Table 7.1 Rural–urban divide in the internet, selected countries, 2013

Country	Non-internet users, millions	Urban	Rural
India	1,063	27%	73%
Pakistan	162	32%	68%
Nigeria	108	37%	63%
Myanmar	53	33%	67%
Rep. of Congo	64	n/a	n/a
Vietnam	50	26%	74%
Thailand	48	31%	69%
Tanzania	47	26%	74%
Indonesia	210	43%	57%
Brazil	97	76%	24%
Ethiopia	92	16%	84%
Mexico	69	75%	25%
Egypt	41	67%	33%
China	736	37%	63%
Philippines	62	57%	43%
Turkey	40	55%	45%

Source: McKinsey (2016) Exhibit 9.

The potential of the internet in rural areas of developing countries

I am not concerned to review all the socio-economic effects of the internet on rural areas in developing countries. For one thing, such studies have already been undertaken (see, for example, the World Bank 2016, Deloitte 2014 and the ITU 2017a). For another, an undertaking of this kind falls outside the purview of this chapter, as this was outlined above. I am concerned, rather, with the specific features of the rural economy that make it especially well-suited to the characteristics of this evolving technology.

Note, however, that the unconnected, who are to be found predominantly in rural areas, are not spread evenly across regions. For example, fully 75% of the population in Africa is offline compared to 55% of those living in the Asia-Pacific regions (ITU 2017a). And in developed countries as a whole, the figure is only 22%.

In order to grasp the potential of the internet in rural areas, it is necessary first to examine the most fundamental problems that afflict such areas. These have much to do with sheer geographical remoteness, an acute lack of information and an inability to communicate effectively with others, be they friends, family, businesses, and government agencies, which is due also to the lack of adequate basic infrastructure such as roads and public transport.

So, for example, firms, farms and individuals may have to confront these distinctively rural problems by undertaking long and expensive journeys to say, distant markets, where this is possible (in fact, it may not be feasible at all in some especially remote areas). A similar fate may befall those needing to communicate with distant friends and family. Or, in yet another prominent example, distance is often a major problem in dealing with government services, such as birth and death certificates or documents confirming that someone is eligible for specific benefits, because, say, he or she is a member of a specific caste.

The Sustainable Access in Rural India (SARI) project is instructive in this regard because the most popular services that it provides are precisely those that have just been mentioned.[3] The Pondicherry project in the same country is also relevant in this context, because it helps to overcome the acute problem of information scarcity in rural areas, as mentioned in the previous paragraph. This project can properly be described as an example of technological blending, in that a modern technology, the internet, is used in combination with a much more traditional technology, namely, the loudspeakers that are used to transmit weather information to local fisherman. In particular, the information thus gleaned from the internet is broadcast four times daily to the fishermen in the village, who are thereby enabled to make more informed decisions and increase their profits.[4]

Telemedicine is yet another area which is well suited to overcoming rural geographical constraints through use of the internet.[5] For one thing, it can promote learning opportunities for medical personnel in these areas. For another, gains will accrue in the form of time and money to patients and doctors, who are saved the task of travelling long distances to dispense or receive medical care. Or again, "it can allow underprivileged rural hospitals to share equipment and human resources within well-equipped hospitals" (Ouma and Herselman 2008, 560).

The opposite side of the coin, though, is that the ability of information technologies such as the mobile phone and the internet, to shrink distance is a defining feature that underlies their particular potential in rural areas. Such potential has not gone unnoticed in the literature. A study by Intel (2010), for example, notes that

> Mobile communications in general, and broadband in particular, have an especially strong impact on the economies of rural areas, which are home to nearby three out of four of the world's poor. Expanding broadband networks to rural areas leads to new opportunities for non-agricultural employment, better-paying agricultural jobs and greater overall productivity. Access to broadband also fosters small-business growth, allows citizens in remote areas to work from home, provides greater access to crop market prices and enables rural businesses to compete more effectively in world markets.
>
> (Intel 2014, 4)

In a similar vein, Correa and Pavez (2016, 47) noted that

> Research has found that internet connectivity would benefit these [rural] areas because it would help to overcome geographic isolation, promote access to resources and opportunities, and encourage social interactions and community attachment, which would lower the possibilities of out-migration and stimulate economic development.
>
> (Figure 7.1)

I now proceed to take the discussion one step further by examining the potential among the unconnected within rural areas of developing countries. What is examined, in particular is the notion that among the unconnected, it is "often the world's poorest and most disadvantaged populations, who could benefit the most from the economic and social inclusion that the internet promotes" (Deloitte 2014, 10). Briefly stated, the idea begins with the recognition that such populations find themselves in these unfortunate conditions, partly because they are often the most isolated and otherwise geographically disadvantaged (in terms, for example, of road and rail infrastructure). And to the extent that it is indeed a lack of information and communication that underlie the severest forms of poverty in rural areas, then it stands to reason that the potential of the internet will tend to be greatest among the poorest inhabitants of rural areas in developing countries. In this sense, it would, then

Figure 7.1 Virginia's BARC electric cooperative installs broadband fiber cable near Lexington.
Source: Wikicommons.

be legitimate to speak of a "pro-poor" component of such technology in those areas. It is worth noting here the empirical finding by Bird et al. (2010) that "there is a strong correlation between isolation and poverty, including chronic poverty" (2010, v) in Uganda. The same authors have also suggested a number of reasons for this statistical finding, namely, that

> Households in more remote areas have lower levels of market participation (including commodity and financial markets), itself associated with poverty; they make less use of public services (which are often more remote); and household members (women and children in remote areas) have to devote more time to fetching wood and water.
>
> (Bird et al. 2010, v)

It is also relevant in this context to consider the results of the Grameen Village Phone project in rural Bangladesh, where all the members of a village share a single phone (which, albeit on a smaller and less elaborate scale than the internet, serves the same basic purpose as that technology of providing information to isolated inhabitants of rural areas). What is especially relevant in regard to the previous discussion is that the main beneficiaries of the project were the group described as "poor" (Bayes et al. 1999).

In what is often described as technological leapfrogging, this group was able to bypass fixed-line phones and progress directly to basic mobiles. There is also an element of leapfrogging in the adoption of smart-phones: that is, that rural inhabitants can circumvent expensive computers and move directly to smart-phones and the internet (on the prices of the latter devices in developing countries see more below). Thus,

> The proliferation of mobile internet subscribers in developing countries is due in part to a lack of fixed-line infrastructure as well as the relatively high price of PCs. As a result, many internet users in developing countries are leapfrogging fixed-line connections and using their mobile phones instead to get online.
>
> (McKinsey 2014, 16)

The constraints on internet use in rural areas of developing countries

If it is the poorest groups in the rural areas that might be the biggest beneficiaries of the internet, as argued in the previous section, it is these same groups that are likely to suffer most from the constraints on the adoption and use of this technology in developing countries. That is why, when I now discuss these constraints, I try to retain the spatial dimension of the argument by continuing (Table 7.1) to distinguish between the rural and urban sectors. In other words, the rural data in themselves do not mean very much without a relevant frame of reference (the urban areas) in the same countries (or conceivably the rural regions in developed countries).

Accordingly, the remainder of this section is devoted to examining the rural-urban divide for each of the major constraints to internet penetration, which comprise incentives, low incomes and affordability, user capabilities, and infrastructure (McKinsey 2014). It is, of course, possible to conceive of other barriers to penetration in developing countries, but those that have just been mentioned exhibit considerable explanatory power. Jointly, that is, they explain more than 50% of the variance in internet penetration across countries selected for a regression analysis (McKinsey 2014).

Consider next a brief discussion of the four constraints that were mentioned in the previous paragraph, in which I pay particular attention to the rural dimension of the problems.

Incentives

It is not always recognized that a substantial percentage of the unconnected population has not chosen to go online, because

> they lack awareness of the internet or its use cases or internet use is not considered culturally or socially acceptable. ... To develop the motivation to go online, consumers must understand the benefits and value

of the internet. However, even a basic awareness of the internet can be an issue. Some population segments – for example, rural residents in developing markets – are not aware of the internet's existence and its potential value.

(McKinsey 2014, 31)

For example, a 2013 study of India found that no less than 69 per cent of those surveyed, gave a lack of awareness of the technology as the reason they were not online.[6]

Incentives, however, are not just confined to awareness of the internet. For, what also matters is the availability of local content. This applies, among other things, to local language content. It is not encouraging in this regard to note that according to one estimate, 80% or more of all internet content is in one of only ten languages, mostly (with the exception of China) from developed countries. One may further suggest that where, say, English content is used on the internet in developing countries, it is more easily assimilated in urban areas where the population is more educated, more modern, and more accustomed to speaking the English language.

It is worth noting, finally, that the issue of linguistic relevance described in the previous paragraph can apparently make quite a striking difference in internet penetration between developing countries. For example, "China – which has the advantage of phones that can support the language – has a smartphone penetration rate of approximately 33 percent; by contrast, Vietnam's penetration rate is just 20 percent" (McKinsey 2014, 32).

Affordability

Even more than they did in the case of mobile phones, low relative incomes act as a severe deterrent to internet penetration in rural areas of most developing countries (especially in rural Africa where the percentage of those in poverty is the highest). Moreover, because of this pronounced regional divide in incomes, it is not surprising that there is also a substantial divide in internet penetration between the two areas. According to the ITU, for example, evidence gleaned from 35 countries indicates that internet use in rural areas is sharply lower than in urban locations (ITU 2017a).

Unfortunately, there is a general paucity of cross-country data on the rural-urban income divide. However, Young (2013) indicates that consumption per individual (a widely used proxy for income) is some four times higher in urban areas than in rural areas in the typical developing country. Similarly, Gollin et al. (2011) have shown that value added per worker is at least twice as high in the non-agricultural than in the agricultural sector.

It is true, on the other hand, that the price of smartphones has declined sharply in recent years. In developing countries for example (where the percentage decline was greatest, prices were reduced from 32.4 to 14.1% of GNI per capita (ITU 2017b). In spite of this, however, the affordability constraint will remain a potent one. For, according to the ITU (2017b) "mobile-broadband prices represent more than 5% of GNI per capita in most LDCs and are therefore unaffordable for the large majority of the population" (emphasis added).

In two poor countries, Tanzania and India, the cost of a smartphone is estimated to represent 16% of income for people living on less than two US dollars a day (GSMA 2017a).

What is more, affordability comprises other dimensions than just the price of the phone alone (assuming that the technology is individually owned; for other modes of adoption see the section below on policy). These other costs include maintenance, repairs and charging (McKinsey 2014:37). Especially in rural Africa, where electricity from the grid is in exceptionally short supply, the cost of charging can be severe. For, instead of travelling distances of at least 5 kilometers, African users often pay local bicycle-powered energy suppliers to charge their phones. In this way the

> lack of adjacent infrastructure can raise the cost of charging a device far above the cost of a data plan; consider that a Canadian pays USD 0.03 per month to charge a phone, while a sub-Saharan pays USD 6 per month, a price equal to half their monthly expense for mobile access.
>
> (McKinsey 2014, 37)

On top of this, there are often taxes which misguidedly treat the internet as a luxury good.

It is hardly surprising then, that affordability, especially in low-income rural areas of Africa, is considered by some as "the main constraint on wider internet take-up and the ITU estimates that broadband remains unaffordable for as many as 3 billion people globally" (Deloitte 2014).

User capabilities

Even if there is no relevance gap, as this was described above, potential users (in rural areas) will generally need to confront a severe capabilities gap. For, unlike the mobile phone, which is undemanding in terms of complexity of use, the internet requires not just the ability to read and write, but also the need to be literate in a digital sense, that is, "the ability to effectively and critic-ally navigate, evaluate, and create information using a range of digital devices and technologies" (Deloitte 2014, 41). That digital literacy is a serious con-straint to internet penetration in rural and other areas, is strongly suggested by considering just a few numbers. In particular, a number of studies draw attention to evidence that indicates just how severe this constraint actually is. McKinsey's research, for example

> found that the most-cited reason for why Africans don't access the internet is that they haven't developed the skills to do so. A 2014 survey

of Chinese consumers found that approximately 60 percent of the offline population cited a lack of knowledge of how to use a computer as the primary reason for not accessing the Internet.

(Deloitte 2014, 41–2)

Regrettably, I know of no study that breaks down overall digital literacy data into its component parts and much the same limitation is found in the case of information on literacy. Such data, however, do exist for a small number of countries, as shown in Table 7.2.

In each case, therefore, rural literacy rates fall below those for urban areas, with the greatest divide occurring in African countries (see, for example the case of Burkina Faso). And although no data on the cross-country divide in digital literacy are available, as far as I can tell the pattern may be similar or, in all likelihood even more pronounced.

Recall, for one thing, that income per head may on average be as much as four times higher in urban than rural areas. With other things being equal, there will then be a tendency for spending on education to be higher in the former as opposed to the latter regions. And while not all forms of education expenditure raise educational achievements,[7] there are certainly some that appear to do so. For example, a study of selected Francophone African countries by Michaelowa (2001) finds that there is a strong significant and positive effect of the presence of student textbooks on learning achievements.[8]

Digital illiteracy is an even more pronounced barrier to overcome in rural schools, because it requires a more expensive infrastructure (such as computers and network connections) and more educated students, not to speak of the difficulty of recruiting information technology teachers to distant, outlying areas. Though progress has been made in some countries (such as India), it is difficult to measure given the lack of available data (and the related difficulty of defining what digital literacy actually means).

In the context of mobile networks, the GSMA (2017b) has described the situation confronting the 30% of the global population lacking access to

Table 7.2 Urban and rural literacy rates (total). Selected sample of countries

Country	Urban	Rural	Date
Brazil	92.5	76.5	2008
Burkina Faso	62.9	19.5	2007
Chad	43.7	13.1	2004
India	78.2	53.7	2001
Niger	52.0	23.4	2005
Pakistan	71.1	46.3	2008
Senegal	69.1	33.2	2009
Sierra Leone	55.7	21.5	2004

Source: UNESCO, Institute for Statistics (2011).

the internet at 3G speeds in the following terms: more often than not the institution argues, these uncovered populations consists of those with low income and live in rural regions of Asia and sub-Saharan Africa (these together account for 3.4 of the 4.8 billion people not yet connected on the internet). The major challenge in reaching such uncovered communities is "overcoming an unfavourable cost-benefit equation": that is, that the high fixed costs of building a network infrastructure and maintaining it are divided over thinly spread populations with minimal purchasing power. Under such circumstances, especially in the absence of paved roads or electricity, network investments often become unprofitable. As far as the global population living in rural areas is concerned, the ITU has suggested that more than two-thirds of them remained uncovered by mobile broadband networks in 2015 (ITU 2017). Clearly, this is a massive gap that needs to be filled and as argued below it will not occur readily; quite the contrary, in fact.

As with the other constraints to internet penetration in rural areas of developing countries, low population density and geographical isolation also play an important role with the provision of adequate infrastructural facilities. For one thing, it is generally more profitable for telecommunication companies to serve areas with relatively high population densities. It is the same reason, incidentally, why it is difficult in general to attract multinational corporations to rural areas in developing countries (together, of course, with the relatively low incomes to be found in these areas as noted above).

Of the various geographical difficulties that confront the areas under consideration, I shall focus on the scarcity of electricity, not only because this is often highly pronounced, but also because of the availability of cross-country data on urban vs. rural differences in the supply of this resource.[9] A selected sample of countries appears in Table 7.3, which also shows such differences at the level of regions.

Note first that in all cases the figure for urban areas is greater than that in rural, though in some cases (such as Pakistan) the gap is very small. Note, too, in relation to regional groupings, the stark contrast between Africa and other developing country averages. For example, the urban–rural gap in Africa of 51.2% is considerably higher than any other region in the table. Relatedly, the rate of rural electricity coverage in this region is less than 25%, a very long way below the comparative figures. And looking again at countries rather than regions, it is clear that there are some African countries with rural electricity rates that are well below even the outlying regional estimate of 25%. I am referring here to Burkina Faso and Mali, with the incongruously low percentage figures of 0.8 and 1.8, respectively.

With respect to this and indeed the other constraints discussed in this section, it does appear as if the (rural) areas with most to gain potentially from the internet, are also those which confront the most severe barriers to adoption of this technology. For this reason, there seems to be much to be gained from policy interventions, especially in certain regions such as Africa. It is accordingly to this topic that I now turn.

Table 7.3 Access to electricity, urban vs. rural areas, selected sample, 2016

Country	Country/region	% of population
	Urban	*Rural*
Angola	70.7	16.0
Bangladesh	94.0	68.8
Benin	70.8	18.0
Botswana	77.7	37.5
Burkina Faso	60.7	0.8
Cambodia	100.0	36.5
India	98.4	77.6
Mali	83.6	1.8
Myanmar	89.5	39.8
Nepal	94.5	85.2
Pakistan	99.7	98.8
Sri Lanka	100.0	94.6
Vietnam	100.0	100.0
Region		
World	97.0	77.3
East Asia and Pacific (excl. high income)	98.8	94.0
Latin America and Caribbean	99.5	94.4
Sub-Saharan Africa	76.0	24.8
South Asia	98.1	79.4

Source: World Bank data (2017).

Policies to promote the internet in rural areas of developing countries

My task in this final section is to discuss ways of alleviating or overcoming the individual constraints to rural internet penetration that were noted above, namely, affordability, relevance, education and infrastructure.[10] I try to focus on regions and countries where the barriers appear to be especially acute (i.e., in sub-Saharan Africa and parts of Asia), and to provide examples (where available) of policy successes in these same locations.

Affordability

Beginning with affordability, the main policy imperatives seem to revolve around raising rural incomes and reducing the price of smartphones to prospective buyers. And it is well to note in relation to the former, that the comparative poverty of rural areas is not simply a function of the singular geographical difficulties they have to confront. Much may also have to do, for example, with politics and in particular the notion of a political struggle between urban and rural areas, with the former holding constant sway over the latter. This thesis has been most persuasively argued by Lipton (1977) in his book on "urban-bias" in world development. It seems to conform broadly to the African development strategy[11] with its emphasis on rapid industrial-ization and imports of expensive foreign technology. In the East Asian experi-ence, by contrast, much more attention was generally paid to rural areas, which led in turn to a more equal (and rapid) pattern of development.

For given incomes and poverty levels, the affordability of smartphones depends heavily on the price of which they are sold. Thus, according to the GSMA (2017a, 2),

> Consumer research shows how the cost of an Internet enabled handset is a critical barrier to using mobile Internet for low and middle-income consumers in emerging markets. India is a clear example of this, where over half of the population live in multidimensional poverty and where an average priced smartphone can cost up to 16% of income for poor and low-income groups. We estimate that over 134 million people in India are unable to afford one of the cheapest Internet-enabled on the market, because it exceeds, an affordability threshold at 5% of income.

The same institutions, however, recognize that there are other ways in which the affordability of handsets can be increased or the terms of purchase made more attractive. In the first place, there are often thriving second-hand markets for handsets in poor countries, which serve to lower the price of smartphones, sometimes to an appreciable degree. Low cost devices are sometimes also available from providers who drive down costs by means of "highly efficient supply chains and/or device subsidies" (GSMA 2017a, 4). Then there are cases where the terms of sale, rather than the price itself, are made more favorable. For example, some users can gain access to finance from various poverty-oriented financial institutions. In this way, they can purchase internet-enabled devices although they cannot pay the entire amount all at once.

Even with these policies, however, it is difficult to conceive of any kind of major surge in rural internet penetration in developing countries, in the short-run, especially in Africa and parts of Asia. This is partly because the discussion so far has been limited to the model of individual internet owner-ship. And while this may be a suitable mode of access to the technology in the rich countries, it is not helpful for the majority of impoverished inhabitants in the rural areas of many developing countries, where sharing makes much more sense (not only financially, but arguably from a cultural point of view as well). In practice, communal access to the internet in rural areas takes place in a wide variety of institutions such as hospitals, schools, libraries, post offices, telecentres, universities, and community radio stations.[12] In previous sections I have provided a few examples of how such communal institutions actu-ally work and will refer to others when dealing with policies to address the remaining constraints. Indeed, some of the community-level institutions that have just been mentioned, often play a crucial role in imparting user capabil-ities to the unconnected rural population, a topic to which I next turn.

User capabilities

Under this heading fall both basic and digital literacy, which exhibit some degree of joint causation, in that the former is often required for the latter, while use of the latter can sometimes promote the former. Schools are the primary institution through which basic literacy is acquired, whereas in the case of digital literacy a wider range of communal institutions is involved[13] (note though, that in rural areas, mobile libraries can also promote literacy).

Recall, to begin with, that for all the countries shown in Table 7.2, rates of rural literacy fall below those for urban areas. Apparently, there are particular problems which need to be confronted when seeking to promote literacy (and hence internet penetration) in these areas.

In part, it is a question of the volume of resources and more specifically the share that goes to rural rather than urban areas. For, although the recent literature focuses more on what is done with resources, one cannot entirely rule out the volume itself. And in this regard, there is much evidence that public expenditure on education benefits primarily the urban sector at the expense of the rural (Sechele 2016). Or, what is broadly the same thing, rich rather than poor inhabitants of developing countries. This result, one should note, is very much in line with what would be predicted by the theory of "urban-bias" described above. So too are the long distances that need to be travelled from outlying rural areas to the Ministry of Education, which is usually centrally located.

Yet, there are countervailing forces which mitigate against the negative effects of "urban bias." One derives from the recognition that the concept should not be deterministically interpreted. For, there have indeed been periods in which some governments in poor countries have taken seriously the need to tackle poverty and inequality.[14] Such an egalitarian focus tends to favor the rural areas in which the poor are mostly located. Cuba, for example, is a socialist developing country, which has attained near universal adult literacy, on the basis of an effective educational system which is applied consistently to urban and rural areas alike (Gasperini 2000).

Then there is a growing recognition in policy circles that a given amount of educational resources can be associated with very different learning achievements. Indeed, the major theme of The World Development Report for 2018 is that "schooling is not the same as learning." As evidence of this, the report points out that

> children learn very little in many education systems around the world even after several years in school, millions of students lack basic literacy and numeracy skills. In recent assessments in Ghana and Malawi, more than four-fifths of students at the end of grade 2 were unable to read a single familiar word.
>
> (World Bank 2018, 5)

Cases such as these have produced quite a substantial literature on how best to spend a given amount of resources in rural and urban schools (some of which is contained in the World Development Report just mentioned).

Consider, finally, the implications of the fact that learning is a cumulative process: that is, initial advantages and disadvantages are amplified during the schooling process and beyond.[15] What is most relevant from the point of view of impoverished rural schools is that "students often learn little from year to year, but early learning deficits are magnified over time" (World Bank 2018, 6). The problem usually begins at the very outset. Stricken as they usually are by severe poverty, rural pupils suffer from inadequate nutrition and medical care. As a result, they are ill-equipped to learn effectively in the earliest years of schooling. And this severe initial disadvantage only gets worse over time as more challenging tasks appear. The implication of this cumulative process is that remedial policy needs to be especially active at the very beginning of the schooling process. Indeed, "getting learners to school ready and motivated to learn is a first step to better learning. Without it, other policies and programs will have a minimal effect" (World Bank 2018L, 21, emphasis added).

The "crisis" in lagging literacy scores in Africa and other poor countries is serious enough in itself, but it also has negative implications for the growth of digital literacy[16] (and the penetration of the internet in rural schools). The first effect arises from the heavy reliance of digital literacy on its more basic counterpart. The second is due to the fact that pupils see less value in the internet and this lessens their incentive to adopt it. Then, lastly, a lack of digital literacy impinges negatively on the benefits users derive from their use of the technology.

It should be stressed, however, that policy designed to promote digital learning should not be confined merely to those who have acquired basic literacy. For, there is some evidence that internet capabilities can be acquired even by those who would not be described as literate. Indeed, according to a well-known experiment conducted in India at the end of the last century (known as the "hole in the wall" project). Mitra (2003) installed a computer with an internet connection in a slum in Delhi and observed the results. What he found in essence was that installation drew the attention of a number of illiterate, poor children, who, even by the end of the first day, had managed to teach themselves to surf the internet, in spite of not knowing what a computer or the internet were. Though the experiment was repeated in other parts of India, its replicability elsewhere remains very much an open question.

Also unclear as yet are the results of a recent nation-wide attempt in that country to bring digital literacy to all public primary schools (though the effort itself underscores the point made earlier in this section, that governments of developing countries do at times intervene on behalf of large numbers of the underprivileged in rural areas).[17] The so-called Prime Minister's Rural Digital Literacy Campaign in India for example began in 2016 with the goal of bringing digital literacy to 60 million rural households by March 2019 (*The Hindu* 2017). To this end, more than $350m was allocated to the scheme, which is clearly one of the world's most ambitious (and as such deserves close scrutiny). In Kenya, too, there is a far-flung and commendable policy effort to bring digital literacy to students in primary schools (Ministry of Information, Communications and Technology, Government of Kenya, n.d.).

Relevance/awareness

Under this heading fall both the relevance of internet content in rural areas and an awareness of the technology itself. The former has to do ultimately with an often severe lack of congruence between a technology that is designed for users in rich countries and prospective adopters in poor rural areas. Broadly, it is the same problem that gave rise in the 1970s to the concepts of appropriate and inappropriate technology (Stewart 1977). And a major lesson from the ensuing debate on the two concepts, was that for appropriate technology to be relevant, it needs to adjust not only to the economic, but also the socio-cultural conditions in the host environment. Such adjustments however have tended to be inadequate confined only to a small area (though there are a number of notable exceptions, such as the Grameen Telecom project in rural Bangladesh). In fact, the awareness problem may be even more acute nowadays because while the technology of the internet has become more complex, traditional rural societies have not become discernibly more outward-looking or more sympathetic to modern technology.

One glaring incongruity of the internet in a rural context is a linguistic one, namely, that only ten (mostly developed countries) languages comprise around 80% of the internet, with the majority of it being in English (McKinsey 2014). "As a result, languages in developing nations, particularly Africa and Asia often aren't well represented on the internet. Language fragmentation within a country compounds the challenge. India, for example, has 22 official languages" (McKinsey 2014, 31). In this way, the unconnected become discouraged and the problem perpetuates itself. Indeed, according to the ITU, in developing countries the chief barrier to household internet adoption is relevance (ITU 2017a).

To this extent, "there would seem to be a particular urgency about implementing policies to create and develop local content/local services and apps in local languages" and "to introduce public awareness campaigns to highlight benefits and value of services/apps/contents" (ITU 2017a). A more concrete policy suggestion is to develop more "intuitive" applications, services with basic graphical interfaces and to support local languages in various ways (McKinsey 2014). Finally, social networks are especially valued in developing countries, partly because they enable local languages to be used relatively extensively and partly because they tend to involve local content. As such, these sites should be used to advertise the relevance of the internet in the rural areas which suffer most acutely from the problem.

Even if the above problems are resolved, however, internet penetration will still require an awareness of the technology. And this, according to several surveys, is certainly not something that can be taken for granted. On the contrary, it seems to represent yet another potent reason why the internet has not penetrated at all widely in rural areas of developing countries. And as with several of the issues described in this chapter, lack of awareness has much to do with the sheer remoteness of rural life and its disassociation from the more modern parts of society. In practical terms, there are far too few points at which (traditional) rural inhabitants would encounter or hear about the internet (a classic problem of the "dual economy").

Suitable policy might therefore be focused on raising awareness at locations which are most frequently visited by the rural population. In Bhutan, for example, a consortium of local and foreign institutions set up telekiosks within the normal operations of remote rural post offices. "A powerful rationale for the project is that the post office, more than any other institution, has a presence in people's lives even in the remote corners of Bhutan" (ITU 2008, 1).

Another project, in rural Sri Lanka, is also relevant here because it makes use of a commonly used technology, the radio, to address several of the problems described above. The idea at Kothmale Community Radio was to engage in blending old (the radio) and new technology (the internet), a process that is sometimes referred to as "radio-browsing." By means, of an internet connection in the studio, broadcasters first select reliable sites that they think will be of interest to the local population and to increase relevance further, they will often invite resource persons from the area (such as a doctor for a program on health issues), to discuss the contents of the sites in the regional languages. Importantly, there is also an element of digital literacy involved in that the broadcasters explain how they are browsing from one page to another. Thus, "listeners not only get the information they requested, but they understand how it is made available on the web" (Hughes 2003). The difficulty of replicating this model, however, is that community radio-stations are not often encouraged in developing countries, largely, one suspects, for political reasons (Girard 2003).

Infrastructure

In an earlier section, I described the decision to invest in rural infrastructure in terms of a simple cost-benefit analysis, which, for many years, has been seen as a deterrent rather than an encouragement to such investments. In this section, however, I suggest that at least with respect to one of its key components, electricity, the cost-benefit calculation has turned more positive, from the point of view of enhancing rural internet penetration in rural areas.

More specifically, solar energy has become relatively inexpensive, partly because the price of solar panels has fallen and partly because the efficiency of light bulbs and appliances has increased dramatically.[18] As a result, Western firms now see solar power in Africa as a chance to reach a sizeable market and make a large profit. Note, in regard to the first reason, that the aim of achieving universal energy access, predates the declaration of global development goals, such as the Sustainable Development Goals of the UN (Farooquee 2017). Note, too, that "electricity will be essential to power and expand all stages of providing internet access, from backhaul and base stations to charging consumer devices" (Bloomberg 2018, 1).

Equally recently, the potential of mini-grids in particular has been emphasized in *The Economist* (2018).[19] The article stresses that such potential does not include only the power that is provided to villages, but also the catalytic role that they often play. The installers, that is to say, "advise villagers on irrigation, farming and marketing to help them develop businesses that require electricity, which in turn justifies the expense of installation" (*The Economist* 2018, 59).

With regard to the specifics of the changing economics underlying the provision of electricity to rural areas, consider the attractiveness of major price reductions in small-scale photovoltaics (PV) and storage, especially when they are combined with non-traditional financing methods. In particular, data show an 82% decline in the costs of PV modules and a 76% fall in the costs of Lithium Ion battery packs between 2010 and 2017 (Bloomberg 2018).

There are already several examples of the emerging role played by cheap solar power in new ways of bringing the internet to rural areas of poor developing countries. One of them, in Kenya, is a Microsoft start-up called "Mawingu," which draws on inexpensive wireless technology and solar energy to create fast internet networks in rural areas. To connect, users rely on Wi-Fi. Another example is drawn from India, where the government, plans to electrify every household in the country by 2019. And "solar power-including the use of small local grids – is likely to be a big part of the push, with 60 percent of new connections expected to be renewable power, according to a report by the International Energy Agency" (Jena 2018, 1).

By far the most novel attempt to bring the internet on a large scale to rural areas, however, is Google's Project Loon (Simonite 2015) (Figure 7.2). Indeed, the basic idea of the project is nothing less than an effort to bring billions of rural inhabitants online by means of helium balloons that are sent over areas where cell towers do not reach (Simonite 2015). The balloons float at an altitude of about 20 kilometres, almost twice the height of commercial flights. "Each balloon supports a boxy gondola stuffed with solar-powered electronics. They make a radio link to a telecommunications network on the ground and beam down high-speed cellular internet coverage to smartphones and other devices" (Simonite 2015). Though still in its early stages, this project has already brought connectivity to 120,000 people in Puerto Rico following the destruction there of infrastructure wrought by Hurricane Maria. And only this month, the Kenyan government announced that it would use the Loon technology to promote the internet in rural areas of the country (Sandoval 2018).

Only time will tell whether a sea-change in the provision of rural internet really occurs. What is encouraging though is the involvement of leading

Figure 7.2 Google Balloon.
Source: Wikicommons.

technology companies in the effort and the willingness of some developing countries to support them. Nor should one ignore the catalytic role in this of solar power and more specifically the fall in its price to levels that are now competitive.[20] Cheap storage too may be said to have come of age.

Conclusions

I began this chapter by listing three reasons for singling out rural areas in developing countries, when dealing with the geographies of global internet diffusion. The first is that on a range of basic indicators, such as literacy, income, and electricity, which help to explain internet penetration, the data show that there is a substantial gap between urban and rural locations in the same country (a so-called internal divide). The equivalent finding, of course, holds true with regard to internet penetration itself within developing countries (an internal digital divide). These divides, I suggested, were partly due to the remoteness and isolation of rural areas and partly to a structural political bias against them (an argument, one should note, which is consistent with, Lipton's influential (1977) idea of an "urban bias" in economic development).

The second reason to lay particular emphasis on rural areas, especially in Sub-Saharan Africa and Asia Pacific, is that the conditions prevailing there, foster the possibility of deriving particular and substantial gains from the penetration of the internet. Much of this has to do with the ability of such technology to shrink the vast distances that so typify the rural areas just described. What also needs to be emphasized in this regard is that many such beneficiaries would tend to be drawn from amongst the poor and female sections of rural societies.

In order to derive these potential gains, however, numerous severe constraints, such as affordability, relevance, skills, and infrastructure, need to be overcome. For each such barrier, the nature of the problems was described and some possible ameliorative policies were suggested. Although many of the problems seem more or less intractable in the short to medium term, several positive developments were discerned. These include the falling price of smartphones, the use of communal institutions to improve digital literacy, the involvement of certain global technology companies in generating solutions to infrastructural barriers in rural areas and the eagerness of certain developing countries to support such endeavors.

Some examples from India and Kenya, involving inexpensive solar energy, and the potential of mini-grids provide hope that a sea-change in the provision of the internet to rural areas might be underway, though it is still too early to reach any firm conclusions. The best-known and most advanced of such projects, though, is Google's Project Loon, which uses stratospheric helium balloons and solar energy to achieve connectivity in remote areas. It has already been applied with some success in the aftermath of the recent hurricane in Puerto Rico. One should not however underestimate the role in this of energy storage, which has improved technically and fallen in price as well.

Notes

1 The comparative remoteness of rural areas can be captured by comparing population densities there, as opposed to urban parts of developing countries. (See, for example, World Bank Indicators.)
2 In following these lines of argument, I rely heavily on reports prepared by consultancy firms and UN agencies. This is mainly due to a comparative paucity of academic literature on the subjects.
3 This project is described, among other places, in James (2018).
4 This project is also described in James (2018).
5 See, for example, Combi et al. (2016).
6 As described in McKinsey (2014), which also offers more data on the topic. See also Surman, Gardner and Ascher (2014).
7 For a detailed discussion see the World Bank (2018).
8 There are also more recent studies on the same topic by this author and her associates. (See, for example, Fehrler, Michaelowa and Wechtler 2009).
9 Indeed, according to one influential recent report, if current trends continue, the lack of available electricity will continue to serve as a significant obstacle to universal connectivity (World Bank 2018). The same report, moreover, makes the telling point that in the absence of electricity access in the home, connecting to the Internet becomes a significant challenge, due partly to the logistics and costs of mobile charging. Indeed, though people might live in communities with 3G networks, their use of a smartphone will be limited without electricity at home (World Bank 2018).
10 It is important to note in this regard that the constraints are not independent of one another.

11 Lipton's thesis was supported by Bates (2014) in his influential study on states and markets in Africa.

12 In addition to cost reductions achieved by a design that excludes certain features of smartphones, such as those that are more relevant in developed countries. Such designs fall readily under the heading of appropriate technology. They often emanate from countries such as China and India.

13 James (2004) covers many examples of how these communal institutions operate in developing countries to bring the internet to rural areas in such countries. Not all of them, however, are successful in this endeavor (Correa and Pavez 2016) and some are drawn from earlier periods. For some recent examples, see the ITU (2017a) and Breitenbach (2013). See also the World Bank sponsored program in Peru, which brought electrical power to more than 105,000 low-income rural households. The program involved communal institutions such as schools, health clinics, and community centres (World Bank 2014).

14 For a detailed, recent discussion see Latchem (2018).

15 As noted in the text, moreover, India has launched a major attempt to bring Internet connectivity to rural areas.

16 This is a process that Myrdal (1968) described as "cumulative causation." A related concept is a vicious circle (see Warren 2007).

17 There are, however, exceptions, such as India's "hole in the wall" project described in the text.

18 Though the reasons why this occurs are not always very clear.

19 See Simonite (2015). According to McKibben (2015) the solar panel is arguably the most "disruptive technology of the era." Its price has fallen, he argues, by 75% in the past six years.

20 A mini-grid is a "bank of batteries charged by solar panels and hooked up to homes, to guarantee round-the-clock power independent of the national network" (*The Economist* 2018, 59).

References

Bates, R. 2014. *Markets and States in Tropical States in Tropical Africa.* Berkeley, CA: University of California Press.

Bayes, A., von Braun, J., and Akhter, R. 1999. Village Pay Phones and Poverty Reduction: Insights from a Grameen Bank Initiative in Bangladesh. Bonn, Center for Development Research, University of Bonn, Discussion Paper No. 8. http://hdl.handle.net/10419/84729.

Bird, K., McKay, A., and Shinyekwa, I. Isolation and Poverty: The Relationship between Spatially Differentiated Access to Goods and Services and Poverty. London, Overseas Development Centre, Working Paper 322. htpps://www.odi.org/sites/odi.org.uk/files/odi-assets/publications-opinion-files/5516.pdf.

Bloomberg New Energy Finance. 2018. *Powering Last-Mile Connectivity.* London and Menlo Park, CA.

Breitenbach, M. 2013. Telecentres for sustainable rural development: Review and case study of a South African rural telecentre. *Development Southern Africa* 30(2):262–278.

Combi, C., Pozzani, G., and Pozzi, G. 2016. Telemedicine for developing countries: A survey and some design issues. *Applied Clinical Information* 7(4):1025–1050.

Correa, T., and I. Pavez. 2016. Digital inclusion in rural areas: A qualitative exploration of challenges faced by people from isolated communities. *Journal of Computer-Mediated Communication* 21(3):247–263.

Dave, L. 2017. Google owner Alphabet balloons connect flood-hit Peru. *BBC News* (May). www.bbc.com/news/technology-39944929.

Deloitte. 2014. Value of connectivity:economic and social benefits of expanding Internet access. file:///C:/Users/Owner/AppData/Local/Microsoft/Windows/INetCache/IE/U7MIK8S8/2014_uk_tmt_value_of_connectivity_deloitte_ireland.pdf.

Farooquee, A. 2017. Revisiting the rural energy challenge in developing countries. *Journal of International Affairs* https://papers.ssm.com/abstract=29.

Fehrler, S., Michaelowa, K., and Wechtler, A. 2009. The effectiveness of inputs in primary education: Insights from recent student surveys for sub-Saharan Africa. *Journal of Development Studies* 45(9):1545–1578.

Gasperini, L. 2000. The Cuban education system: Lessons and dilemmas. *World Bank Education Reform and Management Publication Services* 1(5):1–33.

Girard, B. (ed.) 2003. *The One to Watch: Radio, New ICTs and Interactivity.* Rome: FAO.

Gollin, D., Lagakos, D., and Waugh, M. 2011. The Agricultural Productivity Gap in Developing Countries. London: International Growth Centre, Working paper.

GSMA. 2012. *Connected Society: Unlocking Rural Coverage: Enables for Commercially Sustainable Mobile Network Expansion.* London: GSMA.

GSMA. 2017. *Accelerating Affordable Smartphone Ownership in Emerging Markets.* London: GSMA.

Hughes, S. 2003. Community multimedia centres: Creating digital opportunities for all. In B. Girard (ed.) *The One to Watch: Radio, New ICTs and Interactivity.* pp. 76–89. Geneva: UN Food and Agriculture Organization.

Intel. 2014. *Realizing the Benefits of Broadband.* White paper. Santa Clara, CA: Intel.

International Telecommunications Union (ITU). 2008. *E-Services through Post Offices in Bhutan.* Geneva: International Telecommunications Union.

International Telecommunications Union (ITU). 2017a. *Facts and Figures.* Geneva: International Telecommunications Union.

International Telecommunications Union (ITU). 2017b. *Connecting the Unconnected.* Davos: International Telecommunications Union.

James, J. 2004. *Information Technology and Development: A New Paradigm for Delivering the Internet to Rural Areas in Developing Countries.* New York: Routledge.

James, J. 2018. The internet and rural areas. In B. Warf (ed.). *Encyclopedia of the Internet.* Thousand Oaks, CA: Sage.

Jena, M. 2018. Solar power push lights up options for India's rural women. *Reuters* (Feb.) www.reuters.com/article/us-india-solarpower-women/solar-power-push-lights.

Latchem, C. 2018. *Open and Distance Non-Formal Education in Developing Countries.* Heidelberg: Springer.

Lipton, M. 1977. *Why Poor People Stay Poor: Urban Bias in World Development.* London: Temple Smith.

McKibben, B. 2015. Power to the people. *The New Yorker* (June). www.newyorker.com/magazine/2015/06/29/power-to-the-people.

McKinsey and Company. 2016. Urban World: The Global Consumers to Watch. file:///C:/Users/Owner/AppData/Local/Microsoft/Windows/INetCache/IE/YWLZ8MTP/Urban-World-Global-Consumers-Executive-summary.pdf.

McKinsey Global Institute. 2014. *Offline and Falling Behind: Barriers to Internet Adoption.* London: McKinsey Global Institute.

Michaelowa, K. 2001. Primary education quality in Francophone sub-Saharan Africa: Determinants of learning achievement and efficiency considerations. *World Development* 29(10):1699–1716.

Mitra, S. 2003. Minimally invasive education: A progress report on the "hole-in-the-wall" experiments. *British Journal of Educational Technology* 34(3):367–371.

Myrdal, G. 1968. *Asian Drama: An Inquiry into the Poverty of Nations.* New York: Random House.

Ouma, S., and M. Herselman. 2008. E-Health in rural areas: The case of developing countries. *International Journal of Humanities and Social Sciences* 2(4):304–310.

Sandoval, G. 2018. Kenya will reportedly use Alphabet's cutting-edge balloon project to deliver internet access to rural areas. *Business Insider* (July 6).

Sechele, A. 2016. Urban bias: Economic resource allocation and national development planning in Botswana. *International Journal of Social Science Research* 4(1):44–60.

Simonite, T. 2015. Project Loon. *MIT Technology Review* (March/April).

Stewart, F. 1977. *Technology and Underdevelopment.* London: Macmillan.

Surman, M., Gardner, C., and Ascher, D. 2014. Local content, smartphones and digital inclusion. *Innovations* 9(3–4):63–74.

The Economist. 2018. Empowering villages. (July 14, pp. 59–60). www.economist.com/finance-and-economics/2018/07/12/mini-grids-could-be-a-boon-to-poor-people-in-africa-and-asia.

The Hindu. 2017. Cabinet nod for rural digital literacy campaign. (Feb.) www.thehindu.com/business/Cabinet-nod-for-rural-digital-literacy-programme.

UNESCO, Institute for Statistics. 2011. *Global Education Digest 2011.* file:///C:/Users/Owner/AppData/Local/Microsoft/Windows/INetCache/IE/7TYIA5DU/global-education-digest-2011-comparing-education-statistics-across-the-world-en.pdf.

Warren, M. 2007. The digital vicious cycle. *Telecommunication Policy* 31(6–7):374–388.

World Bank. 2014. *Peru Brings Electricity to Rural Communities.* Washington, DC: World Bank.

World Bank. 2016. *World Development Report: Digital Dividends.* Washington, DC: World Bank.

World Bank. 2017. Access to electricity (% of population). https://data.worldbank.org/indicator/EG.ELC.ACCS.ZS.

World Bank. 2018. *World Development Report.* Washington, DC: World Bank.

Young, A. 2013. Inequality, the urban-rural gap, and migration. *Quarterly Journal of Economics* 128:1727–1785.

8 Geographies of global digital divides

James B. Pick and Avijit Sarkar

Global digital divides are influencing nations and people worldwide by constituting an inequality that influences the knowledge production, decision-making, and economic prosperity of people worldwide as well as geographic units such as nations and continents. This chapter will focus on many dimensions of the divides, including measurement, theories, associated factors of influence, geographic aspects, and implications for people and society. The world in 2018 had 4.2 billion people worldwide using the internet (ITU 2018) or 55% of the planet's population (Population Reference Bureau 2018). In developed countries more than 80% of people use the internet, compared to 45% in developing nations (ITU 2018). Broadband use has also reached record levels, with fixed broadband subscriptions just shy of 1 billion and mobile broadband subscriptions at 5.3 billion in 2018. Mobile broadband subscriptions grew at a yearly rate of 31.1% from 2007 to 2018 (ITU 2018). This overestimates the number of broadband users worldwide, since many users have multiple subscriptions.

The digital divide is defined as differences in access to, and purposeful use of, information technologies by individuals, or by administrative or geographical regions, such as cities, states, countries, and continents. Despite the word "divide," the digital divide is actually a continuum, so a firm cut-off of "haves" and "have nots" is rarely present. Often it is rated on several dimensions; for example, cell phones and broadband, forming an index for digital divide. The different dimensions can also be visualized creatively revealing several measures of digital capability, such as access and uses for two or three specific purposes.

Figure 8.1 shows the dramatic difference between the developed nation of Austria and the developing country Bolivia in 2016 and 2017. The spider diagrams contain a mixture of 11 factors, eight of which are measures of access and use of technologies, while the other three depict education. It is easy to see that from 2016 to 2017 the nations were stable in all dimensions except active mobile broadband subscriptions, which nearly went up over 50% in Bolivia and by about 15% in Austria. In comparing the two nations, Austria is higher in all dimensions but only slightly in secondary enrollment and internet bandwidth per internet user. The comparison illustrates that

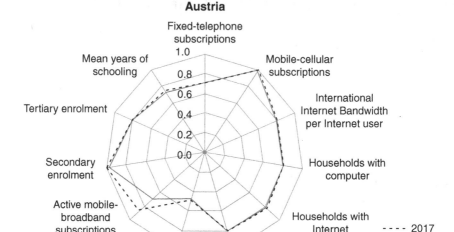

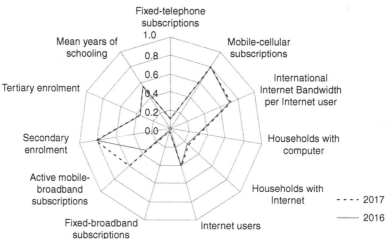

Figure 8.1 Spider diagrams showing dimensions of ICT capability for Austria and Bolivia, 2016 and 2017.

Source: ITU (2017).

including all the dimensions yields a more complex and revealing picture of digital differences.

The digital divide is important because its extent can influence social, economic, cultural, and political phenomena and outcomes. For instance, a high level of technology use can influence educational outcomes. In a meta-analysis of the effects of technology use on learning by over 36,000 students in mathematics, overwhelmingly students learned more if technology use was higher (Li and Ma 2010). The economic benefits of technology are well known. At the level of business, functional as well as enterprise information systems have contributed to the overall competitive capabilities of firms. For individuals, technology knowledge and skills are sought after in the marketplace, generally helping individuals gain employment and advance their careers. For nations, those with high levels of technology adoption and use are associated with strong economies. For example, in a study of the telecommunication industry's value-added contribution to the economy of BRIC nations in 2015, the Russian Federation had 50% value added compared to the entire information and communications technology (ICT) sector, with similar value adds for China of 48%, Brazil of 39% and India of 23% (World Economic Forum 2018). The post- recessionary economic growth in the US from 2007 to 2018 was partly led by the five FAANG firms of Facebook, Amazon, Apple, Netflix, and Google, all exemplars of technology use.

The digital divide is also important since it allows tracking of areas of the world with the lowest levels of technology use and suggests ways that can assist low technology nations bridge their digital gap compared to high technology countries. Indexes also include many dimensions and are discussed in the next section.

In measuring the digital divide, several other factors include unit of analysis, data collection method, and purpose of the tool (Barzilai-Nahon 2006). The unit might be individual, county, province, or nation and should be viewed differently, an issue in geography referred to as the modifiable area unit problem (MAUP). The MAUP problem is that the larger units of analysis are often aggregated from smaller units in an arbitrary way; therefore, they are not a random reflection of the smaller unit level. Another problem is that analysis and evaluation of relationships and effects is different between variables at the different units of analysis. For instance, the relationship between educational level and personal computer (PC) usage at the individual level concerns skills and behaviors learned and how they help encourage or discourage PC use. At the national level, the relationship is between education of a large population and the total computer use of the population, so interpretation might be that certain government educational programs influenced mass use of PCs.

Another aspect is selection of the data source (Barzilai-Nahon 2003). How was the technology data collected and what does that method imply? For instance, did broadband subscription data for a state come from all the internet providers in the state? If so, did they define broadband the same way? Did they count subscribers the same way?

Two well-known indexes for the digital divide, the ICT Development Index (IDI) of the International Telecommunications Union (ITU) and Network Readiness Index (NRI) of the World Economic Forum (WEF), provide overall measures that can be used for worldwide comparisons of nations. The components of the indices consist of variables for which data has been quality-checked and are collected from these two and other leading international organizations such as the World Bank, as well as national governments' statistical agencies. This section briefly discusses the IDI and NRI, and a reader is referred to the extensive methodology and global reports that are published annually or semi-annually (see ITU 2017, 2018; WEF 2016).

The ITU's IDI, first compiled in 2009, consists of 11 indicator variables that are combined into the index and are consistent across countries and over time. There are sub-indexes that constitute the IDI: Access sub-index (ICT readiness including access and infrastructure), Use sub-index (ICT intensity and usage indicators for internet, fixed broadband, and mobile broadband), and Skills sub-Index (capabilities and skills important for ICTs focuses on education). The indicators may change or be weighted differently over time. For instance, active mobile subscriptions per 100 inhabitants, one of the ICT use indicators has had to be weighted less as multiple subscriptions per user has grown. Examples of IDI index components for two countries are shown in Figure 8.1.

The WEF's NRI, first published in 2001, measures the impact of technologies on national development and productivity in order to indicate the level of competitiveness. It includes four categories of indicators: overall environment of technology utilization (innovation, business, regulation and innovation indicators), ICT readiness (ICT infrastructure, skills, and affordability), technology adoption and use (indicators from the three sectors of government, business, and individuals), and economic and social impact of the technologies (WEF 2016). Half of the 57 underlying indicators are drawn from the WEF's Executive Opinion Survey and the other half from government sources at the national level. In 2016 the NRI ranged from the top five nations (Singapore, Finland, Sweden, Norway, and US) with NRIs of 5.8–6.0 to the lowest five nations (Chad, Burundi, Haiti, Mauritania, and Madagascar) with NRIs of 2.2–2.6. The highest NRIs are in advanced economies with high educational levels, while the lowest NRI nations are mostly in Africa, the continent at the low level of the global digital access and use. However, the world's fastest ICT growth rates are currently in Africa.

The digital divide has progressed over time from levels of personal and cultural acceptance of ICTs, to access of ICTs and the internet, to use of ICTs/internet, and then to purposefully making use of ICTs/internet. For instance, two pioneering studies of the digital divide had dependent variables of UNIDO index rating of the most complex technology product produced by a nation (James and Romijn 1997) and the number of internet hosts per 10,000 population for a country (Hargittai 1999). Both dependent variables reflected ICT access. By the 2000s, studies had progressed beyond use to purposeful use, for instance types of uses of computers in households (Shih and Venkatesh 2004), outcomes of use by grade school students (Warschauer and Matuchniak 2010), and quality of use of broadband for nations, reflecting a "quality and capacity divide" rather than a divide of broadband access

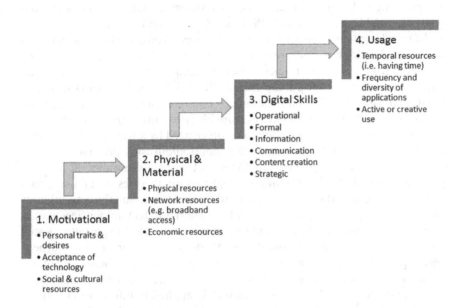

Figure 8.2 Stages of engaging with technologies, from motivation to purposeful usage.
Source: Authors, adapted from Van Dijk, 2012–2013.

(Vicente and Gil-de-Bernabé 2010). This evolution can be summarized as steps (Figure 8.2) in measuring the digital divide of physical and material access, "digital skills" connoting initial uses, and "usage" connoting purposeful, active, and diverse uses (van Dijk 2012, 2013).

As technologies become increasingly accessible and used worldwide, digital divide differences persist in the forms of depth of purpose of use as well as their impacts. If a skilled mobile phone user in Africa can use mobile broadband for agricultural analytics improving yields and quality of production, this would rate high on purposeful use and impact, versus unskilled mobile phone use for texting in Norway would be low on purposeful use and impact. A caveat is that new innovations will continue to appear in the future, for which purpose and impacts are not yet determined, so the initial divide would revert to access, not use.

Theories of the digital divide

As in many aspects of rapid technology change, the rules, laws, and theories have lagged the quick diffusion of the new technology. Unsurprisingly, theories of the digital divide have been few so far. Four that are briefly introduced are Adoption and Diffusion Theory (ADT), Unified Theory of Acceptance and Use of Technology (UTAUT), J. van Dijk's theories, and the Spatially Aware Technology Utilization Model (SATUM). These theories differ on independent and outcome indicators, analysis units, spatial components, and types of variables incorporated (Pick and Sarkar 2016).

ADT is based on an innovation such as a new product, which has a trajectory of early adopters upon release, leading to rapid growth with increasing diffusion from adopters to new ones, peaking at maximum, and finally diminishing with late adoption/diffusion (Rogers 2003). The theory discusses the mechanisms of diffusion including by communications and by geographical proximity, that is, an adopter is more likely to diffuse the innovation to a person located nearby. Innovations are categorized by attributes such as complexity that can be utilized to study the adoption/diffusion process.

UTAUT is a comprehensive theory of user technology acceptance (Venkatesh et al. 2003). It combined eight major theories of IT use into a single unified one and serves as a standard for understanding the determinants and relationships leading to behavioral intention and technology acceptance. The determinants include performance expectancy, effort expectancy, social influence, and facilitating conditions. Moderating factors include experience, age, voluntariness of use, and gender. UTAUT was found to outperform its eight foundational theories and was validated by rigorous testing on four organizations. UTAUT has been widely applied in information systems research on technology acceptance in a wide context. For digital divide research, UTAUT can be applied at the individual level in determining ICT acceptance and use, but not access or purposeful use.

Jan A.G.M van Dijk (2005) argued that inequalities stemming from an individual's position and personal background influences inequalities in the individual's resources, and in turn effect inequalities of ICT access and finally effects disparities by the individual in society and the economy. Those disparities feed back in a loop and influence positional and personal characteristics. The complex theory focuses on the individual unit of analysis but also includes "society," groupings that could range from local community to a nation or continent. An advantage of the theory is its inherent tailoring for digital divide studies and its complexity, as well as its richness of elaboration and posited mechanisms, while a weakness is difficulty in quantitively validating such a complex model (Pick and Sarkar 2016).

The last digital divide theory profiled is SATUM (Pick and Sarkar 2015, 2016). In SATUM, social, economic governmental, and innovation factors influence technology utilization factors, with preliminary screening for spatial autocorrelation of the ICT dependent factors. Spatial autocorrelation diagnoses whether a given dependent factor is randomly distributed or demonstrates significant positive or negative agglomeration. Ordinary least squares regression is applied to test the associations of independent factors with dependent variables, and the regression residuals are tested again for spatial autocorrelation. This process shows the strength of overall regression models as well as individual predictors and diagnoses spatial bias in ICT use (Pick and Sarkar 2015). SATUM is applicable across a wide variety of geographic units, and accounts for geographic proximity. Its weakness is uni-directional linear associations between ICT dependent variables and their independent correlates, rather than more elaborate ones that incorporate complex relationships and feedback.

Factors influencing digital divides

Over the years, a number of theories including the ones discussed in the previous section have been employed to explain ICT disparities between individuals, social groups, communities, provinces, nations, and national agglomerations. From the vast body of prior digital divide research, demographics, geography, social factors, economic variables, infrastructure, regulation, and related policy variables have been consistently found to be associated with ICT adoption, diffusion, and use. In more contemporary studies, social capital, and societal openness have also been important in explaining global and regional digital divides. This section provides a brief overview of factors that explain digital divides in various parts of the world.

Demographic factors: Age structure, race/ethnicity, gender, and educational attainment are key demographic variables known to be associated with disparities in ICT adoption and utilization. Age structure has been found to influence computer use as well as internet and mobile penetration in Japan and the United States (Ono and Zavodny 2007). Race/ethnicity-based discrepancies in internet adoption and use in the United States have been the subject of numerous reports since the early 2000s (NTIA 2002, 2014). Education is among the most well-known correlates of ICT use (Baliamoune-Lutz 2003; Gulati and Yates 2012). Individuals with higher levels of educational attainment are likely to be more skilled at adopting technology making them proficient users of ICTs, often to produce knowledge and related content. Gender has influenced digital divides in parts of the world and women have been found to use more electronic services with increased levels of internet adoption – even in advanced digital societies (Taipale 2013).

Geography: Urban–rural differences in technology adoption and diffusion has transcended extent of economic development, with urban agglomerations such as Tokyo, Beijing, Shanghai, New Delhi, and the Boston-Washington, DC. megalopolitan area often found to be leading hubs of ICT use (Pick and Sarkar 2015). Greater population density often facilitates the deployment of ICT infrastructure, reducing costs, and increasing adoption in urban areas. Urban–rural disparities have been examined in a variety of ways in digital divide studies, yet the role that geography plays in influencing technology levels among neighboring geographies, especially the statistical pitfalls associated with spatial autocorrelation and related spatial bias have not been appropriately tackled in a majority of digital divide theories and studies except SATUM (Pick and Sarkar 2015).

Economic and employment variables: The influence of micro- as well as macroeconomic variables has been studied extensively in the digital divide literature. Gross national income, median household income, and other related variables have been found to be associated with ICT adoption in many parts of the world (Ono and Zavodny 2007; Warf 2013; Pick and Sarkar 2015). The income-education duo has been found to positively influence ICT adoption. Other economic variables associated with greater technology use are openness to international trade (Pick and Azari 2008) and export commodities value (Pick, Nishida, and Zhang 2013). Recent studies have also revealed that higher per capita professional, scientific, and technical services employment – comprised of sectors that demand technology knowledge and skills is associated with higher ICT adoption (Pick, Sarkar, and Johnson 2015).

Policy variables: Governmental policy in education, trade, infrastructure deployment, ICT and telecom policies and regulations have all been the subject of digital divide studies. In developing economies such as India, infrastructure malaise such as unreliable power grids has been found to negatively influence teledensity and hinder broadband adoption (Pick, Nishida, and Sarkar 2014). High mobile-cellular tariffs, often a function of national ICT policies and regulations (Bouali 2017) have been found to be detrimental to ICT adoption and usage, in impoverished economies in Africa as well as in advanced digital societies (for example, the United States).

Social capital and societal openness: The importance of digital skills as part of a construct on skills access has been highlighted in a comprehensive causal and sequential model of digital technology access (van Dijk 2005). Studies have shown that a lack of ICT skills and experience will widen the digital gap (Mendonça, Crespo, and Simões 2015). As argued by Agarwal, Animesh, and Prasad (2009), individual choice to access the internet and other ICTs is subject to social influence ("peer effects") that often emanates from geographic proximity. Social capital in tightknit social communities, that manifests itself in the form of social interactions among peer users of ICTs provides material access and imparts digital skills thereby bridging the access and skills divide and facilitating ICT adoption (Chen 2013; Pick et al. 2014, 2015).

Internet censorship, especially perpetrated by authoritarian regimes in impoverished nations has been argued to be an impediment to technology adoption and societal inclusion by virtue of ICT usage (Warf 2013). Recent studies have shown that open, free, and democratic societies that encourage ICT-enabled communication and transfer of information and knowledge facilitate narrowing of the digital divide (Pick and Sarkar 2015). This in turn deepens democratic engagement among citizens, encourages political participation, and helps to combat corruption due to greater enforcement of laws, often aided by ICT usage (Sassi and Ali 2017; Tettey 2017).

Overall, this section thematically organizes factors that have traditionally influenced global and local digital divides. Readers are encouraged to consult the aforementioned and related studies for greater understanding of some these factors as they relate to the digital divide.

Geographies of the digital divide

Since the turn of the 21st century, worldwide adoption of information and communication technologies (ICTs) has increased as depicted in Figure 8.3. With the exception of fixed telephone subscriptions, subscriptions of mobile telephone, fixed- and mobile broadband, and internet penetration have all increased. Globally, the base of internet users expanded from 1.55 billion in 2008 to an estimated 4.5 billion in 2020. Global mobile-cellular subscriptions exploded from 2.2 billion in 2005 to an estimated 8.2 billion in 2018. Interestingly, in contrast to 2008, when internet users were roughly equally distributed between developed and developing nations, close to three-quarters of global internet users are estimated to be in developing nations. For mobile-cellular, 80% of global subscriptions are estimated to be in developing nations in 2018, compared to 55% in 2008.

At first glance, this distribution shift in user base over the past decade indicates a leveling of ICT adoption and digital divide between developing and developed nations. Yet, as evident from Figure 8.4, from a penetration standpoint, the discrepancy between developed and developing nations is significant. For example, in 2018, an estimated 81 individuals per 100 were internet users in developed nations, compared to 45 in developing nations, a penetration gap of 36%. For mobile subscriptions, there were an estimated

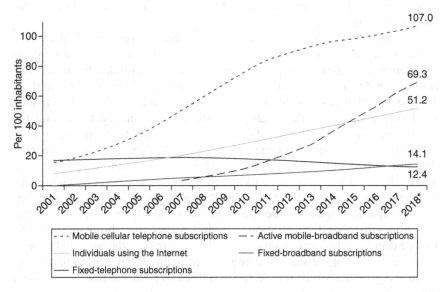

Figure 8.3 Global ICT developments, 2001–2018.

Note: *Estimated

Source: Authors, using data from International Telecommunication Union.

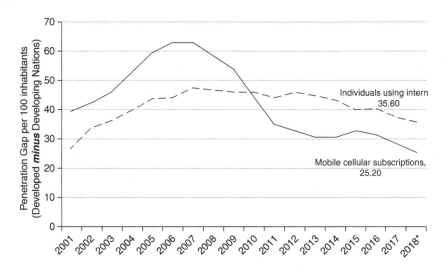

Figure 8.4 Penetration gap for internet and mobile cellular subscriptions, 2001–2018.
Note: *Estimated
Source: Authors, using data from International Telecommunication Union.

128 subscriptions per 100 inhabitants in developed nations against 103 in developing countries, a penetration gap of 25%. Pertinently, the penetration gap (Figure 8.4) has continued to decline over at least the past decade – an encouraging trend.

The next frontier will be closing the broadband access gap (Figure 8.5). Fixed broadband penetration levels remain much lower than mobile broadband, internet, and mobile-cellular penetration rates, with developing nations accounting for only 33 subscriptions per 100 inhabitants, compared to 10 in developing nations. In contrast, mobile broadband has far outpaced fixed broadband. In 2018, developing nations are estimated to account for 111 mobile-broadband subscriptions per 100 inhabitants, compared to only 61 in developing nations, a penetration gap of 50. Smartphones, connected devices and sensors, and the Internet of Things are revolutionizing ICT usage and enabling online social networking, mobile commerce (m-commerce), m-entertainment, m-health, teleworking among other innovations in contemporary, digitally connected societies. Therefore, closing this wide access gap in mobile-broadband penetration – clearly higher than both internet and mobile cellular subscription gaps per capita – represents the next challenge in narrowing the digital divide. Bridging the access divide will lead to narrowing of the usage divide, which is expected to ultimately results in transformational impacts across multiple frontiers.

Geography plays an important role in explaining digital divides, as discussed previously in this chapter. Descriptive mapping reveals important

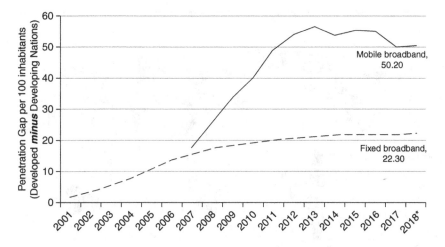

Figure 8.5 Penetration gap for fixed- and mobile-broadband subscriptions, 2001–2018.
Note: *Estimated
Source: Authors, using data from International Telecommunication Union.

visual cues on patterns of ICT access, adoption, and use, shows clusters of high versus low adoption and usage, and provokes thought on exploring underlying reasons for digital divides.

As depicted in Figure 8.6, Europe and the Americas led the world in internet penetration in 2016 with an average penetration of 80 and 71 users per 100 inhabitants. Latin and Central America's internet penetration lagged North America's by almost 28% in 2018 (Internet World Stats 2018). They are followed by the Arab states and the Asia-Pacific region. However, the penetration gaps between these world regions with Europe and Americas is significant, with average penetration at 55 and 47 internet users per 100 inhabitants in 2016. Exceptions are Bahrain, Qatar, and the United Arab Emirates in the Arab region, and Japan and South Korea in Asia-Pacific, where internet penetrations rival globally leading nations. Among Arab states, Bahrain recorded the biggest improvement in internet penetration with a gain of 43 users between 2000 and 2016. Also among the top 20 nations with the biggest gains in internet penetration between 2000 and 2016 were Qatar, Oman, Iran, Jordan, and Saudi Arabia – all in the Middle East. Much of this improvement can be attributed to an increasing focus on societal reform and openness, and an increasing reliance on ICTs for economic reform around knowledge services. Asia has the biggest base of global internet users – exceeding 2 billion in number, many of them in China and India. Yet, penetration in 2016 in both nations remained lower than the global average. Among major world regions, Africa is clearly a laggard in internet penetration with nearly 17% of the estimated world population in 2018, but only 11% of global internet users

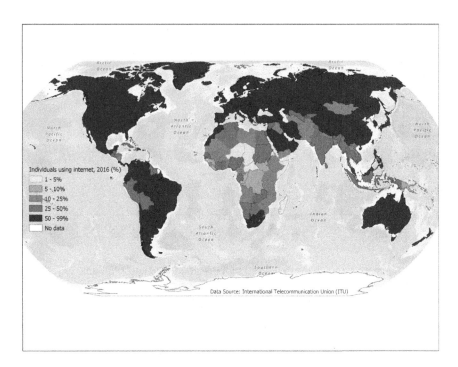

Figure 8.6 Global internet penetration rates, 2016.

(Internet World Stats 2018). Large parts of central and sub-Saharan Africa, long marred by domestic strife, wars, and low levels of human development have struggled to deploy ICT infrastructure. Yet, amidst a demographic transition and an anticipated burgeoning workforce (Ponelis and Holmner 2015), internet connectivity holds the key for transformational human development. For this, some of the leading African nations in terms of internet connectivity such as Morocco and South Africa can show the way. Explosive growth in the use of mobile-cellular devices in Africa holds another key for improving overall internet connectivity in impoverished parts of Africa.

Compared to a base of 4.2 billion internet users, the base of mobile subscribers – at 8.16 billion – is much bigger. Also, mobile subscriptions are estimated to average 107 subscriptions per 100 inhabitants, compared to global internet penetration of 51 users per 100 inhabitants in 2018. Despite the vast differences in subscription base and higher overall penetration, geographic disparities in mobile penetration are largely similar to internet penetration with Europe and the Americas as world leaders, way, closely followed by Asia-Pacific and the Arab states, and Africa as a laggard (Figure 8.7). Unsurprisingly, populous nations – China, India, the United States, Indonesia, Brazil, and Russia have some of the biggest bases of mobile subscribers. Since

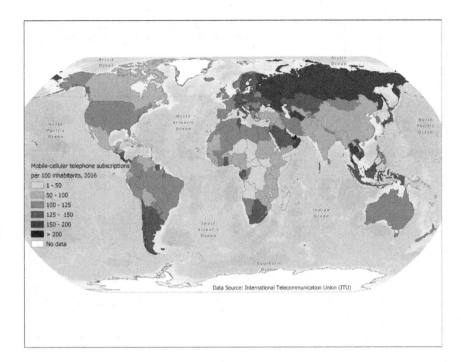

Figure 8.7 Global mobile-cellular penetration, 2016.

2001, the largest gains in mobile cellular penetration have occurred in disparate world regions such as Bahrain, Oman, United Arab Emirates, and Saudi Arabia in the Middle East and Arab world, Costa Rica, El Salvador, and Antigua and Barbuda in Central America and the Caribbean, populous nations such as Indonesia and Russia, Gabon, Gambia, and Ghana in sub-Saharan Africa, and finally Turkmenistan, Kazakhstan, and Ukraine among the Commonwealth of Independent States.

We conclude this section by observing that discourse on measuring the digital divide often centers on dimensions of access, with metrics such as ICT penetration levels. It has been argued that this traditional approach of measuring digital development in terms of telecom subscriptions is becoming increasingly obsolete, especially as mobile phone penetration reaches population saturation levels. Alternative approaches of measuring and tracking the digital divide, using installed bandwidth and effective traffic (fraction of bandwidth effectively used) have been suggested (Hilbert 2016). Recent reports from major telecommunications organizations such as the International Telecommunication Union (ITU) have started to report internet traffic – specifically fixed- and mobile-broadband use – in terms of international bandwidth usage by the percentage of the population using the internet in

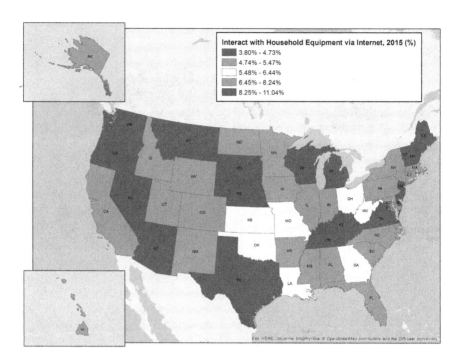

Figure 8.8 Internet use for interacting with household equipment (age 15+ internet
users,%), United States, 2015.

the country (ITU 2018). Pertinently, in mature and developed ICT economies, measuring and tracking effective and purposeful use of the internet
is becoming as prominent as measuring access itself. For example, Figure 8.8
depicts purposeful use of the internet for interacting with household devices
in U.S. states.

With rapid deployment of sensors, connected smart devices, and innovations
such as the Internet of Things (IoT) reshaping contemporary digital societies, tracking and measuring novel as well as traditional dimensions of purposeful internet utilization is becoming more mainstream and effective in
estimating digital divides.

The state of the digital divide in the United States, China, India,
Japan, and Africa

This section provides a brief overview of the state of the digital divide in select
world regions spanning Asia, Europe, North America, and Africa. First, we
focus on advanced economies and the mature digital societies in Japan and
the United States.

As an ICT leader, Japan has always been at the forefront of technology-driven innovation. ICT adoption, diffusion, and use in Japan has been guided by its multi-year plans dating back to the early 2000s, with telecommunications policy planning stretching back to the 1980s. As expected, ICT penetration levels in Japan are much higher compared to global and Asia-Pacific levels. Despite having an advanced information society, potential impediments for further growth in ICT adoption and utilization stem from the nation's age structure and language (Akiyoshi and Ono 2008). A rapidly aging demographic with an age dependency ratio, old (ratio of population older than 64 years to the working-age population, age 15–64 years) of 45.03 in 2017 – the highest in the world (World Bank 2018), coupled with a declining population, poses a unique challenge and is likely to exacerbate the digital divide in Japan. Urban–rural differences also exist; for instance, a 16-prefecture (province) largely rural cluster, located in the north of the main island and in southeastern coastal areas, was found to have the lowest levels of ICT adoption, two to five times lower than Tokyo – a megacity of 38 million inhabitants in 2016 (Nishida, Pick, and Sarkar 2016). Despite these impediments, recent research has shown that a focus on research and innovation, household educational expenditures, along with societal openness that encourages democratic norms and free flowing exchange of information, facilitated by ICTs are key to successfully bridging the internal digital divide in Japan (Nishida et al. 2014).

The United States, a highly developed, industrialized, western nation, has long been at the top of the ICT pyramid worldwide. In 2016, 76.2% of Americans were internet users, and differences in household use of the internet have long been reported by the NTIA to stem from differences in demographics and socio-economic status. Lower reaches of the digital divide in the U.S. have been concentrated among older, less educated, and less affluent populations, as well as in rural parts of the country that tend to have fewer choices and slower connections (Obama White House 2015). Recent studies have also revealed that social capital (Putnam 2001) – the extent of social connectedness between individuals in communities – plays a significant role in explaining the digital divide in the U.S. (Chen 2013; Pick, Sarkar, and Johnson 2016). It has been argued that an individual's decision to go online and use the internet is stronger in the communities with stronger social interaction. In other words, people who live in close proximity and maintain interactions with peers with a propensity to go online will also end up going online. However, social capital itself is not a panacea for bridging the digital divide in the United States. Similar to other advanced information societies, mobile-broadband usage in the U.S. is also on the rise, facilitating mobile commerce (m-commerce), m-entertainment, m-health, and various innovative uses of mobile broadband. Recent evidence suggests that closing the digital divide in the U.S. will require further efforts to reduce barriers in affordability, relevance, and computer literacy (Obama White House 2015).

As the world's most populous nations, China and India together accounted for 1.26 billion internet users (802 million in China and 462 million in India) in 2018, compared to 1.05 billion users in Europe and North America combined. Despite this enormous base of users, internet penetration levels remain low at 57% and 34% in 2018 in China and India respectively, against the global penetration level of 55%, and much higher levels in Europe (85%) and North America (95%) (Internet World Stats 2018). Like Japan and the United States, megacities in China (Beijing and Shanghai) and in India (New Delhi) have been found to be the ICT leaders in the two nations. In Beijing and Shanghai, internet, broadband, and mobile-cellular penetration rates have been at least five times higher than provinces in the central, western, and northern parts of the country (Pick, Nishida, and Zhang 2013). Similarly, in India, Delhi and its adjoining areas have been found to have ICT (broadband and internet) penetrations levels 13–16 times higher than laggard states in more rural parts of the country (Pick, Nishida, and Sarkar 2014). While mobile-cellular subscription penetrations in both countries approach saturation, mobile-broadband subscriptions per capita remain low, particularly in India (17 subscriptions per 100 population, compared to 70 in China in 2016). Recent research has revealed that infrastructural issues including unreliable power grids in rural India as well as broadband infrastructural malaise continue to impede rural teledensity in India (Pick et al. 2014). Yet, mobile penetration has been crucial for the nationwide implementation of India's biometric identification system *Aadhar* that aspires to promote social and financial inclusion – especially among the disenfranchised in remote, rural parts of India (Ghosh 2019).

Similar to the United States, social capital has been found to be key in bridging the digital divide in India. Indian cooperative societies have often been at the forefront of ICT adoption and utilization – especially in rural communities, by fostering ties, providing access to ICT knowledge and skills, spreading the ICT cost burden, and providing a platform of ICT-based cooperation among peers in agriculture, farming, housing, banking, and microfinancing, to name a few sectors (Pick and Sarkar 2015). In China, export commodities value has been found to be significantly associated with technology adoption and utilization (Pick, Nishida, and Zhang 2013). With ICTs playing an increasingly key role in Chinese manufacturing, governmental policies that expedite the optimization and upgrading of broadband networks, increase the level of broadband network applications, and promote digital aspects of continuous improvement of industrial supply chains have been recommended to bridge the gap between large megacities and rural Chinese provinces (ITU 2018).

Lastly, as a major world region, Africa's digital divide represents an extraordinary and evolving case study. Since the turn of the century, Africa's digital transformation has been marked by explosive growth in the base of internet users and mobile-cellular subscribers. Internet usage in Africa has grown from a base of 15 million in 2005 to 213 million in 2017 – a growth of 1353%. Equally impressive is the expansion in mobile-cellular subscriptions which has grown from 87 million in 2005 to 744 million in 2017 – a growth of approximately 750% (ITU 2018). Despite this expansion in user base, penetration levels remain dismally low, with only 22 internet users per 100 people in 2017. Within Africa, disparities in ICT adoption and utilization are stark, with leaders such as Egypt, Libya, Mauritius, Morocco, South Africa, and Tunisia having ICT levels 5–25 times higher than laggards in sub-Saharan and other parts of Africa such as Algeria, Chad, Ethiopia, Gambia, Ghana, Mozambique, Sudan, Tanzania, and Uganda (Pick and Sarkar 2015). Gross national income and industrial output have been found to be key in explaining ICT disparities in Africa along with societal openness. Low levels of human development especially among sub-Saharan African nations, impacts educational attainment lowering research and knowledge production in many African countries, which inhibits ICT adoption and use (Ponelis and Holmner 2015). Internet censorship in varying levels (Warf 2013) remains a key impediment to free and transparent exchange of communication and information using ICTs in Africa (Pick and Sarkar 2015). ICT-enabled technology entrepreneurship is spurring innovation, enhancing e-participation, and accelerating rural development in Africa (Ochara and Mawela 2015; Mwangi and Brown 2015). As a result of regional ICT for development (ICT4D) initiatives, the focus of the early 2000s – bridging the African digital divide by overcoming connectivity and access barriers, is gradually shifting to uptake and impact of ICTs to transform African economies and societies (Ponelis and Holmner 2015).

In summary, a remarkable tapestry of complex underlying issues – evolving demographics, human development, infrastructural malaise, persistent urban–rural gaps, internet censorship, social interactions, and connectedness characterize digital divides and pose policy challenges in various world regions. Many of these issues are geography-specific, while a few are location invariant. As global internet and mobile-cellular penetrations approach saturation, and focus shifts from measuring and bridging the access divide to examining disparities in purposeful usage of ICTs, understanding the roles and impacts of ICTs relative to local geographies towards fulfillment of development goals represents the next frontier for academic research and policy conversations in the digital divide arena.

Implications of the digital divide

Digital divides are influencing education, politics, labor markets, knowledge distribution, and business worldwide. For individuals, the exposure to a level of technology can open opportunities, influence success in career paths, change social relationships, and effect the breadth of information and knowledge. Since the focus of this chapter is on global digital divides, this section will concentrate on the world level and put the implications of the digital divide in the context of its impact on achieving goals for the planet. The United Nations Sustainable Development Goals (SDGs) are considered and those being particularly impacted by the digital divide are examined. Of the 17 UN SDGs, the one most related to the digital divide is Goal 9 (Build resilient infrastructure, promote inclusive and sustainable industrialization and foster innovation) (UN 2018). Technology is assisting in designing and building infrastructure, for instance software is used to design, model, build, and maintain buildings, factories, highways, airports, and products. Accordingly, people's skills and the technology environment in certain locations can favor their participation and their capability to design and build infrastructure.

Finally, innovation is becoming a major driver of economies (UN 2016), and it depends on an environment of high tech, technology educated and trained workforce, and the presence of advanced technology platforms. In the sequence discussed earlier of access to purposeful use to impact, innovation is at the far end of this sequence, using ICT in a creative, impactful way. Hence the digital divide status is strongly linked with the ability to have innovation in a nation or province/state.

Given the relationship of digital divide to the UN's SDC Goal 9, impoverished nations with low economic levels, should consider multiple initiatives to boost their ICT status up at least to the mid-level global standards. These might be changes in government policies, education, investment, and public-private partnerships. An example of a nation that advanced from a lower-middle status in ICT to a high level is Estonia. After gaining independence from the Soviet Union in 1992, Estonia pursued strong central government policies in successive federal administrations, as well as partnerships with leading technology firms in Finland and other neighboring nations. Its education was also overhauled that led to the fostering of highly computer-skilled graduates, enabling its rise in ICT status (Dutta 2007).

Concluding remarks

This chapter has described the background, definitions, and measurement of the digital divide, and pointed to a progression sequence in the digital-divide level indicator from access to use to purposeful use. The theoretical background on the digital divide is limited compared to its magnitude as a global issue. Four theories of ADT, UTAUT, van Dijk's theory, and SATUM, are discussed, each with its strengths and weaknesses.

Much is known about factors influencing the digital divide, which include demographic, economic, social, societal-openness, geographic, and regulatory factors. These factors can be a solid starting base for future digital divide research, and also inform policy and decision makers.

A geographic overview of the global digital divide reveals strong technology zones such as the US Northeastern megalopolis, Silicon Valley, Western and Northern Europe, and advanced East Asian countries including Singapore, Hong Kong, South Korea, and Japan. The lowest technology regions are in Africa and parts of South America. The chapter further explores more detailed geographies at the state and provincial levels for China, India, Japan, and the United States. The growing knowledge of the geography of the global digital divide can be useful to international organizations, trading blocs, alliances of nations, and multi-national companies or small enterprises with digital markets worldwide.

The chapter concludes by considering the implications of the global digital divide in terms of the UN Sustainable Development Goals, and in particular that of Goal 9 which includes digital impacts on infrastructure, manufacturing, and innovation.

References

Agarwal, R., Animesh, A., and Prasad, K. 2009. Research note – Social interactions and the "digital divide": Explaining variations in internet use. *Information Systems Research 20*(2):277–294.

Baliamoune-Lutz, M. 2003. An analysis of the determinants and effects of ICT diffusion in developing countries. *Information Technology for Development 10*(3):151–169.

Barzilai-Nahon, K. 2006. Gaps and bits: Conceptualizing measurements for digital divide/s. *Information Society 22*(5):269–278.

Bouali, S. 2017. Regulated termination rates and competition among Tunisian mobile network operators. Barriers, bias, and incentives. *Telecommunications Policy 41*(7–8):573–586.

Chen, W. 2013. The implications of social capital for the digital divides in America. *Information Society 29*(1):13–25.

Dutta, S. 2007. Estonia: a sustainable success in networked readiness? In Dutta, S. and Mia, I. (eds.). *The Global Information Technology Report 2006–2007*. pp. 81–92. New York: Palgrave Macmillan.

Ghosh, S. 2019. Biometric identification, financial inclusion and economic growth in India: Does mobile penetration matter? *Information Technology for Development 25*(4).

Gulati, G., and Yates, D. 2012. Different paths to universal access: The impact of policy and regulation on broadband diffusion in the developed and developing worlds. *Telecommunications Policy 36*(9):749–761.

Hargittai, E. 1999. Weaving the Western Web: Explaining differences in internet connectivity among OECD countries. *Telecommunications Policy 23*:701–718.

Hilbert, M. 2016. The bad news is that the digital access divide is here to stay: Domestically installed bandwidths among 172 countries for 1986–2014. *Telecommunications Policy 40*(6):567–581.

Internet World Stats. 2018. Usage and Population Statistics. www.internetworldstats.com.

ITU. 2017. Measuring the Information Society Report 2017. www.itu.int/en/ITU-D/Statistics/Pages/publications/mis2017.aspx.

ITU. 2018. ICT Statistics. www.itu.int/en/ITU-D/Statistics/Pages/stat/default.aspx.

ITU. 2018. *Measuring the Information Society Report*, Volume 1. Geneva: International Telecommunications Union.

James, J., and Romijn, H. 1997. The determinants of technological capability: A cross-country analysis. *Oxford Development Studies 25*(2):189–207.

Li, Q., and Ma, X. 2010. A meta-analysis of the effects of computer technology on school students' mathematics learning. *Education Psychology Review 3*:215–243.

Mendonça, S., Crespo, N., and Simões, N. 2015. Inequality in the network society: An integrated approach to ICT access, basic skills, and complex capabilities. *Telecommunications Policy 39*(3–4):192–207.

Mwangi, B., and Brown, I. 2015. A Decision Model of Kenyan SMEs' consumer choice behavior in relation to registration for a mobile banking service: A contextual perspective. *Information Technology for Development 21*(2):229–252.

Nishida, T., Pick, J., and Sarkar, A. 2014. Japan's prefectural digital divide: A multivariate and spatial analysis. *Telecommunications Policy 38*(11):992–1010.

NTIA. 2002. A Nation Online: Internet Use in America. Washington, DC: National Telecommunications and Information Administration.

Obama White House. 2015. Mapping the digital divide. https://obamawhitehouse. archives.gov/sites/default/files/wh_digital_divide_issue_brief.pdf.

NTIA. 2014. *Exploring the Digital Nation: Embracing the Mobile Internet.* Washington, DC: National Telecommunications and Information Administration.

Ochara, N., and Mawela, T. 2015. Enabling social sustainability of e-participation through mobile technology. *Information Technology for Development* 21(2):205–228.

Ono, H., and Zavodny, M. 2007. Digital inequality: A five country comparison using microdata. *Social Science Research 36*(3):1135–1155.

Pick, J., and Sarkar, A. 2015. The global digital divide. In J. Pick and A. Sarkar (eds.) *The Global Digital Divides.* pp. 83–111. Dordrecht: Springer.

Pick, J., and Sarkar, A. 2016. Theories of the digital divide: critical comparison. *Proceedings of the 49th Hawaiian International Conference on System Sciences IEEE*, pp. 3888–3897.

Pick, J., Nishida, T., and Sarkar, A. 2014. Broadband utilization in the Indian states: Socio-economic correlates and geographic aspects. In J. Choudrie and C. Middleton (eds.) *Management of Broadband Technology Innovation*, pp. 269–296. Oxford: Routledge.

Pick, J., Nishida, T., and Zhang, X. 2013. Determinants of China's technology availability and utilization 2006–2009: A spatial analysis. *The Information Society 29*(1):26–48.

Pick, J., Sarkar, A., and Johnson, J. 2015. United States digital divide: State level analysis of spatial clustering and multivariate determinants of ICT utilization. *Socio-Economic Planning Sciences 49*:16–32.

Ponelis, S. R., and Holmner, M. 2015. ICT in Africa: Building a better life for all. *Information Technology for Development 21*(2):163–177.

Putnam, R. 2001. *Bowling Alone: The Collapse and Revival of American Community.* New York: Simon and Schuster.

Rogers, E. 2003. *Diffusion of Innovations*, 5th ed. New York: Free Press.

Sassi, S., and Ali, M. 2017. Corruption in Africa: What role does ICT diffusion play. *Telecommunications Policy 41*(7–8):662–669.

Shih, C. and Venkatesh, A. 2004. Beyond adoption: Development and application of a use-diffusion model. *Journal of Marketing 68*(1):59–72.

Taipale, S. 2013. The use of e-government services and the internet: The role of socio-demographic, economic and geographical predictors. *Telecommunications Policy 37*(4–5):413–422.

Tettey, W. 2017. Mobile telephony and democracy in Ghana: Interrogating the changing ecology of citizen engagement and political communication. *Telecommunications Policy 41*(7–8):685–694.

U.N. 2018 *The Sustainable Development Goals Report*. New York: United Nations.

Van Dijk, J. 2005. *The Deepening Divide: Inequality in the Information Society*. Thousand Oaks, CA: Sage.

Van Dijk, J. 2012. The evolution of the digital divide: The digital divide turns to inequality of skills and usage. In J. Bus, M. Crompton, M. Hildebrandt, and G. Metakides (eds.) *Digital Enlightenment Yearbook*. pp. 57–75. Amsterdam: IOS Press.

Van Dijk, J. 2013. A theory of the digital divide. In M. Ragnedda and G. Muschert (eds.) *The Digital Divide: The Internet and Social Inequality in International Perspective*. pp. 29–51: London: Routledge.

Venkatesh, V., Morris, M., Davis, G., and Davis, F. 2003. User acceptance of information technology: Toward a unified view. *MIS Quarterly 27*(3):425–478.

Vicente, M., and Gil-de-Bernabé, F. 2010. Assessing the broadband gap: From the penetration divide to the quality divide. *Technological Forecasting and Social Change 77*(5):816–822.

Warf, B. 2012. *Global Geographies of the Internet*. New York: Springer.

Warf, B. 2013. Contemporary digital divides in the United States. *Tijdschrift voor Economische en Sociale Geografie 104*(1):1–17.

Warschauer, M., and Matuchniak, T. 2010. New technology and digital worlds: Analyzing evidence of equity in access, use, and outcomes. *Review of Research in Education 34*(1):179–225.

WEF. 2016. The Global Information Technology Report 2016. Geneva, Switzerland: World Economic Forum.

World Bank. 2018. World Bank Data. https://data.worldbank.org/indicator/SP.POP. DPND.OL.

Part II

Political economy of the internet

9 The geography of e-commerce

Bruno Moriset

Electronic commerce has grown today to enormous proportions, disrupting entire industries such as book and music publishing and retail, travel agencies, taxis, and hospitality. On September 4, 2018, Amazon.com became the second company in history after Apple to hit one trillion US dollars of market capitalization (DePillis 2018). In China, e-commerce may have captured about 28% of the global retail market (Statista 2019). From its beginning, e-commerce has attracted the attention of geographers (Leinbach and Brunn 2001; Wilson 2000; Couclelis 2004; Murphy 2007), although its geographic dimension is not easy to capture. For centuries, stores, street markets, shopping malls, and "big box" supermarkets, have been instrumental in the design, growth, and social life of cities. However, since goods and services can be purchased on computers and smartphones without regard to time and location, and delivered at home a few days or even a few hours later, the value chain of e-commerce has lost most of its geographic significance in the eyes of the public. To some extent, e-commerce is subjected to the "death of distance" or "end of geography" fallacies that have, from the beginning, biased the comprehension of the digital economy (Malecki and Moriset 2008). Therefore, the primary purpose of this chapter is to show that e-commerce does have a geography, and to explore this geography.

To understand the geographic features of e-commerce, one needs to comprehend the fundamentals of its business models, presented in the first section. The second section explores the regional idiosyncrasies of e-commerce. The world map of e-commerce diffusion is uneven. It is dictated by purely geographic features such as distance to market and population density, and by economic, social, and political factors, such as age structure, purchasing power, tax regimes, and regulation of commerce, and the degree of information technology diffusion (IT-readiness). The third section is dedicated to the critical link between e-commerce and logistics. The delivery of goods to customers represents most of the physical presence of e-commerce, and the major part of its environmental footprint. Therefore, it is the overarching geographic issue of e-commerce. The fourth section investigates the local implication of e-commerce growth, in urban and rural environments.

Definitions and business models of e-commerce

E-commerce describes

> sales of goods and services where the buyer places an order, or the price
> and terms of the sale are negotiated over an internet, mobile device
> (m-commerce), extranet, Electronic Data Interchange (EDI) network,
> electronic mail, or other comparable online system. Payment may or may
> not be made online.
>
> (U.S. Census Bureau 2018)

In the broadest meaning, e-commerce may span a wide spectrum of IT-
intensive activities such as finance, telecommunications, software, and computer services (Leinbach and Brunn 2001). The present paper focuses on
activities that are close to "commerce" in the common sense, in particular
wholesale and retail trade.

The main categorization of e-commerce used in the literature is based on the
legal statuses of vendors and buyers (Figure 9.1). Trade among professionals
or firms, notably wholesale operations, is said to be "business-to-business"
(B2B). Sales by professional vendors or commercial enterprises to individual,
end-customers fall in the category of "business to customer" (B2C), or
e-retail. The intermediation power of web platforms has allowed millions of
individual demands and offers to meet together, giving birth to the concept of
"consumer-to-consumer" (C2C) e-commerce. "Consumer-to-business" (C2B)
is a less common occurrence which is considered when individuals sell goods
and services to a firm, by posting them on blogs, social networks, or dedicated
portals. For example, people can post photos on Adobe Stock, where they
can be purchased by professional users such as advertisement agencies and
magazines.

Buyer Vendor	Business *(company,* *professional)*	Customer *(individual)*
Business (company, professional)	**B2B** *Amazon Alibaba.com* *Made-in-China* *Tradeindia*	**B2C** *Amazon* *AliExpress* *T-Mall* *JD.com*
Customer (individual)	**C2B** *Google AdSense* *Adobe Stock* *Jobster*	**C2C** *Taobao* *eBay* *Airbnb*

Figure 9.1 E-commerce categorization and examples of firms (in italics).

Other classifications are sometimes used. "Business-to-government" (B2G) describes transactions among private firms and public agencies or departments. The internet is now widely used by governments throughout the procurement cycle at both national and local levels and requests for proposal are commonly posted on public agencies' websites. "M-commerce" describes online transactions made through smartphones and other mobile devices.

Logically, the attention of the public and scholars has been caught by B2C e-commerce, or e-retail, which has entered our everyday lives. Meanwhile, B2B has remained in the shadow (Lilien 2016), because people usually ignore value chain operations that occur between the factory and the store. The procurement chain that is necessary to the production of final goods and services is an increasingly globalized flow of intermediary goods, parts, and business services, which requires an endless series of commercial contracts, mostly hidden to consumers. Because the total of sales in a given value chain exceeds by far the final value-added and the price charged to end-customers, it is unsurprising that global B2B sales are much bigger than B2C. Statista (2017) estimates that the worldwide B2B e-commerce market in 2017 was worth 7.7 trillion U.S. dollars, to compare with B2C (e-retail) global sales of US$2.3 trillion. It is often uneasy to disentangle B2B and B2C e-commerce. For example, major online retailers such as Amazon and Alibaba also operate as neutral platforms for inter-firm transactions.

Electronic platforms and business models of e-commerce

The critical role of web-based platforms derives from their limitless capacity of switching and disintermediation. The "miracle" of Airbnb, eBay and the likes, is that billions of people in the world can get information such as that an apartment in Paris will be available for rent during the next holidays, that a lawn mower is for sale, or that someone can offer seats in their car for a given trip for a modest fee. Disintermediation occurs when incumbent firms and institutions such as hotel chains, travel agencies, taxi companies, and music labels, are squeezed out or "uberized" by web platforms through which individuals and small firms do business directly. Similarly, traditional retailers may be passed over by commercial websites, which play the role of the sole intermediary between manufacturers and consumers.

In the early years of e-commerce, online stores were trading like traditional merchants, buying goods wholesale and reselling them to customers with a profit margin. Amazon and its followers such as Alibaba later implemented a dramatic business innovation, the opening of their platforms to third-party merchants and sellers. In this "pure marketplace" model, the platform does not get ownership of the goods at any point of the trading process, but charges a fee on the transaction. This model is dominant within Alibaba's ecosystem (T-Mall, Taobao, AliExpress). The presence on leading marketplaces carries two main advantages for small and medium-sized enterprises: increased commercial visibility, and the access to global logistical systems (I elaborate on this topic below).

Table 9.1 Amazon, Alibaba Group, and JD.com main data in 2017

	*Amazon.com, Inc.**	*Alibaba Group*	*JD.com, Inc.*
Gross merchandise volume (GMV)	300? (estimation)	768.4**	199.0
Annual revenue	177.9	39.9	55.7
Cost of sales	111.9	17.1	47.9
Income (loss) from operations	4.1	11.0	(0.128)
Net income (loss)	3.0	9.8	(0.02)
Active consumers	-	552 million	292 million
Growth of revenue 2016–2017	30.8%	58.1%	40.3%
Brands	Amazon, Amazon Fresh Amazon Go, Prime Pantry, Prime Now, Whole Foods	Taobao, Tmall, AliExpress, Alibaba.com	JD.com, Yihaodian

Source: Amazon.com, Inc. 2018, Alibaba Group Holding Ltd. 2018, JD.com, Inc. 2018.). Data for the fiscal year ended December 31, 2017 (Amazon, JD.com) and March 31, 2018 (Alibaba Group). Financial data in billion U.S. dollars.

Notes: * Includes results of Whole Foods Market from the date of acquisition by Amazon, August 28, 2017. ** Chinese market only.

The difference between online merchants and mere marketplaces (Amazon is both) makes the interpretation of sales data a tricky exercise. There is a huge difference between the "revenue" (total of sales) of a given company and the "gross merchandise volume" (GMV) actually exchanged on its marketplaces. The Alibaba Group's revenue (about 40 billion US dollars in 2017) is mostly made of fees charged on transactions made by third-party sellers. It pales in comparison to its global GMV, which surpasses those of Amazon by a wide margin (Table 9.1). The comparison with JD.com, China's second largest e-retailer, is revealing: like a supermarket chain, JD.com buys wholesale and resells in retail. Bluntly, JD.com sells merchandise while Alibaba sells trading services. To add to the complexity, some firms have a proportion of non-trade activities: for example, Amazon's subsidiary Amazon Web Services is dedicated to cloud computing and data storage services.

Economies of scale and network effects: a "winner-take-all" economy

The economy of digital platforms shows "winner-take-all" features (Kenney and Zysman 2016) to a degree unseen since Andrew Carnegie and John D. Rockefeller, the founders, respectively, of U.S. Steel and Standard Oil, which essentially held monopolies over the American steel and oil industries in the early 20th century. It was first epitomized by the dominance of Microsoft and Intel. It is today illustrated by the meteoric rise to near-monopoly status of Google and Facebook. The supremacy of Amazon and Alibaba, the successes of Uber and Airbnb, tell a similar story. This tendency stems from two intertwined features of web-based economic models: economies of scale and network effects.

Economy of scale – a fundamental concept in microeconomics – predates the digital economy. It occurs when costs of operation within a company are constant or rise at a slower pace than the final output and the volume of sales, leading to increasing margins and bigger returns on capital. That explains the race for size common in all industries. In the sphere of e-commerce, the rule of economies of scale leads to a Darwinian process of selection and concentration. For example, a commercial website at the startup stage should "burn" huge amount of money in advertisement campaigns, while its competitors do the same. Only a few will sustain the process, achieve commercial success, and capture the market share of those who have failed (Noe and Parker 2005). For surviving firms, the sales/cost of ads ratio will increase and profitability will be in sight. Other effects of scale are seen in fulfillment operations such as warehousing and delivery.

Network effects are a complement of economies of scale. "Metcalfe's Law" says that the value of a network is theoretically proportionate to the square of the number of its users (Metcalfe 1995). Getting more visitors and shoppers on a commercial website makes it possible to offer a larger catalog and to attract more third-party vendors who offer new goods and services, and fuel the attractiveness of the platform, with self-reinforcing, systemic effects.

The intermediation power of the internet has given birth to a "sharing economy," sometimes called a "wikinomics" (Tapscott and Williams 2006), which can be identified in all four categories of e-commerce. These concepts are linked with the idea of "open innovation" (Chesbrough 2003). Rising complexity and the shortening of product lifecycles and time-to-market delays have become definitive features of the digital economy. In this context, large firms must look outside the silos of their internal creative resources (for example R&D and design labs). In diverse industries such as aircraft and automobile manufacturing, or software, major companies have open dedicated platforms to third-party firms, private and public labs, and even individuals. For example, software applications ("apps") for Apple's iPhones are sourced to millions of firms and individuals worldwide. On June 1, 2017, Apple announced that its App Store, the platform on which iPhone users connect and download applications, has generated global revenues of $70 billion since 2008 (Business Wire 2017).

The uneven diffusion of e-commerce

A retrospective view is necessary to comprehend the magnitude of change driven in recent years by the growth of e-commerce. E-commerce made timid debuts in the pre-internet era with the French Minitel in 1982 and Thomson

Holidays, a travel agency, which experimented with electronic sales through a TV screen and a mini-computer interface (Palmer 1988). Modern e-commerce began in the early 1990s in parallel with the diffusion of the internet. Amazon was founded in 1994. Much hype followed. The promise of the "death of geography" attracted thousands of entrepreneurs and inflated the "dot.com bubble," notably in the U.S. However, most of the web-based ventures had underestimated the requirement of fulfillment – customer support and logistics – and collapsed in the "dot.com bubble bust" of the early 2000s (Day et al. 2003; Indergaard 2004; Zook 2005). In 2005, B2C (retailing) e-commerce market share in the U.S. amounted to a mere 2.4%, up from 1.1% in 2001 (U.S. Census Bureau 2006).

Today, e-retailing can by no means be regarded as a marginal phenomenon, with market shares exceeding 10% in many countries, and reaching an astonishing 28% in China (Table 9.2). Worth noting is the dramatic growth of sales made through mobile devices, mainly smartphones, or M-commerce. According to eMarketer (2018), it reached 40% of online sales in the U.S. in 2018, and 75% in China – worth one trillion dollars.

The recent growth of e-commerce has been driven by the coming to maturity of the digital society, or "IT-readiness": mass diffusion of the internet, powerful connected devices, reliable online payment services, and digitally-enhanced logistics solutions. The uneven development of e-commerce is explained by other local idiosyncrasies such as geography (density, percentage of urban population, transportation infrastructures), age structure (the

Table 9.2 Retail e-commerce figures in selected economies, estimates for 2018

	Sales 2018 (billion dollars)	*Percentage of total retail sales*	*Main local e-commerce brands*
China	1530	28.6	Tmall, Taobao, JD.com
USA	526	10.0	Amazon
UK	126	21.5	Tesco, Asos, Argos
Japan	122	9	Rakuten
Germany	71		Otto, Zalando
South Korea	50.5	11.3	Gmarket, 11Street
France	44	7.0	Cdiscount, Vente privée,
India	32.7	2.9	Flipkart, Snapdeal

Sources: www.emarketer.com/content/retail-and-ecommerce-sales-in-china-2018; www. statista.com/statistics/379087/e-commerce-share-of-retail-sales-in-china; www.emarketer. com/Chart/Top-10-US-Companies-Ranked-by-Retail-Ecommerce-Sales-Share-2018-of-US-retail-ecommerce-sales/220521; www.emarketer.com/Report/UK-Retail-Ecommerce-eMarketers-Updated-Estimates-Forecast-20162021/2002188; www.statista.com/statistics/289736/japan-retail-e-commerce-sales-figures; www.emarketer.com/Article/Retail-Ecommerce-Germany-Top-65-Billion/1016261; www.statista.com/statistics/289745/south-korea-retail-e-commerce-sales; www.emarketer.com/Report/Retail-Ecommerce-France-Europes-Third-Largest-Market-Posts-Steady-Gains/2002099; www.emarketer.com/content/india-s-ecommerce-market-continues-to-surge

elderly usually trade less online), and, above all, purchasing power: in 2018, B2C e-commerce sales in India (1.3 billion people) were smaller than France's (67 million).

Local regulations and commerce history matter. For example, the UK has a well-established tradition of mail-order sales, with a high degree of customer protection. That explains, in part, the high market share of e-commerce over global retail sales, which is second only to China.

However, every country pales in comparison to China in terms of e-commerce volume, market share, and speed of growth. In the last decade, China's online sales surged from near zero to worth 1.5 trillion US dollars, about three times those of the U.S. This achievement has been driven by a bundle of factors: the meteoric economic growth of the 2000s and 2010s, the tech-oriented consumption appetite of a young and newly enriched, urban middle class, and the innovative effort of local IT companies such as Alibaba Group and Tencent which have developed versatile mobile payment solutions such as Alipay and WeChat Pay. The rise of China's e-commerce ecosystem is a textbook example of "leapfrogging." In Western countries, people have long had a car, and are used to living near a shopping mall well before having a PC and an internet connection, not to mention a smartphone. At the startup stage, Amazon and other e-merchants have had to compete with a dense physical retail infrastructure, and a tradition of going shopping to the mall strongly anchored in the society. For its part, China jumped straight from mass poverty into the internet era. Chinese citizens shop massively online because so many of them do not own a car to load goods at the nearest supermarket. And they use M-commerce because they often do not have a PC or a fixed line internet connection, although nearly everybody, urban or rural, does have a smartphone.

Tax regimes and the geography of e-commerce

With the growth and rising sophistication of e-commerce, sales, and profit taxation has become a serious issue for governments, notably in decentralized systems such as the U.S. or the European Union. E-commerce and other digital industries do not have the monopoly of tax avoidance and tax sheltering practices. But the fuzziness of e-commerce's geography and its ability to ignore political borders compound the problem governments are facing. Tax regimes and jurisdictions are slow to follow the fast pace of growth and innovation in technology and marketing. As wrote Tarantola (2015, 316), "this new electronic frontier has grown from its infancy to an economic heavyweight in such a relatively short period of time that the legal framework simply has not had time to catch up."

This evolution has resulted in huge tax losses for governments. Estimates of sales tax loss for the U.S. states in 2017 ranged from 8 to 33 billion dollars (SCOTUS 2018). The impotence of governments to collect properly corporate taxes and sales taxes on e-commerce operations results in market distortion and penalizes traditional businesses.

In the European Union, Ireland and Luxemburg notoriously act as tax havens for IT-companies. Apple and Facebook have headquartered in Ireland, where the corporate tax rate of 12.5% is half the average of the other European countries. Amazon has headquartered its European operations in Luxemburg, where the standard tax rate on commercial activities is about 29% while the tax on intellectual property income and royalties amounts to a mere 5.7%. Therefore, Amazon has created in Luxemburg a financial holding compay that charges its main operating business with more or less fictitious intellectual services. In the words of M. Vestager, the European Commissioner for Competition, Amazon's holding is "an empty shell." In the end, Amazon was ordered by the European Commission to repay Luxemburg 250 million euros in back taxes (European Commission 2017).

In the U.S., the decision of the Supreme Court in *South Dakota v. Wayfair Inc.* (June 21, 2018) should be regarded as a decisive turn in the history of e-commerce taxation (Liptak, Casselman, and Creswell 2018). Up to now, the U.S. jurisdiction of e-commerce taxes was stating that sales in a given State should not be submitted to taxes until the vendor shows locally some "significant" physical presence, such as stores, supermarkets, and warehouses. In the 2018 decision, the Court considered that under the current conditions of e-commerce operations, "the physical presence rule is artificial in its entirety" (SCOTUS 2018, 14).

The majority in the Court said that the existing tax regime based upon the 1992 decision was creating "market distortions" (10), "a tax shelter," and "an incentive to avoid physical presence in multiple States" which would deprive local communities of "storefronts, distribution points, and employment centers that otherwise would be efficient or desirable" (13).

From the geographer's point of view, the most interesting point in the 2018 decision is the smart and pragmatic analysis of today's e-commerce geographic reality. The Court stated that consumers' PC and smartphones that host data and cookies resulting from transactions with online merchants are today as much significant as warehouses for the definition of business presence in a given area.

> A company with a website accessible in South Dakota may be said to have a physical presence in the State via the customers' computers. A website may leave cookies saved to the customers' hard drives, or customers may download the company's app onto their phones. Or a company may lease data storage that is permanently, or even occasionally, located in South Dakota.
>
> (SCOTUS 2018, 15)

Actually, Amazon will not be hurt by the recent decision, given it already collects sales taxes in the 45 states that have such taxes in place (Alaska, Delaware, Montana, New Hampshire, and Oregon, do not have sales taxes). The company has been quarrelling for decades with state governments over the sales tax issue, sometimes threatening to close fulfillment facilities in states which tried to levy taxes on Amazon sales. However, since Amazon has turned into a behemoth, with over a hundred physical locations (office, contact centers, and warehouses) throughout the U.S., it has had to comply with local taxing jurisdictions.

Dissenting voices on the Court say that the burden of new sale taxes will fall mainly on "mom and pop" businesses, which could not meet the requirements of the thousands of different taxing jurisdictions in the U.S. However, to relieve the tax pressure on small businesses, many states have implemented a no-tax floor.

Logistics reigns supreme

Scholars have early recognized that logistics is the critical link within the value chain of e-commerce, and the most visible part of its geographical dimension (Anderson et al. 2003; Murphy 2003, 2007). Jeff Bezos, founder and CEO of Amazon and the world's richest person, understood from the beginning that in the absence of physical sale outlets, the key challenge was the management of "the flow" to customers – hence the name of the company (Dodge 2001).

The absence of physical stores has huge implications on the retailing process, notably on the creation of trust. Consumers cannot touch and try their purchases, and cannot immediately take and carry them back at home. That does explain why e-commerce, when in its infancy, turned to standardized goods such as books and CDs from which few bad surprises are expected. Once the payment is processed, the commercial chain is invisible and out of touch: wait for days, hope that the good will not be damaged, lost, or stolen, pray that it will work or fit your body, with nobody and nowhere to complain to face-to-face. Sophisticated web interfaces, efficient customer support centers, and trustworthy payment systems, have been instrumental in the growth of e-commerce. But fast and secure delivery is the corner stone of e-commerce. Said simply, it is a battle against time and space.

The importance of logistics can be read in data on shipping costs and warehousing capacity disclosed by major online retailers. In a tremendous effort to reduce time of delivery, Amazon has so far prioritized investment over profitability. In the recent years, warehousing, transportation, and delivery costs have grown faster than sales. Between 2015 and 2017, Amazon's fulfillment expenses grew from 12.5 to 14.2% of net sales. Shipping costs nearly doubled from $11.5 billion to $21.7 billion, while sales rose "only" by 62%, from 99 to $160 billion (Amazon.com 2018). By the end of 2017, Amazon was operating 19.4 million square meters of fulfillment facilities wordwide, a floor area equivalent to 800 Walmart superstores. JD.com, China's second largest e-retailer, covers the country with 486 warehouses in 78 cities, with an aggregated area of 10 million square meters (JD.com, Inc. 2018). Wang and Xiao (2015) show the much differentiated geography of e-commerce logistics in China. Customers in coastal China and main cities in the interior benefit from free and fast delivery solutions, but in the other parts of the country, accessibility to market is much more unequal, and customers have to choose between high delivery fees or longer delays.

The results of Amazon's logistics effort are impressive. Sisson (2017) reports that half the U.S. population is now living within 20 miles of an Amazon warehouse. In a letter to shareholders (April 18, 2018), the company says it has exceeded 100 million Prime subscribers who are eligible to free two-day shipping on 100 million items. The service Prime Now offers same-day delivery on one millions items in 50 cities and nine countries worldwide.

Last generation warehouses are huge and highly automated, hosting hundreds of robots and miles of conveyor belts. However, warehouses and sorting facilities do not solve the "last mile" issue, or the last leg of delivery to customers' home. In "physical" commerce, customers use to carry their shopping basket back home by their own means. Therefore, shelf replenishment materializes in huge packages and full pallets of goods like soft drink bottles, Nutella jars, or the last best-seller, which are delivered from central warehouses at the back of supermarkets by heavy trucks. And "big box" stores or suburban shopping malls are often located close to an expressway. On the contrary, online retailers must deliver small packages and parcels, directly to the consumer door, in the narrow – sometimes pedestrian – streets of downtown, or in the labyrinthic residential suburbs. Allen et al. (2018) explain that last mile delivery materializes in a growing flow of vans and light goods vehicles such as UPS' iconic "package cars," which adds to the existing urban road congestion and pollution. The issue is made worse by the frequent stops made at the curbside. This trend conflicts with urban planning policies seen today in city centers which favor public transportation, cycling, and walking.

The use of individual mail boxes is possible for very small parcels only. During opening hours, couriers cannot expect the presence of customers and their relatives at home. To solve this problem, e-retailers and third-party parcel express carriers such as FedEx, UPS, and DHL resort to pick-up solutions. For example, DPD has deployed in Europe a network of 37,000 pick-up stations – often managed by small neighborhood stores – and 2260 lockers, used by 5600 online merchants. Amazon is implementing its own network of lockers located in easily accessible spots such as transportation stations or supermarket entries (Figure 9.2). On June 20, 2018, the French Railway Company announced it had contracted with Amazon to place 1000 lockers in its stations.

When e-commerce shapes local economic landscapes

The disruption of retail trade

The fate of Randall Park Mall in North Randall (near Cleveland, Ohio) epitomizes the cross destinies of physical and electronic retails. Once one of the largest shopping malls in the U.S. (opening August 11, 1976), it was

Figure 9.2 Amazon lockers at the door of a French supermarket. Photo by the author, October 19, 2018.

demolished in 2015 after a long decline and the closure of tenants such as Sears and Macy's. In fall 2018, Amazon opened on the site one of its largest and most advanced fulfillment centers. Despite being highly automated with robots and 26 miles of conveyor belts, the 200,000 square meter facility is likely to employ near 2000 people, a recruitment that is much welcomed in a community severely hit by the industrial decline. Amazon's location choice is not fortuitous. Thanks to its central location within the country's transportation system, Ohio hosts so many warehouses that it can be said to be the U.S. "e-commerce capital" (Reagan and Korber 2016).

In a paper on online travel sales, Wilson (2000) wrote about the "fall of the mall." As a matter of fact, travel agencies have almost entirely been "uberized" by e-commerce and have disappeared from the urban landscape. So, the impressive growth of e-commerce rises a critical issue: should we forecast the complete collapse of "main street" and shopping malls, with dramatic consequence on cities urbanity and liveliness, without mention of the loss or displacement of millions of jobs?

The traditional retail industry is actually in deep trouble, notably in the U.S. In the recent years, chains like Macy's, Office Depot, J.C. Penneys, Hancock Fabrics, American Apparel, Kmart, and Sears, among others, have closed stores by the thousands. Toys "R" Us went bankrupt in September 2017 and ceased operations during the summer of 2018. Sears, the country's most iconic department store, filed for bankruptcy on October 15, 2018, after having shuttered 2,800 stores over the past 13 years (Corkery 2018). The rout of these famous brands leads pundits to question the very survival of shopping malls (Guedim 2018). According to a report published in 2017 by Crédit Suisse (a major international bank), one out of four shopping malls in the U.S. could close by 2022 (Isidore 2017). "Retail apocalypse" and "dead mall" are now entries in Wikipedia.

Toward multichannel and omnichannel commerce

E-commerce is often the scapegoat for the crisis of the retail industry. But the glut of commercial property and the competition with low-cost supermarket chains and warehouse clubs must be mentioned as majors factors. In the end, traditional retail will not disappear, but it is submitted to a high degree of disruption (Grossman 2016) and must reinvent itself.

To compete with e-commerce, physical stores and shopping malls must capitalize on their main asset: the potential for providing customers with an enjoyable living and socializing experience. Small and medium-sized malls and main streets in small, remote communities are likely to get into trouble. Meanwhile, shopping malls in tier-one cities are getting bigger and smarter, offering a wide range of integrated services, entertainment and sport facilities, free wifi and dedicated apps, and cultural events such as live music or dance performances, wine testing or culinary demonstrations.

There are several indications that e-commerce and physical retailing, rather than colliding, have become complements in a "brick and click," multichannel commerce (Currah 2002; Steinfield 2004). Evidence of the convergence is given by the tendency of major online merchants to seek for physical stores. In November 2017, Alibaba Group invested $2.9 billion to purchase a stake in Sun Art Retail Group, a major hypermarket chain in China under the brands Auchan and RT-Mart. In January 2018, Tencent, China's largest internet-based company, and JD.com, participated in a $5.4 billion investment in Wanda Commercial, China's leading operator of shopping malls (Bloomberg News 2018). Amazon has shown a similar tendency to go physical with the acquisition in 2017 of Whole Foods Market, a grocery store chain, and with the development of Amazon Go, a concept of automated grocery stores. Competition with e-commerce pushes traditional supermarket chains to a symmetrical move: in 2016, Walmart (the world biggest retail company by revenue) acquired Jet.com, an online seller, in order to improve its e-commerce operations.

To put emphasis on complementarity and combination, rather than mere juxtaposition, Verhoef, Kannan, and Inman (2015) consider the existence of omnichannel forms of retailing, or versatile combinations in time and space of electronic and physical operations. "Click and collect" commerce has become increasingly popular, creating a new marketing and planning challenge for retailers (Kirby-Hawkins et al. 2019). Auchan, a French hypermarket chain pioneered the "drive" service in 2000; Walmart followed in 2014. Customers order on the web and later get their basket of goods loaded in their car at the nearest store. This recombination of shopping practices is much suitable to the life in large cities, where time may be more precious than money. Pouyleau (2018) reports the example of FNAC (France's leading chain store of cultural goods and consumer electronics) whose outlet in La Gare du Nord, a major railway station in Paris, makes about 40% of its sales through articles purchased on www.fnac.com. New combinations of commerce practices are also found in the blogshop phenomenon, which has become much popular in Singapore. Blogshops are "niche markets of fast fashion" (Yeung and Ang 2015, 93) led by young entrepreneurs who have started a commerce – often of apparel and accessories – on a blog, with minimal initial funding. Once they have achieved some commercial success on the web, many blogshops have established a physical store. Blogshops are a much informal model of commerce, driven by a community of "millennial" shoppers and sellers, auto-regulated through social networks.

E-commerce and rural areas: from digital divide to economic inclusion

When the digital economy was in its infancy, many had thought that information technology would act as the great leveller of geographic inequities in terms of accessibility to goods and services. Bluntly, this utopian vision did not materialize. On the contrary, an enormous literature shows that the domination of large cities over the surrounding areas has increased over the past three decades. In particular, cities play a unique role in creative and innovative activities (Florida et al. 2017).

In the field of e-commerce, the domination of cities over rural areas is driven by several factors. The persistence of a digital divide (Warf 2010; Salemink et al. 2017) is often mentioned, as are lack of an adequate internet connection and the lower acceptance of digital devices. However, it is difficult to say if rural dwellers, all other things being equal, shop online more or less than people living in large cities, because social and economic factors (income, education, age) are more important than geography per se (Clarke et al. 2015). De Blasio (2008) on Italy and Farag et al. (2006) on the Netherlands come to the conclusion that rural dwellers are intensive online buyers of much standardized goods such as books and CDs because of the remoteness of stores.

In China, the remoteness from urban shopping malls and supermarkets may be an incentive to shop online. Fan (2016) finds evidence that people in small cities spend online a share of their revenue which is slightly bigger than those living in tier-one and tier-two cities. The access to a larger variety of goods is an important factor of the growth of e-commerce in rural China, which lacks dense networks of retail outlets. Major online retailers such as Alibaba and JD.com have contracted with rural communities to develop the production of high value-added products – for example organic food – and to sell it online to urban customers through their marketplaces. Most of these new agricultural ecosystems are known as Taobao Villages, from the name of Alibaba's main retail platform (Lin et al. 2016). As of 2017, Alibaba had developed a network of 70,000 clusters of farmers and local online vendors throughout the country.

In rural areas, low density and the lack of adequate transportation infrastructure may inflate delivery costs. In rural China, where 600 million people are living, online merchants are facing a big challenge. In the mountainous countryside of Sichuan, Shaanxi and Jiangsu provinces, JD.com is now routinely operating a flotilla of drones that supply isolated settlements (*The Economist* 2018). According to the company, each drone can carry 15 kg of packages over 50 kilometers. Aircraft land in a dedicated area close to the houses; "village promoters" pick-up the payload and distribute the packages to customers. As of May 2017, JD.com said it had about 300,000 village promoters nationwide, and was planning the implementation of one-ton load heavy drones (JD.com 2016).

Conclusion

This chapter provides evidence that electronic commerce, far from being purely virtual, does have an original physical imprint, and features a mountainous geographic and economic landscape. As Wrigley and Currah (2006, 340) write, "place, space and embeddedness continue to matter in the 'new e-conomy' of multi-channel distribution."

The rise of B2B e-commerce has been instrumental in the emergence of a more global economy, based on fast supply chains and global outsourcing (Peck 2017). Some e-commerce firms have become corporate behemoths, whose office buildings, warehouses, and van flotillas are the tangible manifesto of their contribution to local economies. In this regard, the obsession of media and local policymakers over Amazon's 2017 request for proposals for the location of a new headquarters (said HQ2) has been revealing. The decision issued in November 2018 to locate in New York and Arlington, VA, a suburb of Washington, DC, shows that urban buzz and face-to-face contact remain critical to business operations, even in a world of e-commerce (Weise 2018).

One cannot neglect the dark side of e-commerce. The flow of goods and packages adds to urban congestion, pollution, and greenhouse gas emission. Warehouse workers and online food couriers are often described as the galley slaves of the digital economy. In major tourist cities, the dramatic growth of Airbnb lodging has become controversial. It is accused of accelerating the displacement of middle-class households from the center to the periphery.

To tackle this problem, many municipalities like Paris or New York have implemented restrictive regulations.

However, e-commerce potential for growth remains enormous, even in rich countries. In an interview given to *Forbes* (2018), Bezos, Amazon's CEO, said that the market size of the company "is unconstrained." He was arguing on the fact that Amazon has so far captured a mere 5% of the U.S. retail market. Such a grand vision, as we write above, does not predict the "fall of the mall." Apple Stores are getting larger, smarter, and are regarded among the most profitable stores in the world's retail industry because they provide customers with discovery and learning experience through human interaction (Gallo 2016). From the geographer's and planner's point of view, the main problem could occur in small towns and rural regions, where Apple and its fellows will never locate stores. As a key component of the digital economy, e-commerce fuels the obsolescence process which is threatening some local business ecosystems and their communities. But e-commerce may also have some inclusive virtue, for example when its helps communities of Chinese farmers to connect to the global economy.

References

Alibaba Group Holding. 2018. Annual report for the fiscal year ended March 31, 2018. Washington, DC: U.S. Securities and Exchange Commission.

Allen, J., M. Piecyk, M. Piotrowska, F. Mcleod, T. Cherrett, T. Nguyen, T. Bektas, O. Bates, A. Friday, S. Wise, and M. Austwick. 2018. Understanding the impact of e-commerce on last-mile light goods vehicle activity in urban areas: The case of London. *Transportation Research Part D: Transport and Environment* 61(B):325–338.

Amazon.com. 2018. Annual report for the fiscal year ended December 31, 2017. Washington, DC.: U.S. Securities and Exchange Commission.

Anderson, W., L. Chatterjee, and T. Lakshmanan. 2003. E-commerce, transportation, and economic geography. *Growth and Change* 34(4):415–432.

Bloomberg News. 2018 (Jan. 29). Tencent leads $5.4 billion investment in Wanda commercial.

Business Wire. 2017. Developer earnings from the app store top $70 billion. www.businesswire.com/news/home/20170601005532/en/Developer-Earnings-App-Store-Top-70-Billion/.

Chesbrough, H. 2003. *Open Innovation: The New Imperative for Creating and Profiting from Technology*. Cambridge, MA: Harvard Business School Press.

Clarke, G., C. Thompson, and M. Birkin. 2015. The emerging geography of e-commerce in British retailing. *Regional Studies* 2(1):371–391.

Corkery, M. 2018. Sears, the original everything store, files for bankruptcy. *New York Times* (Oct. 14). www.nytimes.com/2018/10/14/business/sears-bankruptcy-filing-chapter-11.html.

Couclelis, H. 2004. Pizza over the internet: e-commerce, the fragmentation of activity and the tyranny of the region. *Entrepreneurship & Regional Development* 16(1):41–54.

Currah, A. 2002. Behind the web store: the organisational and spatial evolution of multichannel retailing in Toronto. *Environment and Planning A* 34(8):1411–1441.

Day, G., A. Fein, and G. Ruppersberger. 2003. Shakeouts in digital markets: lessons from B2B exchanges. *California Management Review* 45(2):131–150.

De Blasio, G. 2008. Urban–rural differences in internet usage, e-commerce, and e-banking: Evidence from Italy. *Growth and Change* 39(2): 341–367.

DePillis, L. 2018. Amazon is now worth 1,000,000,000,000. CNN Business. https://money.cnn.com/2018/09/04/technology/amazon-1-trillion/index.html.

Dodge, M. 2001. Finding the source of Amazon.com. In T. Leinbach and S. Brunn (eds.) *Worlds of Electronic Commerce: Economic, Geographical and Social Dimensions.* pp. 167–180. New York: John Wiley.

e-Marketer. 2018. www.emarketer.com/articles/topics/retail-ecommerce.

European Commission. 2017. Statement by Commissioner Vestager on illegal tax benefits to Amazon in Luxembourg. Brussels, Oct. 4, Statement no. 17/ 3714. http://europa.eu/rapid/press-release_STATEMENT-17-3714_en.pdf.

Fan, J., L. Tang, W. Zhu, and B. Zou. 2016. The Alibaba effect: spatial consumption inequality and the welfare gains from e-commerce. 10th Meeting of the Urban Economics Association, Nov. 11–14, 2015, Portland, OR. http://pubdocs.worldbank.org/en/186601466184172643/Jingting-Fan.pdf.

Farag S., Weltevreden J., van Rietbergen T., and Dijst M. 2006. E-shopping in the Netherlands: does geography matter? *Environment and Planning B* 33:59–74.

Florida, R., P. Adler, and C. Mellander. 2017. The city as innovation machine. *Journal of Regional Studies* 5(1):86–96.

Forbes. 2018. 30 under 30: Retail and ecommerce. (Sept. 30). www.forbes.com/30-under-30/2018/retail-ecommerce/#53c365891682

Gallo, C. 2016. Ten reasons why the Apple Store was never a store. *Forbes* (Aug. 25). www.forbes.com/sites/carminegallo/2016/08/25/ten-reasons-why-the-apple-store-was-never-a-store/#7c3c330276b5.

Grossman, R. 2016. The industries that are being disrupted the most by digital. *Harvard Business Review* (March 21). https://hbr.org/2016/03/the-industries-that-are-being-disrupted-the-most-by-digital.

Guedim, Z. 2018. Will shopping malls survive the decade? Edgy Labs. June 28, https://edgylabs.com/would-shopping-malls-survive-the-crisis   www.bloomberg.com/news/articles/2018-01-29/tencent-leads-5-4-billion-investment-in-wanda-commercial.

Indergaard, M. 2004. *Silicon Alley: The Rise and Fall of a New Media District.* London: Routledge.

Isidore, C. 2017. Malls are doomed: 25% will be gone in 5 years. CNN Business. https://money.cnn.com/2017/06/02/news/economy/doomed-malls/index.html.

J.D.com. 2016 (Nov. 11). J.D.com's drone delivery program takes flight in rural China. http://corporate.jd.com/whatIsNewDetail?contentCode=6IhXLeeSAFLjLLlyuZatDA.

JD.com. 2018. Annual report for the fiscal year ended December 31, 2017. Washington, DC: U.S. Securities and Exchange Commission.

Kenney, M., and J. Zysman. 2016. The rise of the platform economy. *Issues in Science and Technology* 32(3):61–69.

Kirby-Hawkins, E., M. Birkin, and G. Clarke. 2019. An investigation into the geography of corporate e-commerce sales in the UK grocery market. *Environment and Planning B* 46(6):1148–1164.

Leinbach, T., and S. Brunn, eds. 2001. *Worlds of Electronic Commerce: Economic, Geographical and Social Dimensions.* New York: John Wiley.

Lilien, G. 2016. The B2B knowledge gap. *International Journal of Research in Marketing* 33(3):543–556.

Lin, G., X. Xie, and Z. Lv. 2016. Taobao practices, everyday life and emerging hybrid rurality in contemporary China. *Journal of Rural Studies* 47(B):514–523.

Liptak, A., B. Casselman, and J. Creswell. 2018. Supreme Court clears way to collect sales tax from web retailers. *New York Times* (June 22).

Malecki, E., and B. Moriset. 2008. *The Digital Economy: Business Organization, Production Processes and Regional Developments.* Abingdon, UK: Routledge.

Metcalfe, R. 1995. Metcalfe's Law: a network becomes more valuable as it reaches more users. *Infoworld* 17(40):53–54.

Murphy, A. 2003. (Re)solving space and time: fulfilment issues in online grocery retailing. *Environment and Planning A* 35(7):1173–1200.

Murphy, A. 2007. Grounding the virtual: the material effects of electronic grocery shopping. *Geoforum* 38(5):941–953.

Noe, T., and G. Parker. 2005. Winner take all: competition, strategy, and the structure of returns in the internet economy. *Journal of Economics & Management Strategy* 14(1):141–164.

Palmer, C. 1988. Using IT for competitive advantage at Thomson Holidays. *Long range Planning* 21(6):26–29.

Peck, J. 2017. *Offshore: Exploring the Worlds of Global Outsourcing.* Oxford: Oxford University Press.

Pouyleau, A., and V. Houzé. 2018. E-commerce. Destructeur ou disrupteur d'immobilier physique? Jones Lang Lasalle. www.jll.fr/france/fr-fr/Documents/JLL_France_-_E-commerce_destructeur_ou_disrupteur_immobilier_Physique_Juin_2018.pdf.

Reagan, C., and S. Korber. 2016. This is why Ohio is becoming the e-commerce capital. CNBC.com, April 11, www.cnbc.com/2016/04/08/this-is-why-ohio-is-becoming-the-e-commerce-capital.html.

Salemink, K., D. Strijker, and G. Bosworth. 2017. Rural development in the digital age: A systematic literature review on unequal ICT availability, adoption, and use in rural areas. *Journal of Rural Studies* 54: 360–371.

SCOTUS (Supreme Court of the United States). 2018. South Dakota v. Wayfair, Inc. 21 June, www.supremecourt.gov/opinions/17pdf/17-494_j4el.pdf.

Sisson, P. 2017. 9 facts about Amazon's unprecedented warehouse empire. *Curbed. com* (Nov. 21). www.curbed.com/2017/11/21/16686150/amazons-warehouse-fulfillment-black-friday.

Statista. 2017.B2B e-Commerce. www.statista.com/study/44442/statista-report-b2b-e-commerce/.

Statista. 2019. E-commerce share of total retail sales in China from 2014 to 2019. www.statista.com/statistics/379087/e-commerce-share-of-retail-sales-in-china.

Steinfield, C. 2004. Situated electronic commerce: Toward a view as complement rather than substitute for offline commerce. *Urban Geography* 25(4):353–371.

Storper, M., and A. Venables. 2004. Buzz: Face-to-face contact and the urban economy. *Journal of Economic Geography* 4(4):351–370.

Tapscott, D., and A. Williams. 2006. *Wikinomics: How Mass Collaboration Changes Everything.* New York: Portfolio.

Tarantola, D. 2015. Equal footing: Correcting the e-commerce tax haven. *Brooklyn Journal of Corporate, Financial & Commercial Law* 10(9):277–317.

The Economist. 2018. How e-commerce with drone delivery is taking flight in China. (June 9). www.economist.com/business/2018/06/09/how-e-commerce-with-drone-delivery-is-taking-flight-in-china.

U.S. Census Bureau. 2006. Quarterly retail e-commerce sales 1st quarter 2006. www.census.gov/mrts/www/data/pdf/06Q1.pdf.

U.S. Census Bureau. 2018. Quarterly retail e-commerce sales. 2nd quarter 2018. Washington DC: U.S. Department of Commerce. www.census.gov/retail/mrts/www/data/pdf/ec_current.pdf.

Verhoef, P., P. Kannan, and J. Inman. 2015. From multi-channel retailing to omni-channel retailing. *Journal of Retailing* 91(2):174–181.

Wang, J. and Z. Xiao. 2015. Co-evolution between etailing and parcel express industry and its geographical imprints: The case of China. *Journal of Transport Geography* 46(20):20–34.

Warf, B. 2010. Segueways into cyberspace: Multiple geographies of the digital divide. *Environment and Planning B* 28(1):3–19.

Weise, K. 2018. The week in tech: Amazon finally makes an HQ2 decision. *New York Times* (Nov. 16). www.nytimes.com/2018/11/16/technology/week-in-tech-amazon-hq2.html.

Wilson, M. 2000. The fall of the mall? The impact of online travel sales on travel agencies. *Journal of Urban Technology* 7(2):43–58.

Wrigley, N., and A. Currah. 2006. Globalizing retail and the new e-conomy: The organizational challenge of e-commerce for the retail TNCs. *Geoforum* 37(3):340–351.

Yeung, G. and K. Leng Ang. 2015. Online fashion retailing and retail geography: The blogshop phenomenon in Singapore. *Tijdschrift voor Economische en Sociale Geografie* 107(1):81–99.

Zook, M. 2005. *The Geography of the Internet Industry: Venture Capital, Dot-Coms, and Local Knowledge*. Oxford: Blackwell.

10 Online retailing

Emily Fekete

Online retailing, also sometimes referred to as e-tailing or ecommerce, has altered the spaces in which retail activity takes place, bringing shopping to an online platform. Goods and services are now frequently purchased and sold on the internet, bypassing traditional offline storefronts. Tied up in the buying and selling of products online are notions of class, gender, ethnicity, geopolitics, and individual identity. Making online purchases is not a static, disembodied practice, but instead takes on a geography of its own as different websites are used by disparate populations in various locations across the globe. Despite the fact that customers can purchase goods online, most items bought online must be consumed offline. Therefore, not only is there a geography to the online sites of retail themselves, but also to the offline realms associated with this activity. Online retail is increasingly bound to the geographies of distribution, shipping, and transit as physical goods must be delivered to the customer. Online retail is not geographically or socially even, varying based on the types of products being purchased as well as the geographic locations of consumers and retailers, and entails aspects of both the internet and the offline geographies of the physical world.

Though online retail has been growing at higher rates than offline retail over the last two decades, it still does not represent a significant portion of all retail trade. 2014 was the first year that online sales in the United States surpassed $300 billion, an increase of 15.4% from 2013 sales. However, 2014 also saw $4.7 trillion in all retail sales exclusive of food service (Enright 2015). By 2016, adjusted online retail only accounted for 14.4% of total retail sales in the United States (Figure 10.1) (US Census 2018). Nonetheless, online retail sales have continued to grow. At the end of the second quarter of 2018 online retail increased 15.2% over online sales at the end of the second quarter of 2017, while all retail only increased 5.3% during the same time period. Overall, these numbers represent the fact that online retail is still not a popular or common method for obtaining goods and services as opposed to traditional retail outlets, despite its sustained growth.

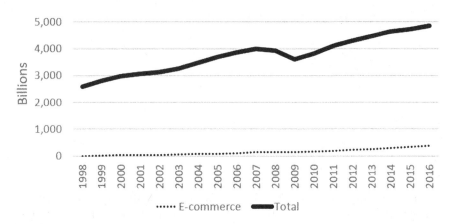

Figure 10.1 Annual U.S. retail sales.

Author, using data from US Census Bureau Monthly Retail Trade Survey Historical Data Reports. www.census.gov/retail/mrts/ historic_ releases.html.

Certain goods and services are more likely to be purchased online over others. Statistics from the U.S. Census Bureau show that clothing and clothing accessories have the highest percentage of online sales as compared to total retail sales. This category is followed by motor vehicles and parts dealers, and books, music, and sporting goods. Categories of goods with the lowest percentage of online purchases compared to total retail purchases include food and beverage as well as building materials and gardening equipment and supplies dealers.

The internet has not necessarily changed the concept of retail, but rather has had an effect on the process of how retail is conducted by creating new spaces of sale and altering the established processes of shipping and offline business. While the internet may open up new retail spaces, it also serves to buttress consumption patterns already in place. This entry examines the development of online retail with an emphasis on its ties to the geography of distribution and shipping, how social media have influenced online shopping, and how the internet has opened up spaces of alternative consumer activities.

The development of online retailing

Online retail trade, defined as "a retail format in which the retailer and customer communicate with each other through an interactive electronic network," represents another step along the evolutionary chain of non-store retail establishments (Williams 2009). Put simply, online retailing is any instance where a retail transaction occurs through the medium of the internet. Retailing has frequently taken place in non-store spaces such as catalogues and television shopping channels (Williams 2009). Therefore, retailing in online spaces serves as the next venture coinciding with the growth and expansion of communications technologies. There are few retail corporations who exist completely online (known as a pure-play establishment), however. Instead, most online stores follow a click-and-mortar, otherwise known as bricks-and-clicks, model, meaning that they are hybrids of traditional retailing in physical storefronts (brick-and-mortar stores) and online institutions.

The dot com bubble of the late 1990s and early 2000s ushered in a wave of online retailers. The rapid establishment of online retail sites led some analysts to predict that online retail development would continue to rise as much as 75% annually (Wrigley et al. 2002). Hoping to capitalize on the perceived revolutionary growth of online retail, investors quickly bought into numerous sites that ultimately could not attract the number of customers necessary to keep them afloat and sustain their business model. Unable to withstand the initial heavy losses in start-up costs, Pets.com and their sock puppet mascot was perhaps the most famous of these failures, eventually encapsulating the entirety of the dot com bubble.

Despite the burst of the dot com bubble in the early 2000s, online retailing has continued to grow. Amazon.com represents one of the first major "success" stories of online retailing. Launched in 1995, in the 20 years it has been online the internet retailer has grown into a company that reports annual sales of $178 billion in 2017, making it the largest online retailer in the United States (Statistica A). The company became the fifth company in the world to be valued at over $1 trillion when its stock prices rose for five weeks in late 2018. Though these numbers are staggering, the revenues have only recently translated into profitability, boasting a net income of $3.03 billion in 2017. Amazon did not make a profit in the first five years of its existence. Despite reporting a negative net income of $241 million as recently as 2014, investors continued to pump money into the corporation. Profits for the company are stunted because of the continued goal of decreasing shipping and fulfillment costs and time (Mangalindan 2015). A recent innovation from Amazon.com is their Amazon Prime membership, a subscription service where users pay an annual fee to receive free two-day shipping on products as well as access to free streaming movies and television shows. Amazon has admitted that the service loses them money, but maintains it to increase brand loyalty among their user base (Mangalindan 2015).

Many early online retail enterprises failed because of the lack of infrastructure required to be able to ship items directly to consumers at a relatively quick pace (Wrigley et al. 2002). Brick-and-mortar establishments often already have supply chains in place, whereas online retailers need to ship directly to consumers and must rely on established shipping companies. Success in the world of online retail is driven by the ability to develop fulfillment centers and distribution networks. Amazon has worked to develop additional new distribution methods aside from the traditional business-to-customer's home model. Innovations such as Amazon lockers where customers go to a central location such as a local convenience store or college campus to pick up and drop off packages is one attempt to consolidate shipping and distribution, making it more efficient for the delivery person to drop off purchases.

Recently, due to rising shipping costs incurred by Amazon as it continues to offer free or discounted shipping to its members, the company has been testing out their own shipping service called Amazon Flex in an attempt to undercut traditional shipping services like FedEx, UPS, and the U.S. Postal Service. The Flex service would work similarly to ridesharing companies, relying on flexible employees in the gig economy – plainclothes contract workers would pick up and drop off packages using their own vehicles. Amazon has even been investing in the possible use of drones to deliver packages as a quicker and more efficient way to reach customer's homes. The future of online retail is largely tied up in the offline geographic networks of shipping and distribution.

While Amazon.com is the largest online retailer in the US, the title for largest global online retail establishment goes to Alibaba Group. Alibaba launched in China several years after Amazon.com, in 1999 as a business-to-business ecommerce site. Presently, Alibaba Group has expanded their ventures to include a business-to-customer site as well as a customer-to-customer online marketplace. These developments have led the corporation to report an annual revenue of $36 billion during their fiscal year ending March 31, 2018. Though reporting lower revenues than Amazon.com, these transactions have translated into much higher profits with a reported annual net income of $11.06 billion in 2016 (statistica B). Alibaba also beats Amazon when it comes to online shopping sprees. The famed Amazon Prime Day event features special deals on products found on Amazon.com throughout a 36-hour event. In 2018, Prime Day was lauded as Amazon's "biggest shopping event in history" and generated approximately $4.2 billion in sales. While impressive, that total was eclipsed in less than ten minutes during Alibaba's Singles Day event on November 11, 2018. Singles Day has become a worldwide retail phenomenon, generating $25 billion in sales over the 24-hour event in 2017. In 2018, sales reached $4.68 billion ten minutes into the event and $6.5 billion at the 20-minute mark (Taylor 2018).

The existence of completely online retail outlets is rare, however. After the initial dot com frenzy, many traditional retail establishments created their own web stores to be more attractive and inclusive to their customer base (Williams 2009; Wrigley et al. 2002). Brick-and-mortar establishments now offer services such as in store pick up, where a customer can purchase an item online and pick it up directly from a store instead of having it shipped to their home, allowing the business to take advantage of both online sales and consolidated, pre-established shipping networks. Big box store giant Walmart has been increasing its efforts to directly compete with online retailers such as Amazon by investing in the development of their online and mobile platforms. Posting a 40% growth in online sales during the second quarter of 2018, Walmart now holds third place for most online sales in the U.S. In an effort to compete with Amazon, Walmart has offered free two-day shipping without having to purchase a membership, free in store pick up, and has been testing the use of driverless cars for delivery services. Online retail has largely supplemented, rather than replaced, traditional methods of retail activity.

A trend in opposition to the click-and-mortar model has developed where online retail giants are seeking to ground themselves in physical space in an effort to compete with established retail corporations. In a reverse clicks-to-bricks move, Amazon, which existed wholly online from its inception, has begun to open physical storefronts in some major U.S. cities, replacing large bookstore chains such as Borders and Barnes and Noble. As other offline retail establishments have closed their doors, Amazon has also stepped in to fill the space. The bankruptcy of Toys 'R' Us prompted Amazon to physically mail 2018 holiday toy catalogs complete with QR codes next to each item to the homes of U.S. customers in a move attempting to play on the nostalgia associated with department store advertisements. Amazon has also worked to root themselves more squarely in the grocery business with the acquisition of Whole Foods, an offline organic grocery store chain.

The grocery business has been one area of retail that online venues have found difficult to crack. Olive oil appears to be one of the few foodstuffs easily sold online to date. Due to the nature of fulfillment and shipping, it is difficult as of now to establish online grocery businesses. Complications with food spoilage and unknown delivery times make delivering groceries problematic outside of urban areas. Distribution networks and the need for extensive warehouses also have prevented many pure-play grocery establishments from succeeding thus far, whereas several grocery stores have successfully adopted a bricks-and-clicks model available in large cities (Murphy 2007). Groceries is one area where Amazon competitor Walmart has dominated, ousting Amazon as the top seller of groceries over the internet. To maintain its position, Walmart has introduced curbside grocery pick up at its stores and same day delivery in some cities, a service it hopes to expand to about 40% of the U.S. population by the end of 2018.

Despite the difficulties of selling groceries over the internet, food retail in general has been altered by the online presence of stores as well as online restaurant delivery services. People are increasingly looking to services such as Yelp and Foursquare for restaurant reviews before they venture out to the restaurant, making it more pertinent than ever for restauranteurs to maintain an online presence (Fekete 2016). Some online retailers allow customers to skip the grocery store all together, providing the option to buy pre-packaged and pre-portioned meal prep kits that are delivered to a customer's home and cooked and prepared on the customer's own time. While Walmart has started to break into the grocery delivery market, middleman companies such as Instacart allow for users to create and purchase a grocery list of items from their local supermarket which are then shopped for and delivered by the employees of the shopping service, not of the grocery store. Similarly, restaurant food delivery services have grown alongside ridesharing services where employees of the food delivery service go to various restaurants and pick up and deliver food to customers who have purchased the food at home online through their mobile phone. Food delivery services have affected the offline restaurant market through analysis of search history data. For example, one service, UberEats, noticed that their users were looking for hamburgers in a location where there were no restaurants selling hamburgers. The company approached some local establishments, encouraging them to start offering hamburgers online only, to which the restaurants complied. The only way to order a hamburger from these physical restaurants is through the use of an online sale and delivery service (Garsd 2018).

There is an extensive literature on the nature of online retail and its' social and economic effects (Currah 2002; Murphy 2002; Wrigley 2002; Weltevreden and Atzema 2006; Weltevreden and van Rietbergen 2009). For instance, the impact online shopping has had on society and geography has been comprehensively explored (Ren and Kwan 2009; Hjorthol 2009). Online shopping is still largely an urban phenomenon, reflecting internet use patterns and availability of distribution networks (Farag et al. 2006). Online purchasing is also highly dependent on proximity to physical retail outlets and whether people are purchasing a highly available product or a specialty item (Forman et al. 2009). Online shopping decreases as access to large retail chains increases. Widely available products are also more likely to be purchased offline and specialized products online (Forman et al. 2009). Some products' desirability, such as extra virgin olive oil, is dependent on the geographic location from which it originates. However, as in the example of extra virgin olive oil, products considered to be enough of a specialty good can benefit from online sales where customers can purchase the product directly from sources in Italy or Spain to verify authenticity (Carlucci 2014).

Gender and socio-economic class also have an effect on online retailing. In the US, women are more likely than men to adopt internet-based shopping as a method of commercial activity (Ren and Kwan 2009). Access to cars has been shown to have a large impact on whether or not people will shop online (Ren and Kwan 2009). Online shopping is often most utilized to save time, money, and travel and is also largely driven by familiarity with the internet (Ren and Kwan 2007; Ren and Kwan 2009). Ren and Kwan (2009, 276) conclude "access to local stores matters and that the spatial distribution of shopping opportunities influences people's e-shopping patterns." Where people shop is still largely a reflection of income and geographic proximity. The internet does not appear to alter consumption patterns significantly in that people are generally not *substituting* their shopping behavior with e-commerce, but rather *supplementing* it (Ren and Kwan 2009).

Retailing has additionally been changed due to the nature of software and coded spaces (Kitchin and Dodge 2011). The increased use of credit cards, ATMs, and store loyalty programs are several examples of the nature of coded consumption today. In some cases, customer experiences would be much altered from what they currently are without the use of computerized management systems that calculate everything from food preferences to daily schedules. Online gambling, an entertainment-based type of retail, has grown rapidly due to the less costly startup funds associated with creating an online casino rather than a brick-and-mortar gambling hall. These sites of online gaming are often located in countries where regulations on gambling are more limited such as Central America and the Caribbean (Wilson 2003).

The adult entertainment industry has also been experiencing a dramatic shift as it switches to online sites of sale. Though the internet has allowed for more people to both produce and purchase pornography from various locations, the industry continues to be largely concentrated in the United States with a few major cities existing as sites of production (Zook 2003). The internet has expanded and altered the way people conduct their retail business, however physical locations are still important for the purposes of production and regulation of said goods and services. Online retail enterprises largely reflect offline societal actions, laws, and fulfillment processing centers.

Coverage of online retail in the academic literature has declined significantly from the height of online retail growth in the mid-2000s (Enright 2015). While online shopping has not become the wave of the future some people suspected it would in the late 1990s, it does not nullify the ways in which online spaces are in fact coloring consumption patterns. Corporations are engaging with location-based services and GeoWeb content (online content that includes location information data) in an effort to capitalize on the fact that the internet is omnipresent in people's lives, yet goods and services are consumed in physical space. Businesses take advantage of the individual behaviors embedded in GeoWeb content to push their goods and services to markets most likely to purchase them, using the internet to encourage customers to visit brick-and-mortar establishments.

Social media applications such as Google Places, Yelp, and Foursquare not only include online customer reviews and ratings of offline retail locations, but sometimes also encourage visits through marketing promotions. Businesses can pay to appear at the top of a Google or Yelp search or with coupons or other promotions for social media users who make a purchase at a particular venue (Fekete 2015). Looking at the offline retail establishments listed on these online applications reveals the continued presence of the urban hierarchy within cities in the United States in terms of the availability of goods and services to urban populations (Fekete 2014). These applications can also point to the locations offline where consumers go most frequently for retail services as well as areas of cities that are underdeveloped from a retailing perspective (Fekete 2015).

The effects of internet use on retail trade also vary depending on the language one uses in online searches for goods and services. For example, Graham and Zook (2011) find that a search for restaurants in Tel Aviv yields drastically different results depending on whether the user is searching in English and Hebrew or in Arabic. Gender bias also exists in the creation of online maps of and navigation to retail establishments. In Open Street Map, (Stephens 2013) finds a wholly gendered space where there are several categories for places that commodify women (e.g., strip clubs, escort services), but suggested categories for traditionally feminized spaces of care (e.g., child care, hospices) have been routinely rejected as necessary additions to the mapping platform by other users. The omission of these types of businesses poses a concern as many applications including Apple products, are abandoning paying for mapping services in favor of the free Open Street Map application.

Searching on Google Maps for services also varies depending on the keywords used in the search as well as on the overall web presence of a retail establishment (Zook and Graham 2007). Those establishments with less of an internet presence will show up less frequently or towards the bottom of a search on Google Maps regardless of whether or not the venue is physically located nearer to the person conducting the search than other retail outlets (Zook and Graham 2007). Similarly, national chain businesses are more likely to appear near the top of search results as opposed to local retailers because of the ability of these large corporations to distort their online presence through marketing techniques and monetary means (Zook and Graham 2007). While Google Maps and Open Street Map are not sites of online retail activity per say, similar to smartphone applications like Yelp and Foursquare, they do drive people to particular brick-and-mortar retail establishments, thus causing the internet to have an effect on retailing in general and making them worthy of inclusion in a discussion of online retail.

Though online shopping has not come close to surpassing traditional consumption patterns, the internet has an ever-increasing effect on the types of businesses people frequent and where these establishments are located. One of the reasons that GeoWeb content is significant is because it represents services that are neither completely online because of their reliance on physical stores for content creation, but not completely off-line either. There are a multitude of retail services that require an offline space because they deal in provisions that are embodied and necessitate a physical presence, restaurants or movie theaters for example. Certain types of goods and services cannot be wholly consumed online, but still need to maintain an online presence in order to compete with other retailers. Location-based applications allow for those companies that sell goods and services to participate in online consumption by creating a hybrid online space in which to advertise and create a forum for consumers to meet and share their experiences (Fekete 2015).

Alternative online spaces of retail activity

The internet has also opened up alternative spaces of sale or ways of exchanging goods and services not associated with offline major corporations. Websites such as Craigslist, Facebook, and Etsy provide online marketplaces for individuals to sell products and goods. While transactions with individuals not associated with corporations existed before the internet, primarily through newspaper classifieds or local craft fairs, the internet has further popularized these alternative options for economic activities in ways that may not have been possible prior to the growth of online sales. Looking further into these internet-based sites of retail could open new ways of understanding second-hand economies, the maker movement, or the retail actions of individuals online.

Craigslist is an early example of a website established for the average person to buy and sell items from other users. Starting in 1995 as an email listserv among friends in San Francisco, California, Craigslist is a mostly free website that allows locals to post classified advertisements to sell or trade goods as well as sell or rent properties or post jobs (craigslist.org). Boasting over 50 million page views each month, Craigslist has grown from a humble listserv to a massive corporation that relies on physical geography as its listings are location dependent. There are approximately 700 local sites on Craigslist for cities, towns, or counties found in over 70 countries worldwide (craigslist. com). Craigslist acts as a facilitator of trade between two unknown local parties, coordinating offline economic activity in an online space, and taking over traditional forms of classified advertising such as newspapers (Seamans and Zhu 2014). One study found that the effect of a new Craigslist local site on local newspaper classified advertisements led to newspapers dropping the cost of their advertisement space by 20.7% while increasing their subscription prices by 3.3% (Seamans and Zhu 2014). The presence of Craigslist has also been shown to lead to a reduction in real estate vacancy rates (Kroft and Pope 2014).

Similarly, Facebook has added features to their social networking site to facilitate consumer activities within a particular town or country, often referred to as Buy, Sell Trade; Online Garage Sale; or Online Yard Sale groups. The Facebook interface within such groups allows for people to privately message specific sellers as well as list prices of items and their location on a group's page (Facebook.com). Selling items on Facebook is different than Craigslist because, rather than setting items up in particular categories for sale, all items are posted with photos to the group wall. Sellers can then "bump" their posts from further down the page back to the top by commenting on the post itself, thus forcing the advertisement of a particular good to the top of the page. As users scroll through the Facebook group, they can tag a friend who might be interested in a particular item in the comments of a post, alerting an outsider who may want to consider the purchase.

The growth of the maker movement and the craft economy has also led to the creation of Etsy, an online site dedicated for "crafters, artists and makers" to "sell their handmade and vintage goods and craft supplies" (etsy. com). Etsy began in 2005 in a Brooklyn apartment and has since "evolved into a sophisticated technology platform that connects Etsy sellers and buyers across borders, languages and devices, a company that spans the globe and a business that is committed to creating lasting change in the world" (etsy. com). Etsy is largely dominated by female crafters who are looking to sell their homemade items as a supplement to their incomes as students or stay at home mothers (Luckman 2013). By establishing themselves as an alternative economy that connects buyers directly with the crafter, Etsy exists as a romanticized handmade economy, drawing on personal nostalgia as much as a desire for unique items to make sales.

Moreover, the internet has created a different way for individuals to participate in the consumption of goods and services without monetary transactions – in the form of barter or trade. The power of bartering on Craigslist, for example, has led to several impressive "barter up" stories, a tactic where a person attempts to barter for increasingly valuable items. In 2006 the Associated Press picked up a story about a man who bartered his way from owning an oversized novelty red paperclip to obtaining a year's worth of rent in a duplex in Arizona (Associated Press 2006). A similar story appeared in 2010, when a 17-year-old spent two years conducting 14 trades to turn an old cell phone into a 2000 Porsche Boxster (Wojdyla 2010).

On the extreme end of anti-economic consumer trade, the Buy Nothing Project exclusively relies on online social networking sites to facilitate consumer activity without formal retail transactions. Activist groups expressing concern over the social and environmental impacts of mass consumption in capitalist markets have begun to take an active stand against over production and limited usability of goods (Iyer and Muncy 2009). One such group, the Buy Nothing Project (inspired by Adbuster's Buy Nothing Day), focuses on reducing the number of goods being produced worldwide and reusing those goods that are currently in circulation (buy nothing project 2016). The Buy Nothing Project stared in 2013 when two people on Bainbridge Island, Washington created a local gift economy to encourage those in their community to share or gift unused items to others as opposed to buying new ones. Exclusively utilizing Facebook Groups, the Buy Nothing Project has expanded to include more than 280,000 members in 18 countries with over 1300 Facebook Groups dedicated to the project (buy nothing project 2016). Local Facebook Groups, similar to the Facebook Online Garage Sales, encourage networking and gifting goods and services rather than selling and buying used items or labor. The group inspires people to ask for items from others with games such as "Big Ask, Small Ask" (where members post comments about items they are hoping to be gifted), "Made By Me Monday" (where members gift items such as food, art, or clothing made by hand), and "Thankful Thursday" (where members post their gratitude for gifts that have helped them or people who have made a difference in their lives).

Though trade through Craigslist or the Buy Nothing Project may not resemble online retail in a capitalist sense, transactions from each are still about obtaining and consuming goods and services. Indeed, the Buy Nothing Project even states their mission as "finding new ways to give back to the community that has brought humor, entertainment, *and yes, free stuff into their lives*" (Buy Nothing Project 2016, emphasis added). Obtaining things is a key component to this type of non-monetary online trade, offering a glimpse into online consumer habits. A monetary transaction does not need to be the defining moment for the receipt of goods and services. The internet may be increasing interaction among communities who wish to exist in an alternative capitalist economy.

Conclusion

Online retailing is the act of retail exchange between customer and seller through an electronic format. The internet hype of the 1990s and early 2000s led to claims that physical storefronts would be replaced with online marketplaces; however this scenario has not come to pass. Online retailers have started to supplement, rather than replace, existing buying patterns in certain segments of society, thus creating new spaces of retail and new ways and practices of making purchases. The most successful models of online retail are currently those businesses using the bricks-and-clicks method, indeed recent trends have shown that enterprises attempting to only exist online have started to expand into offline spaces through various business acquisitions. Online retail is highly dependent on the geographies of shipping and distribution, thus having an effect on preexisting offline distribution and shipping networks.

Despite the fact that online retail represents a small segment of overall retail purchases, the internet is affecting the way people purchase and obtain goods and services through the development of online store fronts, influencing offline retailers to change their products, and breaking into the gig economy with employees contracted for delivery. Businesses have capitalized on GeoWeb services and individual, data-driven analyses of consumers to drive more retail trade. Retail is still largely conducted offline, but the internet has started to open up new locations in which people frequent for purchasing goods and services. Social networks and online sites such as Craigslist have also created pockets of communities that are conducting retail transactions amongst themselves outside of traditional spaces of sale. Alternative spaces of consumer-based transactions have also been developed online, allowing for the consumption of goods without the monetary transactions associated with a capitalistic society. The internet has thus created new interactions among those who are buying and selling products, both in traditional and non-traditional retail.

References

Associated Press. 2006. Man trading up from paperclip to house. *NBC News*. www.nbcnews.com/id/12353171/ns/technology_and_science-tech_and_gadgets/t/man-trading-paper-clip-house/#.VykjeFYrLIU.

Buy Nothing Project. 2016. About. https://buynothingproject.org/about/.

Carlucci, D., B. De Gennaro, L. Roselli, and A. Seccia. 2014. E-commerce retail of extra virgin olive oil: An hedonic analysis of Italian SMEs supply. *British Food Journal* 116(10):1617–1600.

Currah, A. 2002. Behind the web store: The organizational and spatial evolution of multichannel retailing in Toronto. *Environment and Planning A* 34(8):1411–1441.

DeNale, R. and D. Weidenhamer. 2016. Quarterly retail e-commerce sales 4th quarter 2015. *US Census Bureau News.* www.census.gov/retail/mrts/www/data/pdf/ec_current.pdf.

Enright, A. 2015. US annual e-retail sales surpass $300 billion for the first time. *Internetretailer.* www.internetretailer.com/2015/02/17/us-annual-e-retail-sales-surpass-300-billion-first-ti.

Etsy.com. 2016. About Etsy. www.etsy.com/about/?ref=ftr.

Farag, S., J. Weltevreden, T. van Rietbergen, and M. Dijst. 2006. E-shopping in the Netherlands: Does geography matter? *Environment and Planning B: Planning and Design* 33:59–74.

Fekete, E. 2014. Consumption and the urban hierarchy in the Southeastern United States. *Southeastern Geographer* 54(3):249–269.

Fekete, E. 2015. Race and (online) sites of consumption. *Geographical Review* 105(4):472–491.

Forman, C., A. Ghose, and A. Goldfarb. 2009. Competition between local and electronic markets: How the benefit of buying online depends on where you live. *Management Science* 55(1):47–57.

Garsd, J. 2018. Uber's online-only restaurants: The future, or the end of dining out? *NPR.* www.npr.org/sections/thesalt/2018/10/23/658436657/ubers-online-only-restaurants-the-future-or-the-end-of-dining-out.

Graham, M. and Zook, M. 2011. Visualizing global cyberscapes: Mapping user-generated placemarks. *Journal of Urban Technology* 18(1):115–132.

Graham, M. and M. Zook. 2013. Augmented realities and uneven geographies: Exploring the geolinguistic contours of the web. *Environment and Planning A* 45:77–99.

Hjorthol, R. 2009. Information searching and buying on the internet: Travel-related activities? *Environment and Planning B: Planning and Design* 36(2):229–244.

Iyer, R., and J. Muncy. 2009. Purpose and object of anti-consumption. *Journal of Business Research* 62(2):160–168.

Kitchin, R. and M. Dodge. 2011. *Code/Space: Software and Everyday Life.* Boston: MIT Press.

Kroft, K. and D. Pope. 2014. Does online search crowd out traditional search and improve matching efficiency? Evidence from Craigslist. *Journal of Labor Economics* 32(2):259–303.

Luckman, S. 2013. The aura of the analogue in a digital age: Women's crafts, creative markets and home-based labour after Etsy. *Cultural Studies Review* 19(1):249–270.

Mangalindan, J. 2015. Inside Amazon Prime. *Fortune.* http://fortune.com/2015/02/03/inside-amazon-prime/.

Murphy, A. 2002. The emergence of online food retailing: A stakeholder perspective. *Tijdschrift voor Economische en Sociale Geografie* 93(1):47–61.

Murphy, A. 2007. Grounding the virtual: The material effects of electronic grocery shopping. *Geoforum* 38(5):941–953.

Ren, F. and M.-P. Kwan. 2007. Geovisualization of human hybrid activity-travel patterns. *Transactions in GIS* 11:721–744.

Ren, F. and M.-P. Kwan. 2009. The impact of geographic context on e-shopping behavior. *Environment and Planning B: Planning and Design* 36(2):262–278.

Seamans, R. and F. Zhu. 2014. Responses to entry in multi-sided markets: The impact of Craigslist on local newspapers. *Management Science* 60(2):476–493.

Statistica A. 2016. Statistics and facts about Amazon. www.statista.com/topics/846/amazon/.

Statistica B. 2016. Statistics and facts about the Alibaba Group. www.statista.com/topics/2187/alibaba-group/.

Stephens, M. 2013. Gender and the geoweb: Divisions in the production of user-generated cartographic information. *GeoJournal* 78(6):981–996.

Taylor, K. 2018. Alibaba just eclipsed Amazon's estimated Prime Day sales in less than 10 minutes. *Business Insider.* www.businessinsider.com/alibabas-singles-day-beats-amazon-prime-day-2018-11.

Weltevreden, J. and O. Atzema. 2006. Cyberspace meets high street: Adoption of click-and-mortar strategies by retail outlets in city centers. *Urban Geography* 27(7):628–650.

Weltevreden, J. and T. van Rietbergen. 2009. The implications of e-shopping for in-store shopping at various shopping locations in the Netherlands. *Environment and Planning B: Planning and Design* 36(2):279–299.

Williams, D. E. 2009. The evolution of e-tailing. *International Review of Retail, Distribution, and Consumer Research* 19(3):219–249.

Wilson, M. 2003. Chips, bits, and the law: An economic geography of internet gambling. *Environment and Planning A* 35(7):1245–1260.

Wojdyla, B. 2010. How a 17-year-old Craigslist-swapped an old phone for a Porsche. *Jalopnik.* http://jalopnik.com/5591644/how-a-17-year-old-craigslist-swapped-an-old-phone-for-a-porsche.

Wrigley, N. 2002. "Food deserts" in British cities: Policy context and research priorities. *Urban Studies* 39(11):2029–2040.

Wrigley, N., M. Lowe, and A. Currah. 2002. Progress report: Retailing and e-tailing. *Urban Geography* 23(2):180–197.

Zook, M. 2003. Underground globalization: mapping the space of flows of the internet adult industry. *Environment and Planning A* 35(7):1261–1286.

Zook, M. and M. Graham. 2007. Mapping DigiPlace: Geocoded internet data and the representation of place. *Environment and Planning B: Planning and Design* 34:466–482.

11 Finance and information technologies

Opposite sides of the same coin

Jayson J. Funke

The geography of finance has changed dramatically over the past century. Advancements in information technologies, ranging from the telegraph and telephone, to micro-processing and computers, to satellites, the internet, mobile phones, and other digital mobile media devices, have been essential drivers of that change. Financial systems are inherently information systems, and financial institutions and investors have not only been directly involved in financing innovative technologies and telecommunications startups, but have also been among the first to develop transnational networks of capital and information flows using those technologies. The emergence of the financial services industry in the 1980s coincided with the Internet Revolution, placing both at the heart of globalization and the service-based information economy and its knowledge-intensive "network society" (Castells 2003).

Finance, territory, information, and technology

The management of modern capitalist finance is historically predicated on the sovereign state and the territorial governance of people, land, and money. Under neoliberalism, capitalist finance has transitioned from highly regulated national and regional systems of capital circulation and financial intermediation to a more liberalized global geography of flows among centers of financial production. Advancements in telecommunication systems helped to network flows of capital and information, and thus financial centers together, into a complex of market territories and hierarchical networks of spatially rooted global cities. These developments effectively de-territorialized money and inaugurated the era of global finance where money, equities, and speculative derivatives are traded *en masse* across regulatory and market borders through electronic exchanges. These trans-boundary flows of information and capital suggested an "end to geography" in financial markets (O'Brien 1992) as investors easily exploit arbitrage opportunities by moving capital across the world's uneven regulatory and market landscapes. In economics and finance, arbitrage is the practice of profit-making by exploiting a price or regulatory difference between two or more markets. However, the places and spaces of financial production remain important constituents to risk management and the capturing of profits in the global circulation of capital (Budd 1995).

This chapter introduces key concepts that are essential to understanding the institutional character of modern money and capitalist financial systems. The explicitly critical concepts of "rentierism," "spatial-temporal fixes," and neoliberal "debtfare states" are used herein to help frame the "financialization of capitalism" as a particular regime of accumulation that attempts to manage the persistent capitalist crises of over-accumulation and class relations through exploitation predicated on the growth of credit/debt, rentierism, and speculation. It describes the territorial, hierarchical, and informational foundations of modern capitalist monetary and financial systems. It explains how information technologies, the deregulation and disintermediation of financial production, and the proliferation of offshore finance helped advance financialization and the evolution of global finance by hyper-mobilizing money and liberating it from territorial and regulatory boundaries. Disintermediation refers to the direct exchange of money, information, or services between parties that involves no intermediary, like a bank, to manage or process the transactions.

Telecommunication systems and the internet helped open and transform once-national and regional financial systems by providing the infrastructure for an integrated global financial system capable of facilitating vast, multi-scalar networks of transnational capital and information flows. Information technologies eliminated many of the risks and transaction costs arising from spatially and temporally based information asymmetries that historically have plagued financial systems and markets. The economies of scale enabled by information technologies helped centralize information flows and decentralize financial production and financial risks, and at the same time they helped increase competition by enabling financial disintermediation. Financial competition and risk management practices emphasize the importance of accessing, processing, managing, and exchanging information, as well as the need for the cost-saving benefits of using information technologies for risk management, marketing, and rentierism. Encrypted software platforms and the cost saving benefits derived from information technologies fostered disintermediation by increasing competition, new financial instruments, and the growth of digitized online finance. This phenomenon is demonstrated in the emergence of the fintech industry, "big data" analytics, and the development of cryptographic blockchain technologies and virtual currencies, like Bitcoin.

The conclusion contextualizes the role of information technologies in the financialization of capitalism by highlighting how flows of information and money are constituent to the production of finance. It warns that financial information technologies like credit scoring, software protocols, "behavioral" marketing etc. function as panoptic technologies of power that enforce market discipline and centralized decision-making and power, which reinforce the financialization of capitalism.

The foundations of modern finance: financialization, rentierism, and debtfare

Capitalist money and modern financial systems evolved in conjunction with the territorial jurisdictions of sovereign states, and were designed to facilitate the circulation of money and sustain economic development. The ability to create money has historically been claimed as a right of sovereign states, and most modern money is created through the extension of credit/debt. Under central banking systems with territorial fiat currencies, money is created either through the monetization of public/sovereign debt as reserves in central banks and the bond market, or money is created as a financial commodity through the extension of credit in the private banking and financial system (Funke 2017; McLeay, Radia, and Thomas 2014). From this perspective, financial profits are tied to expanding credit/debt. The growth of credit/debt relations has been described as constituent to the "financialization of capitalism," a term deployed by critical scholars to refer to, most broadly, the processes and outcomes associated with neoliberal globalization and the increasing rule and power of financial prerogatives, financial markets, and financial actors and institutions over domestic and international economies, politics, and societies (Epstein 2006; Martin 2002; Stockhammer 2004; Krippner 2005; Lapavitsas 2011).

While each aspect of a financial system is also replete with its own geography, financial systems hinge on the spatial clustering of intermediaries (e.g., banks), service providers (e.g., credit-rating agencies), and other institutions (e.g., collateral registries) in financial centers. Centers of financial production are linked through telecommunication networks, making them the infrastructure of global financial production. Financial production includes regulatory institutions (e.g., central banks), services (e.g., tax reporting) and the development, marketing and trading of financial products, like equities, bonds, and derivatives in primary and secondary financial markets. Financialization infers a blurring of the conceptual line between the "real" and "financial" sectors of an economy. Financial market participants increasingly include non-financial firms like retailers and manufacturers, nonprofit and non-governmental organizations, individuals and households, and governments. Traditional financial firms are increasingly accruing revenue through direct ownership and management of non-financial companies and commodities (International Finance Corporation 2009; Wojcik 2017).

Critical debates surrounding credit/debt often draw on the concept of rent (Birch 2017; Kay 2017). Rent in neoclassical economic theory is the amount that can be charged for the use of an asset. Scholars in the Marxist tradition have argued that since money is considered a private commodity under capitalism, financial profits derived from the circulation of capital as credit are forms of rent accrued through interest. Rent, in Marxist theory, extracts surplus value that is derived from exchange in the "real" economy of production and services. Rent plays a coordinating role in allocating and distributing investments and influences uneven development (Harvey 2006a, 330). The contemporary phase of financialization infers the expansion of rentierism and "value capturing" to help offset the crises of over-accumulation in the circuits of production and exchange. Value is captured through rents levied on credit/ debt instruments, speculative investments, forms of market and regulatory arbitrage, and bouts of "capital switching." Rentierism enhances the power and hierarchy inherent to capitalist money and finance. Financial systems and institutions are legally constructed and bound together through complex, interdependent webs of contractual legal obligations that link investors one to another. Law lends institutional authority to financial instruments and contractually defines their enforceability. The legal rules surrounding financial instruments determine which holders of an instrument will or will not be vindicated, and thus law plays an important role in determining the distribution of gains and losses among the contracted parties. The legal nature of financial instruments makes finance innately hierarchical (Pistor 2013; Bieri 2017).

Shifting investments from unprofitable circuits of over-accumulated capital to more profitable investments in underinvested circuits, termed a "spatial-temporal fix" (Harvey 2006b, 64–8), entails inherent risks. The inherently speculative nature of financialization as a regime of accumulation relies on financial engineering to spatially and temporally alter risks and capture value in the form of an asset's monetized future earnings (Labban 2010), or as arbitrage differentials. For example, the length of time between an initial investment and realization of its profit can vary, so more long-term investment profits always involve a higher risk. Fixes allow the risks associated with production and exchange to be offset spatially and temporally, but do not eliminate them. Financial risk management practices inherently work against the poor and low-income groups. One way that financial risks can be offset is through the practices of financial exclusion, which often infers a spatially defined area, or high risk category of people, lacking access to financial services like savings accounts, credit, insurance, financial advice and so forth. Long-term financial exclusion is associated with social exclusion and higher inequalities, and thus lower living standards as a result of low-quality housing, education, and health services (Leyshon and Thrift 1994). For example, red-lining is the practice of identifying, stigmatizing, and depriving low-income neighborhoods, or communities with poor credit scores, from access to prime mortgage rates. The risky and poor also are susceptible to forms of predatory finance, which includes more than deceptive lending practices, but also subprime mortgages and lending, micro-credit, payday lending, credit cards, overdraft loans and fees, and other high interest or fee-based financial products.

The liberalization of capital and the deregulation of finance undermined the Keynesian welfare state in favor of the neoliberal "debtfare state" and its privatized system of "financial production" based on rents, speculation and arbitrage. The terms "debtfare" and "debtfare state" have been used to describe the neoliberal state's governance role and technologies for mediating the capitalist crisis of over-accumulation and its class struggles (Soederberg 2014). As a governance mechanism debtfare involves the ideological normalization of credit and pervasive debt as a rhetorical development strategy that promotes "financial inclusion." State institutions, philanthropic organizations, and Fintech companies have used digital-based financial inclusion efforts as a form of development intervention, offering micro-credit and insurance, and new ways of marketing and managing risks via new forms of profiling poor households (Gabor and Brooks 2016).

Since the 1990s the over-accumulation of financial capital has been partially offset through the creation and extension of credit/debt, especially consumer, to replace a living wage and social welfare benefits. Exploitation (i.e., accumulation) has shifted to the seemingly "apolitical" consumer realm of exchange/reproduction (creditor/ debtor and service/consumer relations) rather than the more political realm of work (capital/labor relations). The growth of credit/debt has become the basis for capitalist accumulation, dispossession, and increasingly a requirement for the social reproduction of more and more of the population. Under debtfare the creditor-debtor relation is a generalized form of class struggle (Lazzarato 2012), which fuels uneven development and social stratification by further consolidating money and power. In this broader and innately social framework, financialization concentrates and centralizes money and power unevenly across classes, space and time, resulting in a form of "monopoly finance capital" that in turn wields hegemonic power over society (Foster and Holleman 2010). Under neoliberalism and its debtfare governance system, information technologies have been used to expand the mechanisms of rentierism, arbitrage, and speculative financial value-capturing, making them endemic to financialization as a regime of credit-led accumulation (Bryan, Rafferty, and Wigan 2017).

Finance, deregulation, and information technologies

A challenge for territorially based monetary systems lies in the risks and transaction costs associated with trans-boundary foreign exchange involving two or more currencies. There have been two international monetary regimes. The first was the fixed exchange rate regime, which tied currencies to a hybrid US dollar/gold standard that was managed by the Bretton Woods institutions. The second, the neoliberal, or free floating (i.e., market-based) exchange rate regime, ties currencies primarily to sovereign debt that is managed, we are told, by liberalized capital through the "all-seeing" panoptic eye of global financial networks (O'Tuathail 1997). Under the neoliberal regime, information technologies have been used to overcome regulatory barriers to production and exchange by "hyper-mobilizing" money, increasing its velocity and volume by digitally transferring it across national and regional boundaries using transnational telecommunications networks and electronic trading.

Political pressure for deregulation was led by bank lobbies and emerging-transnational conglomerates, like Citicorp, over their desire to remove Regulation Q of the U.S. Glass-Steagall Act, which they charged had competitively constrained American banks by fixing the interest rate on demand deposits. The Eurodollar market, the first offshore market, was out-competing American banks (Holly 1987; Wriston 1992). Digitalized exchanges, like NASDAQ (National Associated Automated Dealers Quotation System) connected users worldwide through satellite, internet, and fiber-optic networks, creating a transnational market in which buyers and sellers traded directly without paying rents to intermediaries, like banks or brokers. The development of electronic exchanges undermined the comparative advantages held by many financial centers, thus forcing them to also adopt digital trading platforms. The growth of lightly regulated offshore financial territories, made accessible by EFTs and telecommunication networks, helped make "offshore financial centers" (OFCs), disintermediation (so-called "shadow banking") accessible. Shadow bank institutions (hedge funds, mutual funds, and special purpose vehicles) operate under light regulations and low tax regimes. Many regulated commercial and investment banks engage in shadow banking activities through subsidiaries and holding companies in offshore havens. OFCs originally referred to centers of financial production located in a foreign territory that provided services to non-residents, and offer liberal regulatory environment and low tax rates. While OFCs originally implied the importance of territory to regulating centers of financial production, the term no longer applies to the geographic locality where financial production is located. For example, London's Eurodollar market circulates currencies (originally U.S. dollars) across the world's variegated market and regulatory landscape through an international telecommunications network among financial centers (Warf 1995; Cobb 2003; Leyshon and Thrift 1999; Michie 1997).

Offshore financial holding companies are used to navigate the world's variegated economic and regulatory landscapes (Roberts 1995). Offshore finance is often used to establish funds and trusts that securitize and bundle assets together under a "special purpose vehicle," which is legally designed for the purpose of offshore financing, arbitrage, financial engineering, mergers and acquisitions, and risk management. In this capacity, even non-financial firms can participate in financial production and rentierism, using EFTs to transmit money through offshore holding companies (Holly 1987). Commercial banks have responded to the threat of offshore finance by lobbying for more liberal onshore regulations, adopting new computer-based technologies, developed new financial instruments, and expanding their size and market reach through mergers and acquisitions. Offshore finance and information technologies have given investors the freedom to play territorial arbitrage across world's asymmetric regulatory landscape. A country or locale that doesn't create a good investment climate or liberal regulations risks losing access to capital. The unevenness of the world's regulatory landscape, along with the hierarchy of finance, create the conditions for maintaining profitable asymmetries that can be exploited through arbitrage (Budd 1995).

As the demand for information technologies increased under deregulation, the increasing volume of digital data and trading escalated the need for new management and processing technologies (Warf 1989). Many financial firms merged with or invested in telecomm companies. For example, credit card companies, like VISA and Mastercard, used telecomm networks like SWIFT (Society of Worldwide Interbank Financial Telecommunications), which was first deployed for international verifications and the processing of credit card transactions, as commercial Electronic Funds Transfer systems (EFTs). EFTs transmit digital information, such as financial payments, check processing, and other financial transactions via electronic telecommunication systems. EFT technologies include inter-bank clearing systems, automated teller machines, point of sale terminals, debit and credit card machines, as well as forms of digital banking by computer (Langdale 1985; Laulajainen 2001; Seese, Weinhardt, and Schlottmann 2008).

Venture capital and the Internet Bubble

Finance has long been key to technological revolutions by facilitating investments in new businesses and technologies (Perez 2012). Deregulation enabled traditional financial firms to merge with or own information technology firms, and thus finance has been crucial in the development of global telecommunications networks. The 1990s Dotcom boom gave rise to venture capital firms and epitomized the transformative role that finance and information technologies would have on the world economy. Venture capital is a form of high risk private equity most commonly associated with firms or funds investing in new companies, and has become especially associated with information technology companies and the stock market Internet Bubble (1995–2000). Venture capital takes a direct equity stake in a business, and venture capital funds manage a pool of capital with the same purpose. Heavy venture capital investments in internet startups helped fuel a stock market bubble in tech companies, especially telecommunications firms like Worldcom and Global Crossing. Between 1995 and 2000, the NASDAQ rose 400%. The dotcom boom gave birth to a generation of young techno-Libertarian millionaires, mostly in North American and Northern Europe, who proffered internet services and online e-commerce spaces. These new firms offered alternative, often ludic workplace environments and more informal wage systems often tied to rents from stock options. When the bubble burst in 2000 many new e-commerce and tech companies collapsed, stock values plummeted, jobs were lost and fortunes erased. Since then financial investments have been re-oriented toward the construction of Web 2.0, which is built around information technologies and online social platforms. Rents can often be high for firms adopting these platform technologies, since all users, both private and commercial, benefit from the free labor involved in their development (Terranova 2010; Zook 2005).

Information technologies, such as telecommunication systems and EFTs, helped drive the market and regulatory race-to-the-bottom unleashed by neoliberalism. The cost saving benefits of information technologies changed financial production and its division of labor. In many financial firms, front office operations typically involve the use of information, while back office operations involve routinized clerical work, including the management of information and technologies. Disintermediation and information technologies have altered the division of labor by allowing financial producers to eliminate or displace many jobs and functions and to standardize financial products. Fewer firms offer tailored products based on the specialized and often localized knowledge of analysts or experts, and carefully nurtured relationships with "insiders" have been replaced with consumer relations and standardized products based on data, theories, and econometric models of calculated risks and returns (Clark and Monk 2013). Mimicking changes undertaken by manufacturing firms, higher-wage front office jobs (client services, etc.) and firm headquarters were centralized in financial centers in global cities, while back office jobs have been decentralized away from corporate headquarters to secondary and peripheral lower-wage locations. Thus information technologies used in centers of financial production have helped concentrate well-paying professional jobs in urban areas that in turn influence real estate prices, which fuel gentrification and social displacement (Pryke and Lee 1995; Warf 1995).

The circulation of capital and the production of finance

Rentierism through financial production involves financial engineering and product innovation; marketing, sales, and placement; data management, information processing, and other clerical functions; and fee-based services associated with the trading of information and financial assets. The territoriality of financial production is manifest in the territory of locales where financial institutions cluster and production occurs. Before the advent of electronic and real-time communication systems, financial markets were plagued with information gaps and communication delays, as well as risks of misinformation. Trading typically occurred a day or more *after* the actual real-world events that led to the trading itself. Such high information asymmetries in trading assets could often result in large and sudden price fluctuations. In order to have information access, foreign banks also clustered in major financial centers. The accurate and timely transmission of information among financial agents and institutions has long been a technological and functional imperative for reducing risks involved in the circulation of capital and the production of finance. The historical and spatial formation of stock exchanges was itself an institutional development that enhanced communications and reduced information asymmetries. The limitations of early information technologies and the need for efficient and reliable information in order to maintain trust, status, and exchange among bankers and investors fostered the clustering of financial agents and institutions together (Ó hUallacháin 1994; Michie 1997).

Financial production was spatially clustered and relied on networks and tacit knowledge for reliable information, and communications among distant centers or investors were predicated on physical transportation technologies. The ability to separate information flows from physical transportation was accomplished with the invention of the telegraph (1844). By the 1870s, the international telegraph system reduced transmission time from days to minutes, and linked many geographically distant markets together through near continuous exchange. By the 1890s the telephone superseded the telegraph, which would be superseded a hundred years later by telecommunication systems, including the internet and smart phones. The revolution in microelectronics and computers, and the switch to digital information formats and computerized database management, were central to the advancement of information technologies and financialization. While both world wars seriously disrupted the process, information technology infrastructure underpinned the integration of securities markets, beginning with Europe and North America, and later the global financial system (Michie 1997).

Prior to the advent of 24-hour electronic trading, exchange was based around daytime trading hours within the succeeding time zones for the world's primary financial centers: New York, London, and Tokyo. Globalization and telecom have altered the geography of finance by fostering the contrasting processes of centralization and decentralization, within both financial production and in the circulation of capital, but did not undermine the importance of place, or location in time, to financial production (Warf 1989; Klagge and Martin 2005). International financial centers like New York agglomerate money, information and power, and are said to be the commanding heights of the global economy (Sassen 1999; Peet 2007). Supporting them are a series of second tier of financial center cities, including Los Angeles, Toronto, Singapore, Dubai, Hong Kong, Bahrain, and Frankfurt networked together through telecommunication systems. Challenges to the territoriality and hierarchy of financial centers include disintermediation and arbitrage. Trading that occurs anywhere and anytime can undermine a financial center's territorial control over a specific market, while also dissolving many of their comparative advantages in financial production and information access. In response, many financial centers enhanced their global comparative advantages through mergers, acquisitions, and alliances among complementary financial centers (Budd 1995).

The production of finance is performed by agents and institutions networked together via global flows of capital and information pipelines comprised of information technologies and telecommunications networks. The value of these information pipelines is derived from their width (the volume of information that can be communicated) and speed (real-time access to spatially dispersed market information). These networks in turn

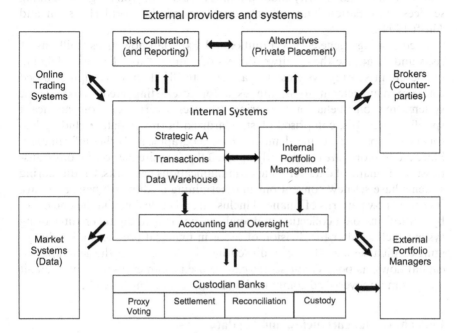

Figure 11.1 Financial institutions and information pipelines.
Source: Author, based on original in Clark and Monk (2013, 1327).

reduce information asymmetries and facilitate communication, enable up-to-date news and data, electronic buying and selling assets, and so on. Communication technologies are conducive to achieving economies of scale by controlling transaction costs for vast volumes of transmitted, processed and managed data. Financial production and risk management practices place a premium on the continuous acquisition, processing, representation, and exchange of information. This includes information flows external and internal to an institution, since risks can arise not only in markets but also internal to a firm from rogue traders, fraud, and other risky activities (Pryke and Lee 1995; Clark and Monk 2013).

The revolution in information technologies allowed for instantaneous trading of information and money. While financial risk management practices vary, the use of spatial framing for analyzing the risks of assets and asset classes is often practiced, albeit in abstract and arbitrary ways. Securitized products, for example mortgages, are often framed regionally in order to conceptually demarcate different assets, markets, and risks (Wainwright 2012). Information technologies also lowered risks through the development of standardized online investment products, service platforms, and credit scoring systems. This includes financial management software, online financial planners, informative websites, and other internet-based financial tools.

Advances in digital security enabled online access to the full range of banking services, information, and markets from around the world (Leyshon and Thrift 1999).

Credit-rating agencies, like Equifax and Experian now assess millions of risks and scrutinize the creditworthiness of entities "at-a-distance." Digital, online credit-scoring systems that are statistically derived, computer-based software management tools, are used for forecasting credit risks. These systems rely on "behavioral" scoring practices ascertained from applicant specific scorecards, comprised of standardized borrower data including history of occupation/income, demographics, geographic and other information. Since credit scores are consistently updated they require the continuous transmission of financial data across telecommunications networks. Credit scoring systems have set new conventions in determining creditworthiness, and have produced new patterns of financial inclusion and exclusion. On the one hand, by centralizing data collection and analyses, automated digital credit-scoring systems helped alter the division of labor in financial production by eliminating credit assessors. Online credit scoring also helped alter the geography of capital flows, as poor credit scores can be used as an excuse to divert credit away from impoverished communities (Leyshon and Thrift 1999).

Fintech, virtual currencies, and big data

While the evolution of financial production has moved in tandem with technological developments in the service economy, electronic exchanges require real-time accurate information in order to execute tactical trades, making them important sites for innovations in information technologies. The proliferation of smartphones and other mobile digital media devices that provide continuous online services and applications enable ever more access to data from the daily lives of users. Personalized services and products, recommended based on the user's demographics and past behaviors, are tailored and delivered in real-time to the user's mobile device. Fintech firms are central to the development of such seamless digital service networks, combining mobile information technologies and big data analytics with digital marketing and services. The term "fintech" has been applied to the use of advanced information technologies to improve efficiencies in financial production, and to the application of information technologies in the design, function, and marketing of financial products and services. FinTech companies use innovative information technologies, especially computers and the internet, to compete with more traditional financial service providers. This includes using smart phones for mobile banking, using cloud storage technologies for data management, or using information technologies for transactions (PayPal, Apple Pay, etc.) as components of financial products or as digital currencies.

Virtual currencies, like Bitcoin, operate through publically accessible internet-based, encrypted, blockchain platforms. Virtual currencies are managed through either a centralized or decentralized system of market control and governance. Centralized virtual currency systems were developed by online gaming companies, and they managed the currency's supply and demand. Decentralized systems were developed by networks of programmers, and the currency is managed via crowd-sourced computations that are based on blockchain cryptographic technologies. Bitcoin, the first decentralized virtual currency, uses digital cryptography to manage the creation of Bitcoins, and the internet to exchange Bitcoins among user's anonymous, and secure electronic "wallets." There are no intermediaries, like banks, or monetary regulations to drive up transaction costs. Bitcoin was developed on Libertarian principles and ideals and the goal of its progenitors was to avoid the high costs associated with financial intermediation. Bitcoin would enable the disintermediation of online digital finance by providing an online value-trading exchange that depends upon peer-to-peer information networks, personal encryption keys, and computational tasks that provide "proof" of transactions between anonymous users (Zook 2018). To date, virtual currencies are not legal tender, and not backed by any commodity or currency and thus avoid the oversight of a centralized monetary authority, like central banks. Lacking monetary legitimacy and liquidity, virtual currencies are volatile, speculative, and risky instruments. While the market for virtual currencies remains relatively small, they consume vast amount of energy and have relatively enormous carbon footprints. The entire Bitcoin network now consumes more energy than some countries, and with most mining activities based in China most of that is dirty energy with high carbon emissions (https://digiconomist.net/bitcoin-energy-consumption).

While information technologies have helped decentralize financial production, their security and cost benefits make them attractive risk management technologies for financial intermediaries, and so they are investing heavily in startups. Cryptographic information technologies like blockchain are being exploited by traditional intermediaries through specialized global networks of financial production, which use private intra- and inter-firm blockchain networks to securely transmit vast flows of capital and information (Zook 2018). The volume of digital data being generated with advancements in information technologies has been tremendous. By 2016 the New York Stock Exchange alone processed around a terabyte of information each day. The growth in the volume of digital information and "big data" analytics has been a gold mine for FinTech firms. Globally, FinTech investments exceeded $22 billion in 2015, growing some 75% compared to 2014 (Accenture 2016).

Algorithmic high-speed trading, based solely on financial models and data, involves computerized automated trading systems that work at super-human speeds and frequencies to execute trades efficiently at the best possible price. The volatility of financial markets has increased as these technologies generate hundreds to thousands of trades within microseconds of each other across global markets, and thus accelerate accumulation by compressing the time and space involved in financial transactions (Grindsted 2016). By incorporating

real-time news, social media, and financial data into one algorithmic platform, the volume and velocity of continuous high-speed algorithmic trading
has helped drive the growth of big data (Zook 2018). Automation intensifies
the power of information technologies over society, and artificial intelligence
working as financial "robo-advisors" and customer service "chatbots" are
playing increasingly essential roles in financial production and risk management practices. The digital revolution in telecommunications helped shift the
costs of financial production toward investments in new information technologies, computerized data management systems, and geo-demographics and
geographical information systems for marketing and branding. Fintech firms
are using big data to capture rents by competing with banks and other traditional intermediaries in providing financial services like payment and check
processing and even savings accounts. Big data refers less to the volume of
information in a data set, but to the use of predictive analytics, user behavior
analytics, and other advanced data analytics methods to extract valuable consumer information. Big data allows for economies of scale applied to financial
production and risk management.

Conclusion

In the electronic ecology of the internet, Web 2.0, and the closely monitored e-
spaces of the social network society, digitized money and information merge
and flow seamlessly together, in effect functioning as "two sides of the same
coin." This virtual economy and online financial system are inherently linked to
the real economy of services and production. Telecommunication systems like
the internet, computerized databases, software and operating platforms, and
personal media devices constitute the digital infrastructure of financial production. They all depend on standardized information practices. Information
technologies, like online credit-scoring and "behavioral" marketing, are said
to constitute new panoptic governance mechanisms that enforce market discipline and the power of finance through persistent surveillance (Neu et al.
2002). They are calculative devices that allow for the standardization, organization, and centralization of information, and can be used by institutions for
centralizing decision-making and power. Power is centralized under control of
those who institutionalize such devices. While the commercial use of platform
information technologies in facilitating financialization has been scrutinized,
hope remains that open source information technologies and other share-
ware systems will be used to provide cheap alternatives to capitalist finance.
Free digital crowd-funding platforms and peer-to-peer networks, often under
democratic control and for the benefit of public participation, are growing in
popularity (Langley and Leyshon 2017). When a firm uses these technologies, it must technically codify and culturally internalize those standards and
protocols. However, when a society adopts them, it too internalizes and codi-
fies those standards, values and norms, and thus effectively transfers power

to financial agents and institutions, their prerogatives, and, ultimately, helps advance the financialization of capitalism.

References

Accenture. 2016. *Global Fintech Investment Growth Continues in 2016 Driven by Europe and Asia, Accentrure Study Finds.* P. Shuttlewood, M. Volin, and L. Wozniak (eds.) London: Accenture Financial Services.

Bieri, D. 2017. Regulatory space and the flow of funds across the hierarchy of money. In R. Martin and J. Pollard (eds.) *Handbook on the Geographies of Money and Finance.* pp. 377–414. Cheltenham: Edward Elgar.

Birch, K. 2017. Financing technoscience: Finance, assetization and rentiership. In D. Tyfield, R. Lave, S. Randalls, and C. Thorpe (eds.) *The Routledge Handbook of the Political Economy of Science.* pp. 169–181. London: Routledge.

Bryan, D., M. Rafferty, and D. Wigan. 2017. The map and the territory: Exploring capital's new financialized spatialities. In R. Martin and J. Pollard (eds.) *Handbook on the Geographies of Money and Finance.* pp. 86–104. Cheltenham: Edward Elgar.

Budd, L. 1995. Globalisation, territory and strategic alliances in different financial centres. *Urban Studies* 32(2):345–360.

Castells, M. 2003. The network society. In D. Held and A. McGrew (eds.) *The Global Transformations Reader.* Malden, MA: Blackwell.

Clark, G., and A. Monk. 2013. Financial institutions, information, and investing-at-a-distance. *Environment and Planning A* 45:1318–1336.

Cobb, S. 2003. Offshore financial services and the internet: Creating confidence in the use of cyberspace? *Growth and Change* 34(2):244–259.

Epstein, G. (ed.) 2006. *Financialization and the World Economy.* Cheltenham: Edward Elgar Publishing.

Foster, J., and H. Holleman. 2010. The financial power elite. *Monthly Review* 62(1):1–19.

Funke, J. 2017. Demystifying money: Fictions of capital and credit. *Human Geography* 10(1):20–35.

Gabor, D., and S. Brooks. 2016. The digital revolution in financial inclusion: International development in the fintech era. *New Political Economy* 22(4):423–436.

Grindsted, T. 2016. Geographies of high frequency trading – Algorithmic capitalism and its contradictory elements. *Geoforum* 68:25–28.

Harvey, D. 2006a. *The Limits to Capital.* 3rd ed. London: Verso.

Harvey, D. 2006b. The "new" imperialism: Accumulation by dispossession. In L. Panitch and C. Leys (eds.) *Socialist Registrar 2004: The New Imperial Challenge.* pp. 63–87. London: Merlin Press.

Holly, B. 1987. Regulation, competition and technology: The restructuring of the US commercial banking system. *Environment and Planning A* 19:633–652.

International Finance Corporation. 2009. Financial infrastructure: Building access through transparent and stable financial systems. In *Financial Infrastructure Policy and Research Series.* Washington, DC: International Finance Corporation.

Kay, K. 2017. Rural rentierism and the financial enclosure of Maine's Open Lands tradition. *Annals of the American Association of Geographers* 107(6):1407–1423.

Klagge, B., and R. Martin. 2005. Decetralised versus centralised financial systems: Is there a case for local capital markets? *Journal of Economic Geography* 5:387–422.

Krippner, G. 2005. The financialization of the American economy. *Socio-Economic Review* 3(2):173–208.

Labban, M. 2010. Oil in parallax: Scarcity, markets, and the financialization of accumulation. *Geoforum* 41:541–552.

Langdale, J. 1985. Electronic funds transfer and the internationalisation of the banking and finance industry. *Geoforum* 16:1–13.

Langley, P., and A. Leyshon. 2017. Capitalizing on the crowd: The monetary and financial ecologies of crowdfunding. *Environment and Planning A* 49(5):1019–1039.

Lapavitsas, C. 2011. Theorizing financialization. *Work, Employment and Society* 25(4):611–626.

Laulajainen, R. 2001. The end of geography at exchanges. *Zeitschrift fur Wirtschaftsgeographie* 45:1–14.

Lazzarato, M. 2012. *The Making of the Indebted Man*. Los Angeles: Semiotext(e).

Leyshon, A., and N. Thrift. 1994. Access to financial services and financial infrastructure withdrawl: Problems and policies. *Area* 26:268–275.

Leyshon, A., and N. Thrift. 1999. Lists come alive: Electronic systems of knowledge and the rise of credit-scoring in retail banking. *Economy and Society* 28:434–466.

Martin, R. 2002. *Financialization of Daily Life*. Philadelphia: Temple University Press.

McLeay, M., A. Radia, and R. Thomas. 2014. Money creation in the modern economy. *Quarterly Bulletin*. London: Bank of England.

Michie, R. 1997. Friend or foe? Information technology and the London Stock Exchange since 1700. *Journal of Historical Geography* 23:304–326.

Neu, D., E. Gomez, O. Garcia Ponce de Leon, and M. Zepeda. 2002. Facilitating globalization processes: Financial technologies and the World Bank. *Accounting Forum* 26(3–4):271–290.

O'Brien, R. 1992. *Global Financial Integration: The End of Geography*. New York: Council on Foreign Relations.

Ó hUalláchain, B. 1994. Foreign banking in the American urban system of financial organization. *Economic Geography* 70(3):206–228.

O'Tuathail, G. 1997. Emerging markets and other simulations: Mexico, the Chiapas revolt, and the geofinancial panopticon. *Ecumene* 4(3):300–317.

Peet, R. 2007. *The Geography of Power: The Making of Global Economic Policy*. London: Zed Press.

Perez, C. 2012. Financial bubbles, crises and the role of government in unleashing golden ages. In A. Pyka and H.P. Burgof (eds.) *FINNOV* Milton Keynes, UK: Open University.

Pistor, K. 2013. A legal theory of finance. *Journal of Comparative Economics* 41(2):315–330.

Pryke, M., and R. Lee. 1995. Place your bets: Towards an understanding of globalisation, socio-financial engineering and competition within a financial centre. *Urban Studies* 32(2):329–344.

Roberts, S. 1995. Small place, big money: The Cayman Islands and the international financial system. *Economic Geography* 71:237–256.

Sassen, S. 1999. Global financial centers. *Foreign Affairs* 78(1):75.

Seese, D., C. Weinhardt, and F. Schlottmann (eds.) 2008. *Handbook on Information Technology in Finance, International Handbooks on Information Systems*. Berlin: Springer.

Soederberg, S. 2014. *Debtfare States and the Poverty Industry: Money, Discipline, and the Surplus Population.* Abingdon: Routledge.

Stockhammer, E. 2004. Financialization and the slowdown of accumulation. *Cambridge Journal of Economics* 28:719–741.

Terranova, T. 2010. New economy, financialization and social production in the Web 2.0. In A. Fumagalli and S. Mezzadra (eds.) *Crisis in the Global Economy: Financial Markets, Social Struggles, and New Political Scenarios.* pp. 153–170. Los Angeles: Semiotext(e) and Ombre Corte.

Wainwright, T. 2012. Number crunching: financialization and spatial strategies of risk organization. *Journal of Economic Geography* 12(6):1267–1291.

Warf, B. 1989. Telecommunications and the globalization of financial services. *Professional Geographer* 41:257–271.

Warf, B. 1995. Telecommunications and the changing geographies of knowledge transmission in the late 20th century. *Urban Studies* 32:361–378.

Wojcik, D. 2017. The global financial networks. In G. Clark, M.P. Feldman, M. Gertler, and D. Wojicik (eds.) *The New Oxford Handbook of Economic Geography.* Oxford: Oxford.

Wriston, W. 1992. *The Twilight of Sovereignty: How the Information Revolution is Transforming Our World.* New York: Charles Scribner's Sons.

Zook, M. 2005. *The Geography of the Internet Industry: Venture Capital, Dot-Coms, and Local Knowledge.* Malden, MA: Blackwell.

Zook, M. 2018. Information flows, global finance and new digital spaces. In G. Clark, M. Feldman, M. Gertler and D. Wojicik (eds.) *The New Oxford Handbook of Economic Geography.* Oxford: Oxford University Press.

12 E-tourism

Irene Cheng Chu Chan and Rob Law

Travel has become one of the most popular leisure activities in modern society. According to the United Nations World Tourism Organization, the number of international tourists increased by 7% in 2017, representing the sharpest growth since 2010 (UNWTO 2018). One out of six people travels to another country annually. The increase in travel is shaped mainly by the surge in demand from emerging tourist source markets, such as Brazil and Russia, as well as the gradual recovery of the global economy after the 2008 financial crisis. The strong and continuous growth of international travel consolidates the role of the tourism industry as a key economic sector in the world, generating jobs, trades, and prosperity in many developing and developed countries.

The significant achievement of tourism goes hand in hand with the rapid development of information and communication technologies (ICTs), including the internet, wireless networks, smartphones, and so on. ICTs enable instant communication and information dissemination at almost no cost and have become an integral part of consumers' lives and businesses. Tourism is one of the fastest growing sectors of the internet, with travel search and purchases being among the most popular activities performed online. According to eMarketer (2018), global online travel sales totaled USD 630 billion in 2017 and was projected to exceed USD 800 billion in 2020 before the outbreak of the coronavirus pandemic. Moreover, an increase has also been observed in the use of smart devices to search for information and make bookings online (eMarketer 2017).

The mutual influence of tourism and the internet has led to the creation of electronic tourism (e-tourism) in the late 1990s. The concept of e-tourism has evolved considerably over the last two decades and has given rise to related concepts, such as mobile tourism (m-tourism) and smart tourism (s-tourism). This entry first provides an overview of the origin and history of e-tourism, followed by a review on the various internet-enabled ICTs adopted by corporate industrial players. Afterward, the present study delineates important conceptual issues and debates related to e-tourism and finally, geographical and spatial implications are outlined.

Origin and history of e-tourism

Tourism refers to the movement of people to places outside their usual environments for various purposes, including leisure and business. Tourism products, such as flights, hotels, tour packages, and attractions, are service oriented. They are highly intangible, and therefore cannot be touched or experienced before actual consumption. While an airline can provide customers with a flight ticket to bring them to the target destination, the ticket itself does not represent the service, but a promise that can only be evaluated during or after use. The production and consumption of tourist products and services cannot be separated. For example, the tour guide and tourists have to be present for the touring services to be delivered. The involvement of the customers in the process of service production, delivery, and consumption makes product standardization almost impossible. Therefore, consumption experiences of tourism products vary from person to person and from time to time even for the same tourists. Tourism products and services are also highly perishable because their inventories cannot be stored. In other words, if a hotel room is not occupied for one night, the revenue for that particular night is lost forever. Therefore, tourism operators act promptly to sell all inventories to maximize profits. The unique characteristics of tourism products render their evaluation before consumption difficult, and thus information plays an important role in reducing risks and uncertainties perceived by customers. As such, the tourism industry is highly susceptible to the influence of the internet and internet-enabled technologies.

The internet refers to a collection of computer networks which allows unlimited numbers of users to share information worldwide. Before the internet, travel often involved extensive efforts. To select a destination, people took inspirations from sources limited to family, friends, travel magazines, and books. No comparison websites existed, and people had to visit different travel agencies to collect information on different deals to determine the best offers. Information on hotels was not available and each decision represented a high risk. Thus, group travel and packaged services offered by travel agencies were popular modes of travel. Flexibility during the trip was very limited. Tourists normally followed what was planned to avoid additional costs and troubles. Moreover, when they returned from their trips, people could only share their experiences with close friends and family.

The situation was not any better for tourism product suppliers, such as airlines, hotels, and travel agencies. Before the internet, airlines and hotels had to manage manually all business processes, such as reservations, guest check-ins and registrations, check-outs, and so on, which were labor-intensive and inefficient. Tourism suppliers had to rely passively on their brick-and-mortar firms, newspapers, and magazines, or other intermediaries as distribution channels to reach out and communicate with their customers, which often represented high costs.

E-tourism refers to the adoption of different ICTs by tourists and businesses, which transforms the business functions, value chains, and relationships between customers and tourism firms (Buhalis 2003). E-tourism interlinks the business, information technology/system, and tourism sectors. From a strategic perspective, e-tourism changes the business landscape and processes of the tourism industry as well as the relationships among tourism organizations. From a tactical perspective, e-tourism empowers tourism firms to maximize efficiency and effectiveness of business processes through the application of ICTs.

E-tourism did not evolve overnight. E-tourism dates back to the 1950s when American Airlines installed the first automated reservation system, later known as the computerized reservation system (CRS). Despite the system needing human operations, it largely reduced the time required to handle customer reservations and communications. In the following decades, other airlines and travel agencies began to adopt the CRS. In the 1980s, the first global distribution system (GDS) was developed to connect the individual systems of different tourism firms and allow travel agencies to book different services through a single platform. At the time, GDS was adopted mainly by large firms because smaller firms could not afford the expensive commissions and membership fees.

The internet opened up for commercial use in the mid-1990s, which largely reconfigured the landscape of tourism operations. The wide adoption of the internet provided new means of distribution, allowing service providers to reach customers directly, bypassing traditional intermediaries. At almost the same time, e-businesses including the Lonely Planet website and Expedia by Microsoft were established. In the 2000s, online social media and mobile technologies played a significant role in fueling the development of e-tourism. The most representative social media sites include TripAdvisor and Facebook. Furthermore, the first flight search engine, Kayak, and the first accommodation sharing platform, Airbnb, were subsequently established. Throughout the years, innovative services were constantly being developed while old ones died out. These newly emerging technologies not only transformed the way customers accessed information, made purchase decisions, and performed transactions but also restructured the power balances between customers and tourism suppliers.

E-tourism technologies

Various technologies and innovations have been developed for business purposes such as tourism. Technologies can be categorized into three major types, namely, distribution, operational, and research and analytical. Distribution and operational technologies are grounded on and largely driven by the technologies that customers adopt before, during, and after their trips, such as mobile devices and social media.

Distribution technologies are tools and channels that facilitate transactions between tourists and businesses. The CRS and the GDS, which are automated reservation systems adopted by tourism suppliers such as airlines, travel agencies, and hotels to facilitate operations and collaborations, are examples of early distribution technologies. With the GDS, travel agencies could easily obtain information from various service providers on the GDS network. As the internet became more common, tourism firms started to set up their own websites to sell directly to their customers. In addition to being a distribution channel, websites also represent the first point-of-contact between suppliers and customers. Thus, tourism firms seek to build their brand images through their websites and put continuous efforts in improving them.

The development of the internet has led to the emergence of a new form of a business model called online travel agencies (OTAs). An OTA is a website that provides one-stop travel planning and reservation services to customers, from airplane ticket, hotel, local tour bookings, to car rentals. Tourists enjoy the freedom to formulate their own travel packages using OTA websites. Examples of OTAs are Expedia, Agoda, and Ctrip. By simply providing the name of the destination, customers can browse a wide variety of available options and perform price, location, and star rating comparisons. The convenience and flexibility of OTAs sustain their competitive position in the distribution process of tourism products. Popular social media sites, such as TripAdvisor and Facebook, have also started to develop their role as distribution technologies by redirecting users to booking sites for various travel products while they enjoy the advantage of interacting with users daily.

Recently, metasearch engines have become a new distribution channel for OTAs and a new marketing opportunity for tourism service providers. Metasearch engines include Skyscanner and Kayak for flights, HotelsCombined and Trivago for accommodations, and CarRentals for automobiles. A metasearch platform compiles all available rates from numerous booking websites and OTAs into a single platform. Users can easily compare all available options and be directed to the booking site with the best offer for their choices. The business model of metasearch engines is different from that of OTAs. Instead of charging commissions to hotels, metasearch engines charge the booking sites directly for cost-per-click.

Operational technologies are the tools and systems adopted by tourism firms to enhance efficiency and competitiveness. Traditionally, property management system (PMS) is software designed for the accommodation sector to assist front-office tasks, such as reservations, guest check-in/outs, room allocations, and billing. Examples of PMS are Oracle's Opera and Maestro PMS. The development of cloud technology has extended the service offerings of PMS to facilitate other hotel operations, such as housekeeping, inventory, revenue, customer–relationship management, and performance analytics. The PMS largely enhances the productivity and personalization of service delivery to tourists.

In addition to PMS, other technologies have been implemented in the hospitality sector to facilitate operations, such as self-service kiosks, smart devices, and robots. These technologies are mainly adopted to replace labor resources and increase operational efficiency. Various chain hotels, such as Hyatt and Marriott, have developed their own mobile applications, which allow guests to check in and out, choose and access their rooms, and also communicate requests. Meanwhile, hotel rooms have been equipped with smart features that allow guests to adjust all in-room settings, such as lighting, temperature, and so on, with just a few clicks on a single device. Customers can even use the virtual concierge service on their smart devices to browse information about destinations, request additional amenities, or customize their itinerary.

Aside from the hospitality sector, other tourism firms have gradually adopted ICTs to enhance tourist experiences. For example, attraction managers have adopted advanced technologies to offer individual and interactive touring services to visitors. Museums, such as the British Museum in London and the Louvre in Paris, have invested in interactive displays and augmented reality technologies. The Audio Guide Louvre-Nintendo 3DS XL provides maps and three-dimensional photos together with audio commentaries. Similarly, visitors to the British Museum can simply place their smartphones over an ancient statue to see its missing parts and how the statue looked before being destroyed. Some museums are testing the iBeacon technology, which alerts visitors through their smartphones to interesting information as they pass a work of art.

Tourism-related firms, such as airlines and hotels, have started to adopt artificial intelligence (AI) to implement dynamic pricing and yield management. By tracking the occupancy rate and customers' demographic characteristics, consumption habits, and loyalty, the pricing software can determine each customer's willingness to pay and their responsiveness to different offers. Accordingly, the system will determine the optimal price for each customer and perform price discrimination, and customers may pay a higher or lower price for the same product. This technology allows tourism firms to maximize the lifetime value of each customer. Moreover, AI has also been developed into chat robots (chatbot) to simulate human conversation. For example, KLM Royal Dutch Airlines uses chatbots in the Facebook Messenger application to offer personalized and efficient customer care. Information such as booking confirmations, check-in notifications, and flight status updates are sent to customers via the chatbot, which is fully automated without human intervention. By handling the routine queries, chatbots save time for customer service agents, allowing them to perform more complex tasks.

Research and analytical technologies are used by tourism firms use to track and analyze tourist behavioral patterns, thereby informing their strategic and operational decisions. The wide adoption of ICTs by tourists, such as social media and mobile devices, has generated a large amount of behavioral data, including information searches, transactions, spatial movement at a destination, social media mentions, photos, online interactions with other customers, and even emotions. The huge amount of data generated from ubiquitous computing devices are known as big data, which are potential indicators of tourists' preferences, motivations, travel planning behaviors as well as actual experiences. Unlike traditional data available for analytics, big data are highly unstructured and fragmented across various platforms, which is beyond the capacity of traditional database systems to handle. As a result, specialized software has been developed to mine the associations and frequencies and to perform predictions. Some hotels try to draw insights into customer experiences from social media websites using data mining software, such as IBM technologies and statistical analysis systems. Hotel owners can now know which specific hotel features satisfy guests the most and which cause customer frustration. Accordingly, management attention and resources can be directed.

Moreover, destination marketing organizations (DMOs) are utilizing web traffic data on Google Analytics and Baidu to forecast tourist arrivals and hotel occupancy rates. A geographical information system (GIS) is used to record, transform, and analyze geographical and spatial data. By monitoring the movement of tourists using GIS, DMOs gain valuable insights into tourists' spatial and temporal travel patterns at the destination. Specifically, DMOs can determine the most popular attractions at certain time periods and the most frequently used transport by tourists, so that they can implement proper policies on crowd management and destination planning.

Key conceptual issues and debates about e-tourism

As a field of study and research, e-tourism has grown significantly since its emergence. The conceptual issues and debates on e-tourism revolve mainly around two domains, namely, (1) consumer behavior and demands and (2) industry and business functions. The following section discusses the major conceptual issues in each domain.

Consumer behavior and demand

E-tourism has tremendously changed the way tourists manage travel. Understanding the role of ICTs in every stage of tourists' decision-making processes is very important. The decision-making processes of tourists range from information search and evaluation of alternatives, to purchase and consumption activities at the destination and post-purchase or post-trip communications. A wide range of studies has sought to understand and predict tourists' acceptance and adoption of different ICTs for each stage in the process. These studies are grounded primarily in the conceptual frameworks of information systems, such as the technology acceptance model (TAM) and unified theory of acceptance and use of technology model (UTAUT). The TAM suggests that when users are presented with a new technology, two factors will affect their decision to use it, namely, perceived usefulness and perceived ease of use. Technologies perceived to be useful and easy to use will generate positive attitudes in the users and drive them to use the technologies (Davis 1989). UTAUT supplemented two other factors, namely, social influence and facilitating conditions, which suggest that in addition to being useful and easy to use, a technology's social acceptability and enjoyability are also important (Venkatesh, Morris, Davis, and Davis 2003). The service environment should also encourage usage. Other factors such as risks and users' privacy concerns as well as their personal characteristics, such as self-confidence, personal innovativeness, and previous experiences are added to the various drivers of technology acceptance and adoption. It is important to note that these factors do not bear the same weight in influencing tourists' adoption, and conflicting arguments exist in determining which factor dominates. Nevertheless, privacy has been demonstrated consistently to be a key concern of tourists and gaining their trust is still a key challenge for tourism firms. Thus, tourists' perceptions of and attitudes toward different technologies and ICT-enabled tourism services, such as location-based advertising, travel mobile applications, and personalization, are also important concepts in e-tourism.

Online reviews represent real consumption experiences and have become a reliable reference for decisions on tourism products which are difficult to evaluate before consumption. Assessing the effects of online reviews on customers has been a popular research avenue. In general, tourism products with more positive reviews than negative reviews are considered more favorable and thus engender higher purchase intention among tourists. However, operators should give specific attention to negative reviews because they are more influential due to negativity bias and prospect theory. However, the congruity of opinions is also important. A hotel with unanimous negative reviews may have to consider lowering its price to mitigate the overall negative effect caused by those online reviews. In actuality, online reviews describing different customer experiences rarely reach consensus. Customers facing ambivalent reviews may rely on other peripheral information, such as the reviewers' identities, expertise, or even similarities, to determine the usefulness and helpfulness of each review (Liu and Park 2015). This behavior suggests that consumers value the information provided by different reviewers differently, and review prioritization and customization are important to help customers determine the most relevant ones for their decisions.

Industry and business functions

The conceptual issues from the industry's perspective revolve mainly around the strategic and tactical levels of businesses. Studies with a strategic view focus on investigating whether the integration of ICTs into businesses enriches their resources, thereby improving the dynamic capabilities of the firm to carry out its business objectives. Scholars have attempted to investigate the readiness and capability of tourism firms to leverage ICTs to increase revenue, reduce cost, and improve quality for their businesses to win over competitors.

However, some studies have focused on examining the impact of the internet and ICTs on the survival of travel agencies or the threat of disintermediation. Tourism distribution refers to a chain of businesses or intermediaries through which a tourism product passes until it reaches its end consumers. Traditionally, tourism products were distributed in a channel approach, which is a linear and static process of commercial exchange. The internet facilitates information access that disturbs the classic distribution process. Instead of selling its rooms through travel agencies, a hotel can reach its customers directly by setting up its own website, allowing customers to make their own reservations. Direct distribution represents significant cost saving and allows hotels to bypass expensive intermediaries. Thus, most hotels introduce guaranteed low prices and personalized services to encourage direct bookings from customers.

Therefore, disintermediation may result from the evolution of e-tourism and threaten the survival of intermediary firms. However, the increasing number of OTAs indicate that online intermediaries are growing continuously against the predicted trend of disintermediation. In contrast to traditional travel agencies, OTAs reach their customers only through the internet. OTAs provide a wide range of tourism products to customers, offering high levels of convenience and flexibility. Therefore, optimistic scholars argue that technologies only transform the distribution process. Instead of reducing the number of intermediaries in the distribution channel, e-tourism has created an increasingly complex network of intermediaries (Kracht and Wang 2010). Travel agents will still remain key players in tourism distribution; however, their role will change from being supplier-centric to consumer-centric. While disintermediation is not likely to take place any time soon, tourism product suppliers must maintain their competitiveness by adopting a multi-channel strategy (Beritelli and Chegg 2016). They should not only strengthen their direct distribution by taking advantage of advanced technologies, but also improve their service offerings and relationship with customers constantly through loyalty programs and enhance their presence and collaboration with key intermediaries.

Studies with a tactical view focus on the role of ICTs in various business functions, such as marketing and management. Websites are one of the most important channels for tourism firms to reach and communicate their brand images with their customers, thus website evaluation has become an important concept to improve website design constantly. During its early stages, scholars were interested in developing evaluation metrics to measure the overall quality of a website (Law 2019). Website quality is assessed based on two key dimensions, namely, functionality and usability. Functionality refers to the information and content on the website (Chung and Law 2003), while usability refers to the structure and presentation of the website (Au Yeung and Law 2004). Each dimension can be evaluated at four levels, from the basic information to the communication, transactional, and relationship levels (Law 2019). However, some scholars stressed the significance of other aspects, such as visual aesthetics and atmospherics, which may evoke emotional responses and attachment among users (e.g., Kirillova and Chan 2018). They suggested that instead of a functional website, a persuasive website that could leave a good impression on users or even convert them into customers is more important.

The rising number of tourists spending an increasing amount of time on different ICTs means digital marketing has become an important means of marketing and advertising for tourism firms. Scholars are interested in examining how tourism operators adopt various ICTs to develop more effective marketing strategies and conduct promotional campaigns. The different types of ICTs being used have led to the emergence of different types of marketing, including website marketing, social media marketing, and mobile marketing. Recently, scholars have begun to investigate an emerging concept of SoCoMo marketing, which refers to "an advanced systematic method of content marketing on smart mobile devices that integrates social media to empower co-creation of value" (Buhalis and Foerste 2015, 155). Marketers can increase their value by revolutionizing market offerings and dynamically co-creating products and services with customers. Instead of being a passive receiver of service offerings provided by tourism firms, tourists have become "prosumers" actively involved in the service provision process. This concept is evident in the rapid development of the sharing economy in tourism, such as Airbnb and BlaBlaCar. In response, tourism firms must shift their focus from merely creating values to co-creating values with tourists. This concept of co-creation has become widespread in the industry, influencing the e-tourism business landscape.

Tourism firms seek to build and maintain lucrative relationships with customers through relationship marketing or customer relationship management (CRM) strategies. The concept of CRM in tourism and hospitality has transcended from a strategic marketing initiative to a multidimensional management philosophy (Law, Fong, Chan, and Fong 2018). The extensive adoption of technologies has extended the relational approach of CRM to e-CRM, which focuses on establishing and enhancing customer relationships through electronic channels. The popularity of social media has led to the further development of the concept into social CRM, which integrates social media applications to engage customers in collaborative interactions to enhance customer relationships (Greenberg 2010; Trainor 2012). Scholars are interested in identifying the level of adoption of social CRM and different strategies adopted by tourism firms as well as the various challenges and outcomes of social CRM implementation (Chan, Fong, Law, and Fong 2018).

In addition, online reputation management has become an important area of research because of the significance of online reviews. Specifically, understanding the appropriate strategies to address negative online reviews and mitigate unfavorable effects is imperative. Prospective customers rely on cues, such as frequency and speed of response to those negative reviews as a way to judge the sincerity of tourism firms. However, as the pressure for maintaining good reputation grows, morally questionable practices by tourism firms to improve online ratings have started to emerge, such as buying positive reviews and suppressing negative ones. This practice may cause customers to become frustrated and suspicious toward online ratings and comments.

Geographic and spatial implications of e-tourism

Tourism has been traditionally defined as much by space as by time. The introduction of the electronic component into tourism has brought about significant geographic and spatial implications. First, technology eliminates the constraints of distance and geographic boundaries among different nations and regions. Most of the time, information flows freely among internet users, buyers, and sellers across borders, resulting in enlarged markets and access to global supply chains. Thus, competitive and cooperative activities among tourism firms largely progress from being on the local or regional level to the national or international level. The distinction between activities based on different geographical scales also becomes ambiguous. Thus, tourism firms should hold a wide viewpoint over the global environment when it comes to the strategic development of the business.

Second, technology empowers and mobilizes individuals. International tourism flows tend to be biased toward more developed destinations as people possess more knowledge and lower risk perceptions on these destinations. E-tourism may remove this psychological distance in tourists' minds as the internet opens its eyes to more unknown places, communities, and cultures around the globe. As technologies enable greater ease in traveling, tourists are able to discover less popular destinations and attractions.

The ability of technologies to remove geographical barriers has significantly changed the original conception and definition of a community, which used to be more geographically confined (e.g., neighborhoods, ethnic group, villages). A community may now exist without public space, which is known as a virtual community. Virtual communities do not have geographical boundaries but boundaries between members and non-members. Instead of being geographically linked, a virtual community consists of a group of like-minded individuals, oftentimes strangers, who share similar interests, values, and traveling lifestyles. For example, various backpacker travel communities have been established for travelers with similar mindsets to share their experiences or even travel together. They resemble real-life communities where members exchange information and provide support and friendship to one another. Virtual communities have strong influences on its members and sometimes extend beyond the virtual platform. For example, CouchSurfing, with over 14 million members, offers a platform for travelers looking for opportunities to stay and interact with local communities when they travel. The social change mediated in virtual communities is an interesting phenomenon that influences and is influenced by e-tourism development. Essentially, virtual communities represent a novel travel culture that goes beyond geographical assumptions.

Tourists are heterogeneous in their background and preferences. Empowered by ICTs to access information anytime and anywhere, tourists are able to make informed decisions, thereby becoming more sophisticated. Segmentation can account for the heterogeneity among tourists by grouping them into market segments with members similar to one another based on certain characteristics. Performing market segmentation allows operators to focus on the needs of a particular group and target their marketing resources accordingly. Traditionally, segmentation of tourists is often based on geographical features, such as country of origin, as evident from statistical reports prepared by DMOs. Tourists can also be segmented according to their travel motives, activities, and expenditure. In the contemporary e-tourism epoch, the segmentation of tourists based merely on geographical dimension is inadequate or even obsolete. E-tourists should be analyzed and grouped according to their frequency of ICT usage and usage behavior. They can also be segmented based on their level of engagement in different virtual communities. For example, depending on the length of time that tourists spend on their mobile phones every day, video advertisements should be displayed to attract their attention more effectively.

Despite the possibility of technology in removing geographical barriers, unequal access to and usage of ICTs still exist among individuals, which is referred to as the digital divide. This concept can be attributed mainly to the disparity in the access to technological infrastructure and knowledge among individuals in different nations. Tourism firms seeking to adopt advanced technology for their businesses must be cautious of the implications of the digital divide and should consider the local infrastructure of their target markets. For example, although virtual reality is a popular way of showcasing the three-dimensional view of a destination, potential markets without access to such technology may not be reached. Hence, the efforts of implementing high-tech solutions will be deemed inappropriate and redundant. Furthermore, applying robots for service delivery may enhance brand image and operational efficiency of the hotel but it may create inconvenience and disturbances to non-tech savvy customers.

Concluding remarks

With only a few decades of presence, e-tourism has benefited tourism firms in various ways and has enhanced tourists' travel experiences. The tremendous impact of technological advancement will continue to fuel the development of e-tourism. In the future, e-tourism will be transformed from a joint sector of business, information technology/system, and tourism into a comprehensive smart eco-system. However, this vision will only be feasible with the collaborative efforts and commitment of different actors in the system.

References

Au Yeung, T., and Law, R. 2004. Extending the modified heuristic usability evaluation technique to chain and independent hotel websites. *International Journal of Hospitality Management 23*(3):307–313.

Beritelli, P., and Schegg, R. 2016. Maximizing online bookings through a multi-channel-strategy: Effects of interdependencies and networks. *International Journal of Contemporary Hospitality Management 28*(1):68–88.

Buhalis, D. 2003. *eTourism: Information Technology for Strategic Tourism Management.* London: Pearson (Financial Times/Prentice Hall).

Buhalis, D., and Foerste, M. 2015. SoCoMo marketing for travel and tourism: Empowering co-creation of value. *Journal of Destination Marketing & Management 4:*151–161.

Chan, I.C.C., Fong, D.K.C., Law, R., and Fong, L.H.N. 2018. State-of-the-art social customer relationship management. *Asia Pacific Journal of Tourism Research 23*(5):423–436.

Chung, T., and Law, R. 2003. Developing a performance indicator for hotel websites. *International Journal of Hospitality Management 22*(1):119–125.

Davis, F. D. 1989. Perceived usefulness, perceived ease of use, and user acceptance of information technology. *MIS Quarterly 13*(3):319–340.

eMarketer. 2017. *Mobile drives growth of online travel bookings.* www.emarketer.com/Article/Mobile-Drives-Growth-of-Online-Travel-Bookings/1016053.

eMarketer. 2018. *Digital travel sales worldwide, 2017–2022 (billions and % change).* www.emarketer.com/Chart/Digital-Travel-Sales-Worldwide-2017-2022-billions-change/221442.

Greenberg, P. 2010. The impact of CRM 2.0 on customer insight. *Journal of Business & Industrial Marketing 25*(6):410–419.

Kirillova, K., and Chan, J. 2018. "What is beautiful we book": Hotel visual appeal and expected service quality. *International Journal of Contemporary Hospitality Management 30*(3):1788–1807.

Kracht, J., and Wang, Y. 2010. Examining the tourism distribution channel: Evolution and transformation. *International Journal of Contemporary Hospitality Management 22*(5):736–757.

Law, R. 2019. Evaluation of hotel websites: Progress and future developments. *International Journal of Hospitality Management 76*:2–9.

Law, R., Fong, D.K.C., Chan, I.C.C., and Fong, L.H.N. 2018. Systematic review of hospitality CRM research. *International Journal of Contemporary Hospitality Management 30*(3):1686–1704.

Liu, Z., and Park, S. 2015. What makes a useful online review? Implication for travel product websites. *Tourism Management 47*:140–151.

Trainor, K. J. 2012. Relating social media technologies to performance: A capabilities-based perspective. *Journal of Personal Selling & Sales Management 32*(3):317–331.

UNWTO. 2018. 2017 international tourism results: The highest in seven years. http://media.unwto.org/press-release/2018-01-15/2017-international-tourism-results-highest-seven-years.

Venkatesh, V., Morris, M. G., Davis, G. B., and Davis, F. D. 2003. User acceptance of information technology: Toward a unified view. *MIS Quarterly 7*:425–478.

Further reading

Buhalis, D., and Law, R. 2008. Progress in information technology and tourism management: 20 years on and 10 years after the internet: The state of eTourism research. *Tourism Management 29:*609–623.

Buhalis, D., and O'Connor, P. 2005. Information communication technology revolutionizing tourism. *Tourism Recreation Research 30*(3):7–16.

Condratov, I. 2013. e-Tourism: Concept and evolution. *Ecoforum* 2(1–2):58–61.

Karanasios, S., and Burgess, S. 2008. Tourism and internet adoption: A developing world perspective. *International Journal of Tourism Research 10*(2):169–182.

Law, R., Buhalis, D., and Cobanoglu, C. 2014. Progress on information and communication technologies in hospitality and tourism. *International Journal of Contemporary Hospitality Management 26*(5):727–750.

Law, R., Chan, I. C. C., and Wang, L. 2018. A comprehensive review of mobile technology use in hospitality and tourism. *Journal of Hospitality Marketing & Management 27*(6):1–23.

Law, R., Leung, R., and Buhalis, D. 2009. Information technology applications in hospitality and tourism: A review of publications from 2005 to 2007. *Journal of Travel & Tourism Marketing 26*(5–6):599–623.

Leung, X.-Y., Xue, L., and Bai, B. 2015. Internet marketing research in hospitality and tourism: A review and journal preferences. *International Journal of Contemporary Hospitality Management 27*(7):1556–1572.

Liang, S., Schuckert, M., Law, R., and Masiero, L. 2017. The relevance of mobile tourism and information technology: an analysis of recent trends and future research directions. *Journal of Travel & Tourism Marketing 34*(6):732–748.

Navío-Marco, J., Ruiz-Gómez, L. M., and Sevilla-Sevilla, C. 2018. Progress in information technology and tourism management: 30 years on and 20 years after the internet – revisiting Buhalis & Law's landmark study about eTourism. *Tourism Management 69*:460–470.

O'Connor, P., and Murphy, J. 2004. Research on information technology in the hospitality industry. *International Journal of Hospitality Management 23*(5):4730–484.

Standing, C., Tang-Taye, J.-P., and Boyer, M. 2014. The impact of the internet in travel and tourism: A research review 2001–2010. *Journal of Travel & Tourism Marketing 31*(1):82–113.

13 The state and cyberspace

E-government geographies

Barney Warf

Among the innumerable changes unleashed by the internet is how governments interact with their citizens. Electronic government, or e-government, is the deployment of web-based applications to provide information and deliver public services more efficiently. There are several definitions of e-government (Yildiz 2007), but they all revolve around the use of the internet to improve administrative procedures and enhance citizen feedback and participation. This process has been accelerated by the adoption of Web 2.0, which allows users to provide input into government agencies rather than just passively receive information. As Burn and Robins (2003, 26) argue, "eGovernment is not just about putting forms and services online. It provides the opportunity to rethink how the government provides services and how it links them in a way that is tailored to the users' needs." E-government has received considerable scholarly attention (for a review see Rocheleau 2007).

This chapter outlines e-government in several steps. It opens with a summary of its forms and applications. The second section summarizes several theoretical interpretations. In the third part, the many and varied obstacles to implementing e-government successfully are reviewed. The fourth section focuses on e-government in practice, as practice by three East Asian cities with extensive programs for delivering public services online, Seoul, Singapore, and Shanghai. The conclusion summarizes the major themes.

Summary of e-government applications

E-government alters how state agencies interact with one another and their interactions with the public. It thus figures prominently in many "reinventing government" discourses. There is considerable variation in the nature and sophistication of e-government implementation, ranging from simple one-way delivery of information to more intensive two-way interactions that incorporated user input and citizen feedback.

A common means of differentiating the types of e-government includes government-to-business (G2B), government-to-government (G2G), and government-to-citizens (G2C) variations (Fountain 2001a,b). G2B e-government subsumes phenomena such as online calls for proposals; submissions of digital bids and bills, and electronic payments; and internet management of supply chains. G2G e-government is heralded as a means to enhance communications among different government agencies through electronic flows of information. It is frequently touted as a means to increase citizen accessibility, improve efficiency, create synergies, and generate economies of scale in the delivery of public services. Some argue that e-government encourages the shift of public bureaucracies from traditional hierarchical forms of control to more horizontal, collaborative models (Ho 2002; Ndou 2004).

The most common type is G2C e-government, which includes the digital collection of taxes; electronic voting; payment of utility bills, fees, and dues; applications for public assistance, permits, and licenses; online registration of companies and automobiles; and access to census and other public data. Some countries such as Estonia and France have experimented with e-voting. Online access to information allows citizens to reduce uncertainty, and minimizes trips to government offices and the long wait times associated with them. In rural areas, e-government applications such as distance education or telemedicine offer services to people who otherwise may not be able to utilize them. Local governments use the internet to entice tourists and foreign investors. Interactive municipal websites give residents access to information about schools, libraries, bus schedules, and hospitals.

E-government has numerous benefits. By improving efficiency in the delivery of public services, it increases satisfaction and trust with the state. Often it is celebrated as a vehicle for reducing corruption; electronic payments limit opportunities for graft. It may minimize the growth of public employment, a key goal of neoliberalism. Online solicitation of citizens' opinions give them a voice in governance, such as urban planning. Periodic concerns over e-government include the potential invasions of privacy, hacking of government files, and the inequality of access generated by digital divides (Belanger and Carter 2008); for those without internet access, e-government is worthless.

E-government is frequently viewed through the lens of technological determinism, a perspective that tends to suggest that its effects are essentially the same across the world. This view lends itself to the notion that there is one, universal model that can be applied everywhere regardless of local or national contexts (e.g., Grant and Chau 2005). Unfortunately, this approach ignores the importance of political, cultural, and economic contexts, which profoundly shape who uses e-government and for what ends. A more realistic notion emphasizes varying institutional environments, leading to the recognition that there are inevitably significant differences in impacts. Even in the allegedly placeless world of cyberspace, place still matters. A sizeable body of literature has revealed profound geographical variations in the nature and consequence of e-government among the world's countries. When assessed from the perspective of social constructivism, e-government comprises a series of diverse practices that vary over time and space. These variations reflect different political climates, cultures, and institutional imperatives.

In order for e-government to be implemented successfully, it must have decisive and consistent leadership, cooperation by government bureaucrats (many of whom fear it will displace them), sufficient funding, clear lines of accountability, and effective mechanisms for feedback (Rose and Grant 2010). Thus, e-government is as much an administrative process as a technological one. These comments serve to illustrate that the adoption of e-government is highly contingent and path-dependent, and is shaped by a variety of cultural, legal, and political forces. The highly political nature of e-government implementation and its effects imply that its usage changes over time and space, and that its consequences are inevitably geographically differentiated.

Theorizing e-government

Studies of e-government typically arise from those working within public administration and political science. Fountain's (2001) *Building the Virtual State: Information Technology and Institutional Change* has been widely acclaimed for its institutional analysis, including virtual agencies and single portals through which citizens can access numerous public services.

Many theorizations emphasize "stages" models ranging from the primitive to sophisticated, or more condescendingly, from the immature to the mature. Holliday and Kwok (2004) differentiate between enhanced, interactive, and seamless forms of e-government along a continuum. Lane and Lee (2001) offer a four-stage model. Davison et al. (2005) similarly emphasize the transition from government to e-government, including three models of "maturity" ranging from simple digital presence to full-scale, interactive service delivery. Often this work deploys metaphors of life stages, evolution, contagion, or learning curves, often with the implicit assumption that of teleological inevitability, as if all countries march mechanically through a predefined set of stages. In fact, variations and skipping stages are common. Again, there is no single trajectory: it is impossible to understand e-government without its spatio-temporal contexts.

Chadwick and May (2003) summarize three approaches to understanding e-government – managerial, consultative, and participatory – that reflect the experiences of the U.S. and the European Union. Managerial approaches to e-government emphasize the speed and efficiency of delivery of government services to citizens. The consultative approach incorporates citizen input via Web 2.0 technologies (e.g., internet voting, polling, and electronic communications with public officials); this line of thought often touts e-government as emancipatory and liberating. Finally, the participatory model allows for input from non-state actors such as corporations and non-governmental organizations. These views fall across a continuum of social access in which the consultative and participatory models are the most socially inclusive forms.

A widely admired view of e-government stems from Layne and Lee (2001), who outline developmental stages ranging from a simple online presence (i.e., a public webpage); interfaces that allow citizen access to data and services; vertical integration in which citizens can actively participate (e.g., for license applications); and horizontal integration, in which government websites offer a broad range of government functions (e.g., payments of fees, applications for permits and licenses). Empirical evaluations of the quality of e-government typically focus on how well websites are designed and operate, including their user-friendliness, missing links, readability, the publications and data displayed, contact information for public officials, languages in which content is provided, sound and video clips, ability to use credit cards and digital signatures, security and privacy policies, and opportunities for feedback.

A hugely influential theorization of e-government that draws from earlier works on technological diffusion is the technology acceptance model (TAM) (Davis 1989, 1993), which emphasizes how users accept and utilize computer-related technologies. This view has the advantage of including social and personal psychological variables such as experience, education, cultural norms, and gender roles, as well as personal considerations such as self-assurance. Potential adopters of e-government judge it based on its perceived usefulness, ease of use, and potential savings in terms of time and cost. The decision to adopt is also made in light of social pressures from peers and supervisors, and reflects the wider context of policies, regulations, and the legal environment. The TAM model implicitly reduces citizens to the role of passive consumers. This approach has often been applied to e-government in order to make sense of its uneven adoption among countries and groups (e.g., Carter and Belanger 2005; Hung, Chang and Yu 2006). The TAM model is not so much wrong as it is self-evident, and pays little attention to the roles of power, conflict, class, and inequality in shaping access to all technologies.

A common claim about e-government is that it facilitates democracy by making governance more transparent (Rose 2004; Prasad 2012). This notion is grounded in utopian views of the internet. As Johnson and Kolko (2010, 17) note, "Many current models of e-Government have an unstated assumption that a desire to fulfil democratic functions motivates the creation of e-Government initiatives and that the ultimate goal of e-Government initiatives is increased transparency and accountability." This notion contains considerable merit, which is why e-government is often touted as a means of improving efficiency in the delivery of public services, trust and accountability, as a means of obtaining citizen feedback, raising public satisfaction, and as a vehicle for combatting corruption (Andersen 2009; Bertot et al. 2010; Kim et al. 2009; Neupane et al. 2014). Globally, the most effective implementations of e-government are situated in wealthy, democratic societies. However, it is also the case that authoritarian governments can use e-government to improve popular opinions without moving toward democracy, as China aptly attests. Uncritical perspectives of e-government tend to overlook this fact.

Obstacles to implementing e-government

Implementing e-government invariably faces a variety of social, political, and technical obstacles. The most severe of these include poverty, illiteracy, and lack of internet access. For the very poor, for whom daily survival is a challenge, e-government is remote and unimportant. In impoverished countries, for many people a personal computer is unaffordable, and many do not have the requisite technical skills. In many developing countries the supply of electricity is often unreliable.

Gender roles are also important in limiting women's access to the internet, and thus e-government. In many developing countries, particularly those with entrenched traditions of patriarchy, women have fewer opportunities to learn minimal computer skills. Women's incomes and literacy rates are typically lower than those of men, and they are often burdened by the demands of childcare and domestic labor. Even visiting government-sponsored telekiosks may invite unwelcome scrutiny and harassment from men.

Governments that attempt to implement e-government may faces resentment by public officials and employees. In states in which patronage systems are widespread, e-government is a challenge for the politically connected. In very corrupt countries, where e-government may reduce the need for intermediaries in obtaining public services, it limits the opportunities for bribe-taking and graft. Moreover, government workers may lack the necessary information technology skills. Those employed in overstaffed public bureaucracies may oppose e-government's potential for streamlining and downsizing, rightly viewing it as a "disruptive technology." Hanna (2008, 128) observes

> IT units and their managers often stand as barriers to change. They have a vested interest in stand-alone systems; coordinated services and shared infrastructure or the more recent Web-based technologies may erode their monopoly on information and technology-oriented services.

In addition, insufficient funding for e-government may lead to shortages of equipment and software. Such technologies quickly become out of date and need continuous updating. Funds intended for this purpose may be siphoned off by corrupt managers. In governments riven by turf battles among different agencies, e-government can fall victim to inter-agency rivalries. Finally, e-government is vulnerable to hacking, viruses, worms, malware, and invasions of privacy.

One significant concern about e-government is the digital divide, or social and spatial barriers to access to the internet (Yigitcanlar and Baum 2006). One-half of the world's population uses the internet, which also means that one-half does not, notably the poor, elderly, and those living in rural areas, where penetration rates tend to be lower than in cities. For those excluded from the information highway, e-government is useless (Helbig et al. 2009). As Fountain (2001, 48) puts it, "An increasingly digital government favors those with access to a computer and the internet and the skills to use these sophisticated tools competently." People who cannot afford a personal computer, or even access at cybercafes, or lack the technical skills, can never apply for services, benefits, permits, licenses and such online. Tragically this population is precisely the one that needs such services the most. Indeed, because e-government improves the lives of information "haves" over the "have-nots," it may amplify existing inequalities (Dugdale et al. 2005; Hossain 2005).

The obstacles to implementing e-government are frequently insurmountable. For this reason, there is a long and tragic history of failed e-government initiatives (Dada 2006). These observations underscore the contingent nature of e-government and refute teleological interpretations that portray it as inevitable. The politics of e-government indicate that its adoption is open-ended, contingent, and always context-dependent.

E-government in practice: three east Asian cities

E-government in theory is one thing; e-government in practice is another. To illustrate the complexities of this phenomenon, this section offers three case studies of e-government implementation drawn from East Asia: Seoul, Singapore, and Shanghai. All three are large, wealthy and highly globalized metropolises in which the internet has been adopted successfully, to one extent or another, to enhance the delivery of public information and services.

Seoul

South Korea has adopted e-government on a massive scale in order to enhance the transparency of public decision-making processes and raise the efficiency of service delivery. For example, the country's citizens can provide input to the national government through the online petition and discussion portal, the presidential Blue House's *shinmoongo* system, which also allows citizens to appeal judicial decisions and report corruption (Lee and Hong 2002). The webpage is used by roughly 75,000 people annually. Similarly, Seoul uses electronic bulletin boards called "Citizen's Agora" to evaluate officials' responses to their comments. Seoul also introduced a digital "Appeal to City Mayor" in 1998, which receives roughly 4,200 comments annually.

Seoul ranks at the top of the world's most well connected "smart cities," with one of the most developed network infrastructure on the planet (Lee et al. 2014). In 2011, the government announced the "Smart Seoul 2015" program to maintain the city's status as a model of e-government. About 10,000 public offices offer free wifi, which serve 13% of the metropolitan region's area. The e-Seoul safety program uses RFID tags to monitor the locations of children and people with Alzheimer's; if the tag wearer leaves a designated zone or pushes an emergency button it notifies authorities immediately. The city also pioneered 10 smart work centers from which employees can work, resulting in shorter commutes. Smart meters in homes have reduced energy consumption. In Seoul's Eunpyong district, completed in 2011, 45,000 people live in smart homes in which they can obtain information on their living room walls. Throughout the city, a one-stop integrated service system allows residents to make appointments for one of the 30,000 public services online. Online three-dimensional virtual tours of the city allow planners to simulate developments. The city's u-Shelter bus stops offer maps, schedules, and routes to a variety of smart devices. A citywide smart payment system allows buyers to purchase goods and services with their smart phones without entering stores. The HomePlus virtual stores allow pedestrians make purchases using their phones and have them delivered via a mobile payments system. Seoul's "telepresence" system allows residents to conduct video chats around the world. Home devices can be programmed via mobile phones. Sensors alert riders waiting for buses as to the arrival time. At the Cisco Innovation Center, a small "IoT (Internet of Things) cube" encourages software startups to develop new applications.

The wealthy Gangnam district adopted e-government services that led it to become an internationally recognized "intelligent community" (Ahn and Bretschneider 2011). The community started a local area network connected to public kiosks in 1997, from which residents could pay taxes and apply for licenses and registrations. Gangnam implemented internet broadcasting of senior staff meetings to allow residents to observe the decision making of senior personnel. The Cyber Local Autonomous Government Management System enables citizens to make suggestions, and an online citizen survey that began in 2001 allows residents to voice their approval or disapproval of how their local government is run. In 2003, it began a service known as Online Publication of Official Documents that put government documents on its website. Its system for providing access to public documents was adopted by the national government in 2002. More than 200,000 residents belong to an email service in which the Gangnam administration solicits their opinion on public policies. The online Movement to Keep Basic Order allows people to report public disturbances. In 2006, Gangnam launched TV GOV, a set of interactive e-government applications on televisions that enables users to access news programs, cultural and arts channels, and specialized information for seniors, women, and children. In 2007, it began to place wireless motion detectors in the homes of the elderly that trigger an alarm if they do not detect motion for an extended period of time.

Singapore

Singapore, a city that is an island that is a country, boasts one of the world's most advanced e-government systems (Chan and Pan 2008). In 2018 the internet penetration rate was 84%, third-highest in Asia (following South Korea and Japan). The Infocomm Development Authority (IDA) claims to have built a "world-class e-government" that enables citizens to "be involved, be empowered, and be a pacesetter."

Singapore's e-government initiatives began in the early 1980s with the launch of the Civil Service Computerisation Programme and the National Computer Board. Successive national plans followed, including the National IT Plan (1986 to 1991), the IT2000 Master Plan (1992–1999), and Infocomm (or Singapore) 21, which began in 2000 with a series of forums and surveys; it proceeded to accelerate the development of government webpage. Subsequent efforts sought to create an "intelligent island" (Ke and Wei 2006). The Singapore One project, which began in 1997, created the world's first nationwide broadband network, and reaches 99% of the population. In 2000 Singapore unveiled an E-government Action Plan, and in 2011, it started the E-government Master Plan, with the explicit aim of providing "government-with-you" (Baum and Mahizhnan 2014). The most recent initiative is eGov 2015. All e-services follow identical security, electronic payment, and data exchange procedures. Today, more than 1700 government services can be delivered online (Ke and Wei 2004; Ha 2013), meaning that "any public services that can be delivered online must be digitized" (Ha and Coghill 2006, p. 107).

Singapore's e-government system reveals the fruits of a well-designed, coordinated, and implemented system. The state was careful to identify potential stakeholders in designing and implementing its e-government system (Tan, Pan, and Lim 2005), including public servants, corporations, academics, and labor unions. From the outset, it took pains to make sure that its plans were inclusive and citizen-centered (Ha 2013; Srivastava and Teo 2009). The Infocomm Development Authority (IDA) invested heavily in training programs and cultivated considerable goodwill (Pan et al. 2006). As Sriramesh and Rivera-Sanchez (2006) note, Singapore is also well educated and has a corporatist, communitarian culture.

Since 1999 a one-stop e-citizen portal (www.ecitizen.gov.sg), the world's first to be provided by a government (Netchaeva 2002), has offered a single point of access through which citizens can check traffic, download publications, register births and deaths, search for jobs, pay fees, fines and taxes, check retirement benefits, register to vote, and obtain health care advice. Sriramesh and Rivera-Sanchez (2006) report that it receives about nine million hits per month; 75% of Singaporeans have used it and 80% of them report satisfaction. It has become the model of many such portals around the world.

Singapore's e-government also includes an advanced Electronic Tax Filing system (Tan et al. 2005), which was implemented in light of significant uncollected revenues. In 1998 e-filing began, allowing Singaporeans (even those overseas) to pay taxes through the internet. The vast bulk of tax returns are filed this way. The digital system not only reduced paperwork, but necessitated fewer staff as well. Corporations were encouraged to submit information about their employees to expedite the process. The system allows taxpayers to communicate with the tax authority; this feedback is used in upgrading and improving the system.

Several other aspects testify to the sophistication of the country's e-government. Singapore was the first county in the world to implement a national e-library and an e-citizen center. The government launched a "Connected Homes" system for home networking and community services and a MySingapore website giving citizens access to a broad array of services. The pubic e-litigation service simplifies legal filings and allows lawyers to appear in court via videophones (Ha and Coghill 2006). The Singapore Immigration & Checkpoints Authority offers discounts to those who apply for visas and passports digitally. The Ministry of Manpower introduced iJOBS, an online job matching site. The Singapore Sports Council implemented iBook, an online service for booking for sports facilities. Singapore's government has been especially innovative in using social media, particularly Facebook, to engage citizens (Soon and Soh 2014). For example, it deployed this medium to alert residents during the SARS outbreak of 2004 (Pan et al. 2005). The Immigration and Checkpoints Authority gives permissions for exit permits to go overseas by SMS, while other agencies send text messages to pay parking tickets and national service obligations (Trimi and Sheng 2008).

E-government has been enthusiastically adopted by Singapore's business community, whose participation was central to its success (Chan and Al-Hawamdeh 2002; Wong 2003; Tung and Rieck 2005). The eBusiness Industry Development Scheme introduced subsidies to encourage firms to enhance their e-commerce capacities. The Cyber-Trader Act, the Electronic Transactions Act further encouraged this trend, notably among venture capitalists, start-ups, and patent holders. In 1997, Singapore established Tradenet, an electronic clearing house that seamlessly integrates firms, the Customs Department, the Trade Development Board, and air and seaport authorities (Teo et al. 1997). The portal gebiz.gov.sg allows firms to conduct many functions online, including electronic registration, tax payment, submission of contracts, and license renewal, update or termination. It has cut red tape considerably, and as a result the time needed to incorporate a company dropped from two days to two hours (Sriramesh and Rivera-Sanchez 2006). Similarly, the One-Stop Public Entertainment Licensing Centre cut processing times "from 6 to 8 weeks to about 2 weeks" (Ha and Coghill 2006, 113).

To address national security and infectious disease concerns, the Singaporean government launched the Risk Assessment and Horizon Scanning (RAHS) program, which collects and analyzes large datasets in the hope of predicting terrorist attacks, epidemics, and financial crises (Kim et al. 2014). Its Experimentation Center, which opened in 2007, focuses on new technological tools to support policy making for RAHS and enhance and maintain RAHS through systematic upgrades of the big-data infrastructure. A notable application is exploration of possible scenarios involving importation of avian influenza into Singapore and assessment of the threat of outbreaks occurring throughout Southeast Asia.

Singapore too faces a digital divide that poses an impediment to e-government (Ke and Wei 2006). To address this issue, the government erected 27 self-service Citizens Connect kiosks offering free internet services. It also pumped S$25 million into a program to encourage late internet adopters to get online (Pan et al. 2006). However, as a former British colony, Singapore provides its e-government services only in English, a linguistic obstacle in a country where the majority of residents speak Mandarin, Tamil, or Malay.

While it is a world leader in information technology and e-government, Singapore is also widely regarded as an authoritarian state, and censors the internet regularly. Its primary vehicle in this regard is the Singapore Media Development Authority (MDA), which has regulated internet content under the guise of monitoring a broadcasting service since 1996. Internet service providers are licensed by the Singapore Broadcasting Authority, which routes all internet connections through government servers. Licensees are required to comply with the 1996 Internet Code of Practice, which restricts "prohibited material" or content deemed "objectionable on the grounds of public interest, public morality, public order, public security, national harmony, or is otherwise prohibited by applicable Singapore laws" (OpenNet Initiative 2007, 3). As a result, Singapore's government has achieved absolute control over its internet contents minimal loss of political legitimacy.

Shanghai

China has aggressively encouraged the adoption of e-government to reduce corruption, improve efficiency, and enhance the legitimacy of the ruling Communist Party. The leading city in this regard is clearly Shanghai, a metropolitan area of 23 million people and the country's largest and most globalized city. Shanghai established an official website early (www.shanghai.gov.cn), which offers news and information about municipal governance. A web of 19 local universities is connected through the Shanghai Science and Education Network, a system to implement information technology as rapidly as possible. Shanghai University established a Smart City Academy. Today Shanghai is China's best-connected city, with an internet penetration rate of 71%; it has been particularly successful with broadband. The city's i-Shanghai project consists of 4000 free wifi points (Wang et al. 2016). The city also established an Internet of Things center in Jiading.

Shanghai's shift towards a smart city date back to the late 1999s. In 1998 it initiated a smart card system with many different uses (Chen and Huang 2015), which also provides the municipal government with a centralized database containing personal information about holders (including fingerprints and medical insurance account numbers) (Lili Cui et al. 2006). Users can use the cards in numerous places, including hospitals and government offices, and through them can do things such as apply for home loans, drivers' licenses, marriages, divorces, and government subsidies; register their employment status in the *houkou* system; and make housing and utility payments. Today the system is used by more than three-quarters of the metro population. E-government services have also become indispensable for firms operating there.

In 2011, the metropolis launched the Shanghai Smart City Development Program (Gil and Zheng 2017), or Smart Shanghai, which was followed by a second plan in 2014. The system has numerous components, including the use of information technology to monitor illegal construction, street peddling, and parking, fine those who evaded paying tickets for the city's metro transit system, electronic toll collection, and dynamic parking guidance systems designed to maximize the efficiency of the city's limited supply of space. It also began implementation of programs for smart lighting, interactive electronic display boards, and a series of smartphone apps that display bus schedules as a means of discouraging driving. Leading the way was the Pudong financial district, whose i-Pudong system utilized the internet for purposes such as congestion alleviation, environmental protection, the bus dispatch system, and emergency responses. It also offers 15 digital services in 450 public places, including tourist centers and transportation hubs. For example, load-bearing sensors now indicate whether trucks going onto bridges carry more than the maximum allowable weight and automatically notify authorities.

In 2013, the Shanghai Government Data Service Portal opened, and provided an enormous array of small but important services online: payments of utility bills, fines, tickets, and taxes; applications for licenses; scheduling of appointments with public officials; registration of sales and purchase of houses; marriage and divorce records; and birth and death certificates. Similarly, the AIRNow-I project, which began in 2010, provides information on air quality and allows citizens to upload data via mobile phones, a form of citizen science.

Shanghai's successes in implementing e-government raised its visibility in national government circles, and it became a model for the rest of the country. Notably, the system was designed to be reproduced elsewhere. As a result, Shanghai's initiatives are being emulated throughout large parts of China. The city provides expertise for other provinces and metropolitan areas through collaborative agreements.

Concluding thoughts

As the internet has grown in the number of users and number of uses, e-government has increasingly become popular – indeed, normalized – throughout large parts of the world. There are several types of e-government, including government-to-business and government-to-citizens applications, but all of them revolve around the use of digital technologies to provide information and services. With the diffusion of Web 2.0 technologies that allow user feedback, many public functions such as applications for permits and payments of bills and fees have moved online. The result has been a gradual improvement in efficiency, often by eliminating the need to visit government

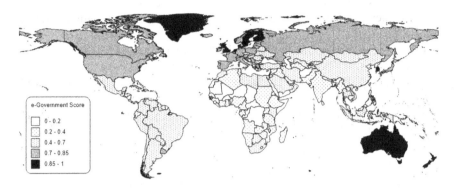

Figure 13.1 Map of U.N. e-Government Readiness Index, 2018.

offices and wait in line. e-Government is often touted as a means to make government more democratic, reduce corruption, and improve transparency.

This process is not without obstacles, however. Firm leadership and adequate funding are mandatory. The adoption of digital technologies presumes the equipment, software, and technical skills are readily available. Suspicious bureaucrats may resent how e-government reduces the demand for labor and reduces the opportunities for graft. In many countries the digital divide leads to vast numbers of people being excluded from this process, notably the poor, elderly, illiterate, and those in rural areas. Indeed, e-government may enhance the advantages of information "haves" at the expense of "have-nots," raising social and spatial inequality.

E-government varies in sophistication around the world (Figure 13.1). Generally, wealthy countries have the most advanced systems, such as in Scandinavia. However, the rapid growth of e-government in Asia indicates new centers of innovation. Seoul, Singapore, and Shanghai offer three such examples, but there are many other emerging centers in Japan, Taiwan, India, and elsewhere. In short, e-government has a geography, and takes different forms in different places, a process that is continuously and rapidly changing over time.

References

Ahn, M. and S. Bretschneider. 2011. Politics of e-government: e-Government and the political control of bureaucracy. *Public Administration Review* 71(3):414–424.

Andersen, T. 2009. E-government as an anti-corruption strategy. *Information Economics and Policy* 21:201–210.

Basu, S. 2004. E-government and developing countries: An overview. *International Review of Law Computers* 18:109–132.

Baum, S. and A. Mahizhnan. 2014. Government-with-you: e-Ggovernment in Singapore. In S. Baum and A. Mahizhnan (eds.) *E-Governance and Social Inclusion: Concepts and Cases.* pp. 229–242. Hershey, PA: IGI Global.

Becker, J., B. Niehaves, P. Bergener, and M. Räckers. 2008. Digital divide in e-government: The e-inclusion gap model. *Lecture Notes in Computer Science* 5184:231–242.

Belanger, F. and L. Carter. 2008. Trust and risk in e-government adoption. *Journal of Strategic Information Systems* 17:165–176.

Bertot, P. Jaeger, and J. Grimes. 2010. Using ICTs to create a culture of transparency: e-Government and social media as openness and anti-corruption tools for societies. *Government Information Quarterly* 27:264–271.

Burn, J. and G. Robins 2003. Moving towards e-government: A case study of organizational change processes. *Logistics Information Management* 16(1):25–35.

Carter, L. and Bélanger, F. 2005. The utilization of e-government services: Citizen trust, innovation and acceptance factors. *Information Systems Journal* 15(1):5–25.

Chadwick, A. and C. May. 2003. Interaction between states and citizens in the age of the internet: "e-Government" in the United States, Britain, and the European Union. *Governance* 16:271–300.

Chan, B. and S. Al-Hawamdeh. 2002. The development of e-commerce in Singapore: The impact of government initiatives. *Business Process Management Journal* 8(3):278–288.

Chan, C. and S. Pan. 2008. User engagement in e-government systems implementation: A comparative case study of two Singaporean e-government initiatives. *Journal of Strategic Information Systems* 17(2):124–139.

Chen, A. and W. Huang. 2015. China's e-government. In S.L. Pan (ed.) *Managing Organizational Complexities with Digital Enablement in China: A Casebook.* pp. 97–106. Singapore: World Scientific Publishing.

Dada, D. 2006. The failure of e-government in developing countries: A literature review. *Electronic Journal of Information Systems in Developing Countries* 26(7):1–10.

Davis, F. 1989. Perceived usefulness, perceived ease of use and user acceptance of information technology. *MIS Quarterly* 13(3):319–340.

Davis, F. 1993. User acceptance of information technology: System characteristics, user perceptions and behavioral impacts. *International Journal of Man-Machine Studies* 38:475–487.

Davison, R., C. Wagner, and L. Ma. 2005. From government to e-government: A transition model. *Information Technology & People* 18(3):280–299.

Dugdale, A., A. Daly, F. Papandrea, and M. Maley. 2005. Accessing e-government: challenges for citizens and organizations. *International Review of Administrative Sciences* 71:109–118.

Fountain, J. 2001a. *Building the Virtual State: Information Technology and Institutional Change.* Washington, DC: Brookings Institution Press.

Fountain, J. 2001b. The virtual state: transforming American government? *National Civic Review* 90:241–251.

Gil, O. and T.-C. Zheng. 2017. The Smart City Plan 2011–2013 in Shanghai. In Y. Jing and S. Osborne (eds.) *Public Service Innovations in China.* pp. 127–149. Singapore: Springer.

Grant, G. and D. Chau. 2005. Developing a generic framework for e-government. *Journal of Global Information Management* 13:1–30.

Ha, H. 2013. E-government in Singapore: Critical success factors. In J. Gil-Garcia (ed.) *E-government Success around the World: Cases, Empirical Studies, and Practical Recommendations.* pp. 176–194. Hershey, PA: IGI Global.

Ha, H. and K. Coghill. 2006. E-government in Singapore: A SWOT and PEST analysis. *Asia-Pacific Social Science Review* 103–130.

Hanna, N. 2008. *Transforming Government and Empowering Communities: The Sri Lankan Experience with e-Development.* Washington, DC: World Bank.

Helbig, N., J. Gil-Garcia, and E. Ferro. 2009. Understanding the complexity of electronic government: Implications from the digital divide literature. *Government Information Quarterly* 26:89–97.

Ho, T. 2002. Reinventing local governments and the e-government initiative. *Public Administration Review* 62:434–444.

Holliday, I. and R. Kwok. 2004. Governance in the information age: building e-government in Hong Kong. *New Media and Society* 6(4):549–570.

Hossain, F. 2005. E-governance initiatives in developing countries: Helping the rich? Or, creating opportunities for the poor? *Asian Affairs* 27:5–23.

Hung, S.Y., C.M. Chang, and T.J. Yu. 2006. Determinants of user acceptance of the e-government services: The case of online tax filing and payment system. *Government Information Quarterly* 23(1):97–122.

Johnson, E. and B. Kolko. 2010. E-government and transparency in authoritarian regimes: Comparison of national- and city-level e-government web sites in Central Asia. *Digital Icons: Studies in Russian, Eurasian and Central European New Media* 4(3):15–48.

Ke, W. and K. Wei. 2006. Understanding e-government project management: A positivist case study of Singapore. *Journal of Global Information Technology Management* 9(2):45–61.

Kim, G.H., S. Trimi, and J.H. Chung. 2014. Big-data applications in the government sector. *Communications of the ACM* 57(3).

Kim, S., H. J. Kim, and H. Lee. 2009. An institutional analysis of an e-government system for anti-corruption: The case of OPEN. *Government Information Quarterly* 26:42–50.

Layne, K. and J. Lee. 2001 Developing fully functional e-government: A four stage model. *Government Information Quarterly* 18:122–136.

Lee, J., M. Hancock, and M. Hu. 2014. Towards an effective framework for building smart cities: Lessons from Seoul and San Francisco. *Technological Forecasting and Social Change* 89:80–99.

Lee, K. and J.H. Hong. 2002. Development of an e-government service model: A business model approach. *International Review of Public Administration* 7(2):109–118.

Lili Cui, L., C. Zhang, C. Zhang, and L. Huang. 2006. Exploring e-government impact on Shanghai firms' information process. *Electronic Markets* 16:312–318.

Ndou, V. 2004. E-government for developing countries: Opportunities and challenges. *Electronic Journal on Information Systems in Developing Countries* 18:1–24.

Netchaeva, I. 2002. E-government and e-democracy: A comparison of opportunities in the North and South. *Gazette: International Journal for Communication Studies* 64(5):467–477.

Neupane, A., J. Soar, and K. Vaidya. 2014. An empirical evaluation of the potential for public e-procurement to reduce corruption. *Australasian Journal of Information Systems* 18(2):21–44.

OpenNet Initiative. 2007. Singapore. https://opennet.net/sites/opennet.net/files/singapore.pdf.

Pan, S.L., C.W. Tan, and E. Lim. 2006. Customer relationship management (CRM) in e-government: A relational perspective. *Decision Support Systems* 42:237–250.

Pan, S.L., G. Pan, and P. Devadoss. 2005. E-government capabilities and crisis management: Lessons from combating SARS in Singapore. *MIS Quarterly Executive* 4(4):385–397.

Prasad, K. 2012. E-governance policy for modernizing government through digital democracy in India. *Journal of Information Policy* 2:183–203.

Rocheleau, B. 2007. Whither e-government? *Public Administration Review* 67:584–588.

Rose, M. 2004. Democratizing information and communication by implementing e-government in Indonesian regional government. *International Information and Library Review* 36:219–226.

Rose, W. and G. Grant. 2010. Critical issues pertaining to the planning and implementation of e-government initiatives. *Government Information Quarterly* 27:26–33.

Soon, C. and Y. Soh. 2014. Engagement@web 2.0 between the government and citizens in Singapore: Dialogic communication on Facebook? *Asian Journal of Communication* 24:42–59.

Sriramesh, K. and M. Rivera-Sanchez. 2006. E-government in a corporatist, communitarian society: The case of Singapore. *New Media and Society* 8(5):707–730.

Srivastava, S. and T. Teo. 2009. Citizen trust development for e-government adoption and usage: Insights from young adults in Singapore. *Communications of the Association for Information Systems 25*(1):31.

Tan, C.W., S. Pan, and E. Lim. 2005. Managing stakeholder interest in e-government implementation: Lessons learned from a Singapore e-government project. *Journal of Global Information Management* 13:31–53.

Teo, H.H., B.C.Y. Tan, and K.K. Wei. 1997. Organizational transformation using electronic data interchange: The case of Tradenet in Singapore. *Journal of Management Information Systems* 13(4):139–165.

Trimi, S. and H. Sheng. 2008. Emerging trends in m-government. *Communications of the ACM* 51(5):53–58.

Tung, L. and O. Rieck. 2005. Adoption of electronic government services among business organizations in Singapore. *Journal of Strategic Information Systems* 14:417–440.

Wang, M., F. Liao, J. Lin, L. Huang, C. Gu, and Y. Wei. 2016. The making of a sustainable wireless city? Mapping public wi-fi access in Shanghai. *Sustainability* 8(2).

Wong, P.K. 2003. Global and national factors affecting e-commerce diffusion in Singapore. *The Information Society* 19:19–32.

Yigitcanlar, T. and S. Baum. 2006. E-government and the digital divide. In M. Khosrow-Pour (ed.) *Encyclopedia of E-commerce, E-government, and Mobile Commerce.* pp. 353–358. Hershey, PA: IGI Global.

Yildiz, M. 2007. E-government research: Reviewing the literature, limitations and ways forward. *Government Information Quarterly* 24:646–665.

14 A geography of the internet in China

Xiang Zhang

The internet has changed the way how we communicate with one another and how we gather and synthesize information (Curran et al. 2016; Reed 2014). It has radically altered social interactions among people, groups and communities (Howard and Jones 2004), and revolutionized how business transactions are conducted (Castells 2002). The internet brought a boom of new theories to understand the function and character of the cyberworld, such as social network formation (Ellison 2007), spatial identities (Papacharissi 2002), political implications (Chadwick 2006), and global cooperation and disputes (Mueller 2010). In the field of geography, the term cyberspace reflects the new structures of digitized social formations (Graham 2010; Warf 2009; 2013). Social activities are increasingly conducted over cyberspace (Schuler and Day 2004; Tai 2007). The spatial structure of interpersonal linkages has been reorganized as virtual reality opened another window for people to rethink the identity of place (Zook 2003; Learner and Storper 2014; Zook and Graham 2007).

The growth of the internet in China has paralleled the rapid expansion of the Chinese economy and the shift of the Chinese economic system from a centrally planning system to neoliberalism. The rapidly growing economy of China has led to many technical and institutional innovations, especially in information and communication technology (ICT), making the internet one of the most important catalysts to sustain the development of the Chinese economy (Yang 2003). For the rapid growth of China, international trade and manufacturing contribute significantly to capital accumulation (Chen and Feng 2000; Liu et al. 2002). Therefore, the expansion of e-commerce became the most noticeable figure of the growth of the internet in China. The introduction of the internet not only removed spatial obstacles to personal and social communications, but facilitated international trade and reduced transaction costs among firms (Freund and Weinhold 2002, 2004). Along with the rapid growth of the internet in China, Chinese society, culture, and economy have all entered a newly reorganized spatial order. Connected by the internet, the information acquisition become a simple task of searching keywords with search engines; digital interpersonal communications can be as vivid as face-to-face real time interactions with online streaming and social media;

the demand from foreign buyers and the supply by Chinese sellers are easily matched. The internet not only enhances the economic prosperity and social integrity of China, but also promotes Chinese integration into the world community and the global market (Sila 2013; Zhou et al. 2013). This relationship grew in accordance with the global trend of the internet and digital economy growth after the turn of the millennium (Wymbs 2000; Kraemer et al. 2005; Damm and Thomas 2006), and continues to grow stimulated by the government policy of promoting information technology development and economic structure transformation.

Growth of the internet and the digital divide in China

China began to communicate with the world in electronic format in the late 1980s. A research team in a foreign language institute in Beijing sent the first Chinese email to its partner department in a German institute on September 20, 1987, which says "Across the Great Wall, we can reach every corner in the world." After traveling in cyberspace for seven days, it was received by Karlsruhe University in Germany. Though this message did not comply with the internet protocol and was not commonly considered as a trial to access the internet, it is still marked as the first Chinese message and greeting in the cyberspace (China Internet Network Information Center [CNNIC] 2007).

The story between the world's second largest economy and the internet formally begun in the early 1990s. China first connected digitally to the world internet in 1994, when the Institute of High Energy Physics of Chinese Academy of Science built a connection to the Stanford Linear Accelerator Center in California (Li and Zorn 2006). The major purpose of the internet connection then was serving research and educational institutes and not the public. The first server allowing people to visit the internet was built in Beijing on September 30, 1995 (Wu 2001). However, due to the high cost and slow speed of the dial-up connections, the number of internet users remained at a very low level until the introduction of broadband into China after 2000. Since then, as the application of broadband technology significantly decreased the cost of using the internet (Hausman et al. 2001; Horrigan 2010), the number of broadband internet users climbed from 4 million in 1999 to 210 million in 2007, which marks the first stage penetration of the internet in China.

Since the 2008 global financial crisis, the Chinese government has utilized a portfolio of fiscal and institutional tools to maintain its economic growth. Transformation and upgrading of industry are one of the major targets in China's economic growth plan. To transform China from a producer and exporter of global product orders to a designer and developer of new technologies, state policy emphasizes the need to develop technological information and communication innovations and applications (National Development and Reform Commission of China 2010).

Figure 14.1 shows the growth of three critical factors in e-commerce during the most recent decade: economic growth, the number of internet users, and

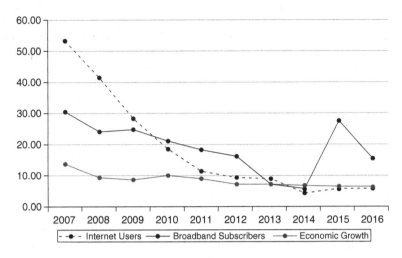

Figure 14.1 The growth in the number of internet users and broadband subscribers in China, 2007–2016.

Source: Author, using data from National Bureau of Statistics, 2017.

broadband subscribers. After the rapid economic growth before the 2008 global financial crisis, the Chinese economy slowed down to one-digit growth at around 7% per annum. Although that growth was quite fast by Western standards, it remains low for the Chinese economic miracle, which, from 1990 to 2007, had experienced an average annual growth of over 10%. The growth in the number of internet users and broadband subscribers has been maintained at a relatively high level in comparison to the country's economic growth. This indicates that the internet and broadband service are expanding faster than China's domestic economy. The rapid growth of the internet and broadband service has stimulated the enormous growth of internet and information technology (IT) companies in China during this period, and it lays the foundation for the transformation of economic transactions from the traditional business mode to the new internet-based pattern.

To illustrate the growth of the internet penetration and the size of user group in China, Figure 14.2 shows the number of internet users from 2005 to 2017 as recorded by the China Internet Network Information Center (CNNIC). During this period, the number of internet users expanded seven-fold, and the penetration rate increase from 10% to 50%. China now has the largest group of internet users in the world (est. 854 million in January, 2020), and the internet has already become a necessary component in people's daily life. Thanks to the growing use of social media, mobile payments, and online shopping services, the internet can provide any essential service for a Chinese family and foreign residents.

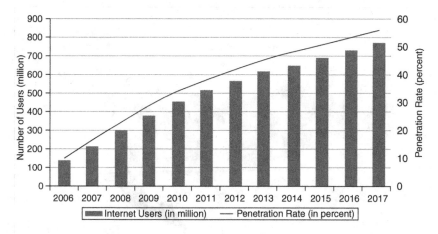

Figure 14.2 The growth of the internet in China, 2006–2017.
Source: Author, using data from CNNIC, 2017.

Geography of the internet in China

As access to the internet requires adequate telecommunication infrastructure and financial investment, internet growth, and development in China have long been highly correlated with regional and local economic development (Lum 2006). Like the uneven growth of the economy in different regions in China, the development of the internet in China reveals a nationwide inequality in terms of the internet penetration rate, which describes the percentage of residents with access to the internet (Pick and Sarkar, 2015). Because fiber optic cables are part of a country's infrastructure, the level of local economic development determines the level of infrastructure that can be provided to residents. The physical geography and terrain also determine the level of difficulty in building a modern telecommunications network (Pick and Nishida 2015). Thus the geography of the internet in China tends to be constrained by two major conditions, the natural environment and the economic development.

Figure 14.3 shows internet penetration rates across Chinese provinces in 2016. The map shows that the It reveals a spatial pattern that is similar to the uneven economic development level in China. Thus, the coastal areas have the highest internet penetration rate, particularly in highly urbanized Beijing (77.8%), Shanghai (74.1%), and Guangdong (74.0%). The inland areas have a relatively low penetration rate in comparison to the coastal ones. According to the data used for this map, 20 of the 31 provincial units in China have an internet penetration rate greater than 50%. The lowest rate was recorded in the southwest mountainous Yunnan Province, where a large group of ethnic

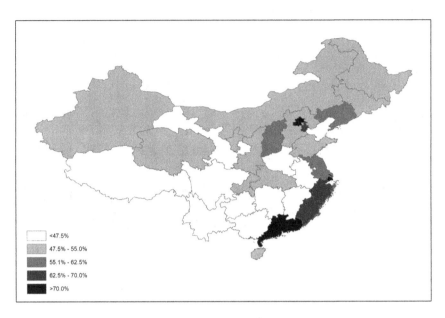

Figure 14.3 Internet penetration rates in Chinese provinces, 2016.
Source: Author, using data from CNNIC, 2017.

minorities are distributed over the mountainous edge of the Tibetan Plateau and the hilly southwestern karst area.

However, a more balanced distribution could be possible in the future due to the nationwide promotion of and investment in a telecommunications infrastructure (State Council of China 2016). Figure 14.4 shows the growth of internet penetration in 2016. The pattern is the reverse of that presented in the previous map (Figure 14.3). The growth rate in coastal areas slowed down, while the growth rate in the inland provinces such as Jiangxi (15.7%), Anhui (13.6%), Guizhou (13.2%), was faster than it was for their more developed neighbors. This trend sends a promising signal for a more balanced spatial pattern of internet penetration across the country. And a catch-up trend of the internet penetration in the inland and underdeveloped provinces can be anticipated.

In addition to the geographical unevenness of internet access, gaps in internet access are affected by demographic factors such as gender, age, and education levels. According to the most recent report (CNNIC 2017), in 2016, 72.6% of all netizens in China lived in an urban area; only 27.4% of all netizens came from rural areas. 52.4% of all netizens are male and 47.6% are female. By comparison, 51.2% of the Chinese population is male and 48.9% is female, so the percentage of males who are netizens is higher than

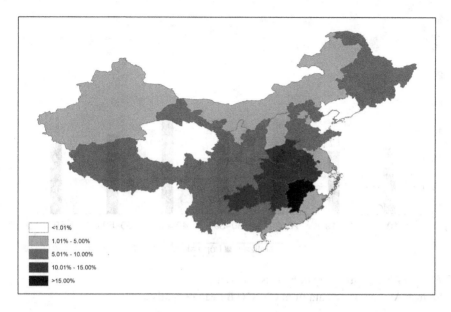

Figure 14.4 Internet penetration growth in China, 2016.
Source: Author, using data from CNNIC, 2017

the percentage of that for the overall population. Figure 14.5 shows the age ranges of Chinese netizens in 2016. In contrast to the total population, this implies that the internet attracts more young people than children and older people. People ranging in age from 10 to 39 comprise nearly three-quarters of all Chinese internet users.

Figure 14.6 presents two more indicators of the demographic structure of Chinese netizens. In terms of the education level among Chinese netizens, 37.3% only completed the nine-year compulsory education, 26.2% had a high school degree, 15.9% only finished elementary school, and only 20% of all Chinese netizens hold an associate's degree, a college degree, or a higher educational degree. This marks an interesting landscape of the demography of Chinese internet users, as the internet in China is a space that is primarily shared by less educated people, which is not consistent with most cases in the rest of the world (Sheehan 2002), but more fitted into an online consumer search model (de los Santos 2018). This phenomenon implies that Chinese netizens tend to be more susceptible to other information and communication online. In terms of the income structure of Chinese netizens, only 7% earned more than 8000 yuan (USD $1250) per month, and 60% of Chinese netizens earn a monthly income less than 3,000 yuan (USD $470).

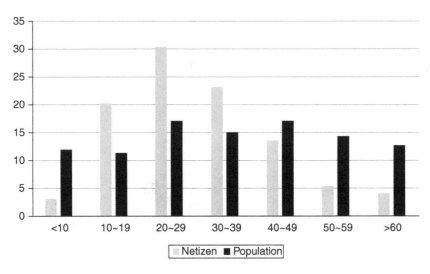

Figure 14.5 Age structure of Chinese netizens.
Source: Author, using data from CNNIC Bureau of Statistics.

The demographical structure of the internet users in China remarks some distinctive characteristics in China and the social landscape in the Chinese cyberspace. Unlike the rest of the world where the majority of internet users is well-educated, the user group in China present a more grass root composition. This demographic pattern could induce several influential social and cultural impacts for the future development path in China. First, such a large group of grass-roots internet users could be the potential beneficiaries by the promotion of business and economic opportunities created by the expansion of e-commerce. Second, the large group of grass root netizens could make significant contributions to waves of new cultural and social elements as the internet has already become the new stage for the Chinese mass media. Third, the grass root feature on the Chinese internet could open up new tasks and opportunities for the future development of the internet and web applications in the education and public service sectors. Therefore, this grass root group with access to the internet could become a distinctive social group who witnesses, participates, and changes the traditional social and cultural spatiality of China.

Besides the above spatial and demographic patterns related to the geography of the internet in China at a nationwide level, the internet in China also has some sophisticated features at the micro level. Figure 14.7 presents the spatial character of internet users in China at a micro scale by showing where they get the access. Based on the data gathered by CNNIC, most Chinese netizens use the internet at home, which indicates that internet access in China

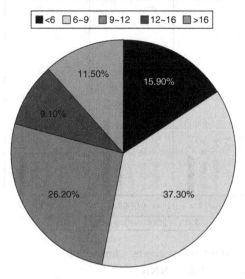

NUMBER OF SCHOOL YEARS OF NETIZENS

■ <6 □ 6~9 ■ 9~12 ■ 12~16 □ >16

15.90%

37.30%

26.20%

9.10%

11.50%

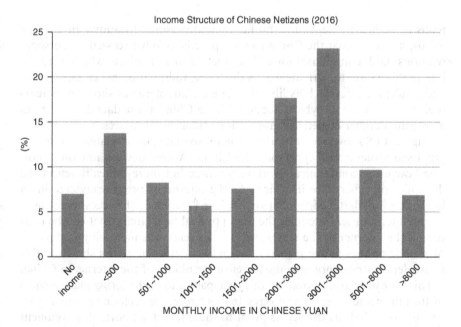

Figure 14.6 Education and average monthly income levels (in yuan) of Chinese netizens.

Source: Author, using data from CNNIC.

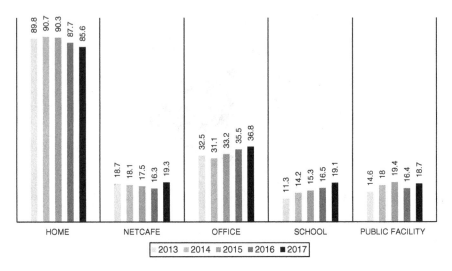

Figure 14.7 Location of internet access by Chinese Netizens.
Source: Author, using data from CNNIC.

tends to be heavily dependent on family and personal locations. In another words, it implies that the Chinese cyberspace is mainly accessed by personal relations and communications. The second major place where Chinese internet users use to surf the web is the office, followed by net cafes, schools, and public facilities such as libraries. These characteristics show an increasingly public nature of cyberspace activities in China as the data demonstrates a growing number of netizens use the internet at public places.

Figure 14.8 shows the second microlevel spatial feature related to the geographical structure of cyberspace in China. Access devices not only indicate how users communicate with each other, but more critically reflect the dynamics of cyberspace in China. As the internet is presumably creating a borderless and flattened "global village," the location of the access point used by the internet users becomes the only physical benchmark that can be used to link the internet to the real world. In this sense, the increasing popularity of mobile phones as the primary device for internet activities shown in this figure demonstrates not only the increasing mobility of the internet in China, but also the increasing dynamics of cyberspace and the fostering placelessness of the internet in geographical ways. In this figure, the declining popularity of desktop terminals as the access point to the internet supports this argument from another perspective. As a fixed access point to the internet, behaviors and activities engaged on a desktop terminal also reflect the locality to the internet user and the corresponding activities in cyberspace. Therefore, the desktop terminal shares the spatiality of traditional fixed geographical coordinates in analyzing the spatial organization of cyberspace. The trend

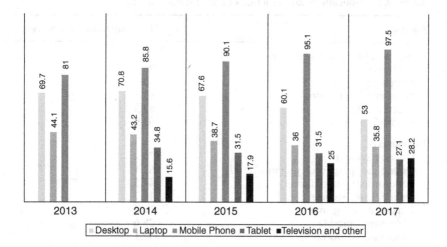

Figure 14.8 Devices used for internet access in China.
Source: Author, using data from CNNIC.

toward mobile internet access is reflected by Figure 14.7, which portrays the future of the spatiality of the internet and the digital geography as spatial locations, relations, and interactions are evolving into a more dynamic state with increasing uncertainty and a more flexible topological positionality.

Application of the internet: e-commerce and its geography

The application of the internet in China covers almost every aspect of Chinese daily life, from social communication to e-governance, from personal shopping to international trade, from online payments to equity trading, etc. Most netizens use the internet as a medium for daily communication and information service. More than half of Chinese internet users engage into different e-commerce activities online. Many other users also use the internet as a platform for leisure and entertainment. Table 14.1 lists the penetration of the internet among Chinese netizens based on a nationwide survey conducted in 2017 (CNNIC).

The application pattern of the internet in China also has significant geographical implications. First, the widespread usage of the internet as an information and communication instrument illustrates a new trend in the social relations in China. Considering the number of netizens in China, such a large group of online social and information users is creating a noticeable change of social and cultural communications. The extensive usage of the internet for communication and information transforms social relations from the traditional spatial pattern based on Euclidean and Cartesian space to the topological and metamorphic pattern exemplified by the internet. Second, the

Table 14.1 Applications and functions of the internet in China

Function	Penetration rate (in percent)
Communication and information	
SMS and email	93.3
Search engine	82.8
News and information	83.8
Social media	87.3
E-government	62.9
E-commerce	
Online payment	70
Online shopping (excl. prepared food)	69.1
Food-ordering and delivery	44.5
Online travel agency	48.7
Financial management	16.7
Rideshare – bicycle	28.6
Rideshare – automobile	37.1
Leisure and entertainment	
Online music service	71
Online leisure novel and reading	48
Online gaming	57.2
Stream video	75
Live streaming	29

internet performs a critical role in the industrial organization in China since the application of e-commerce is now widely used by the Chinese netizens. Such a diverse pattern revises the traditional spatial relation of economic geography as it changes the landscape of production and consumption in many aspects including retail, eatery, leisure, and financial markets. It will produce a profound influence on the economic structure and business performance in China. Third, the internet changes the way people entertain and leisure in daily life, which remarks a shift of cultural life from the traditional pattern which closely connected to the positionality of a physical location, to an intertwined, multifaceted organization between cyberspace and real space. In this manner, the internet is producing a new organization and spatial structure of culture and accelerating the evolution and modification process of culture in the Chinese society. That is, the application of the internet is changing the landscape of cultural geography in China by increasing the interaction and communication between different cultural elements.

The following section offers an overview of the geography of internet applications in China. It focuses on the geography of e-commerce due to its core position in the economy of China and extensive linkages to the social and political development in China and the international market. Therefore, the analysis of the geography of e-commerce provides, on the one hand, a nationwide market feedback to the development of the internet in China, and on the other hand, a review about the socioeconomic impact of the technological progress led by the application of the internet.

E-commerce in China began in the early 1990s, not long after the internet was introduced (Tan and Ouyang 2004). Like its path in the rest of the world, e-commerce in China started when companies utilized internal applications to build a digitized information and database system for their own operations. The growth of e-commerce in China cannot be separated from the rapid growth of internet usage as well as the country's industrial structure. The role of China as the world's largest second largest producer and largest exporter drives the country's rapid e-commerce development in comparison to the rest of the world. As the world's largest importer of materials and the largest exporter of light industrial products, China's economy relies heavily on international trade. Because e-commerce provides the convenience to conduct transactions online without physical barriers, the first wave of rapid expansion of China's e-commerce focused on international trade, and it was used as a platform to link both domestic and global businesses. In 2002, several appliance manufactures in Shandong established the first e-commercial platform to sell their products online. In the same year, a group of textile companies created a shared short message service (SMS) system to conduct business online. In 2002, Alibaba, then a small online business-to-business (B2B) service provider, experienced positive net revenue on its balance sheet for the first time (CNNIC 2003).

In 2005, the Opinion of the General Office of the State Council on Accelerating the Development of Electronic Commerce was officially issued. This document signified that the promotion of e-commerce had become a legitimate, nationwide strategy to help grow the Chinese economy. At the same time, new legislation was passed to ensure the security and regulation of e-commerce over the internet. In the following year, the State Council readdressed its support on promoting the development of e-commerce in the 11th Five-year Plan of China, which aimed to stimulate the rapid growth of e-commerce. This initiative proposes several key points to promote e-commerce such as to increase state investment in building a telecommunication infrastructure, to advocate the application and education of e-commerce in retail and service sectors, to develop more reliable encryption and authentication technologies, etc. (State Council 2007).

In recent years, China has witnessed rapid growth of e-commerce (Zhang et al. 2013; Turban et al. 2017), which helped to fuel its economic boom that began with the Open and Reform policy of the 1980s. The internet became a daily necessity for more than 770 million Chinese netizens by 2018. Increasing productivity and incomes helped to make China into not only the world's second largest economy, but one of the largest consumer markets as well. With the help of the internet, the huge potential in the Chinese domestic market has been triggered, as e-commerce makes it easy for people to buy and sell products without a trip to the department store or the cost to rent a vendor stall. Since Alibaba, the largest e-commerce company in the world, developed its business/customer-to-customer (B/C2C) e-commerce portal, Taobao.com, e-commerce rapidly became a popular method for people to sell and buy products and services as this online business model eliminated the travel cost for consumers and the rent for sellers. By the end of 2016, these two direct-to-customer methods constituted 12.6% of total consumption in China (Ministry of Commerce 2017). Online shopping, e-payment, takeaway ordering, and ride sharing are the most popular practices among Chinese netizens. E-commerce has already become a critical part of almost every aspect of a Chinese daily life with economic, social, cultural, and political significance for the entire country. It became the best object mirroring the development and application of the internet and its reorganization of socioeconomic relations in China.

Online shopping and selling by individual customers made the customer-to-customer (C2C) business mode favorable for Chinese e-commerce. Traditional businesses began to broaden the scope of their websites to sell their products online to individual customers across the country. According to government statistics, the total market value of e-commerce in China increased from 410.8 billion yuan (62.8 billion USD) in 2005 to 12.3 trillion yuan (1.9 trillion USD) in 2014 (CNNIC 2015); the size of the e-commerce market has expanded from less than 2% of China's total GDP in 2008 to over one-tenth of the world's second largest economy (Bureau of Statistics 2015).

In the meantime, China's economic growth is also inseparable from its political institutions and legislative and bureaucratic resources (Allen et al. 2005). With the government's role in economic activities, the policymaking process in Beijing has a significant impact on the industrialization, transformation, and integration of different economic sectors. E-commerce is highly controlled by government agencies at different levels (Zhu et al. 2013). The state engages in internet censorship nationwide and throughout the e-commerce marketplace. The state governs the monetary and fiscal policies for investing in the market of e-commerce. The state endorses regulations over the establishment of new businesses that participate in e-ecommerce. Local authorities enforce different practical policies for zoning and for promoting e-commerce firms. Therefore, e-commerce operates within a triangular stage where the interplay among socioeconomic factors, state governance, and cyberspace is carried out.

To generate a complete picture of the geography of e-commerce in China, a measurement of the e-commerce penetration has been designed based on the data from the official statistical report from the Bureau of Statistics and the company registration database of the Ministry of Commerce of China. This variable, called e-commerce penetration rate, is a normalized variable that calculates the number of e-commercial enterprises divided by the local population.

$$\text{eCommerce Penetration Rate (EPR)} = \frac{\text{Number of E} - \text{commercial Companies}}{\text{Population}}$$

Normalized by population, this ratio compares the frequency of e-commercial entrepreneurship across Chinese provinces and prefectural cities without the influence of disparities in demographic size.

Figure 14.9 is a series of maps showing the spatial pattern of e-commerce penetration rate (EPR), as defined above, from 2013 to 2016. This series of maps clearly demonstrates the spillover of e-commerce from the coastal area to underdeveloped inland China during this four-year period. In 2013, only seven provincial units had a penetration rate higher than 0.25, all located along the coast. The development of e-commerce in China was led by the well-developed units of Beijing, the capital city; Shanghai, the largest economic center; and Zhejiang, the province containing the most private firms. At the same time, e-commerce was almost nonexistent in the inland areas of China. Growth began in 2014. By the end of 2014, provinces along the Yangtze River had hit the 25% penetration rate and had become fertile new land for the development of e-commerce. For coastal areas, e-commerce continued to grow as a popular channel for business in Jiangsu, Fujian, and Guangdong. In 2015, most provinces in central China witnessed an increase in their EPR to the 0.25-level threshold. Coastal provinces had grown into hot spots for e-commerce business, particularly the Yangtze River Delta area, where all three provincial units saw a penetration rate of 150%. In 2016, although the coastal areas still led the development of e-commerce in China, the inland areas were catching up at a rapid rate. The growth of e-commerce in China became a nationwide phenomenon, with all provincial units except Heilongjiang in the Northeast region reaching the 0.25 threshold. The only inland provincial-level municipal city, Chongqing, became a hot spot for e-commerce in the heartland of China.

Figure 14.10 shows the average growth rate of both the economy and e-commerce in China during this four-year period. Both maps exhibit a spatial pattern in which the rapidly growing e-commerce tends to overlap the area of rapid economic growth, particularly in underdeveloped central and western China. It is also noticeable that the average growth rate of e-commerce is much higher than the overall economic growth. This rapid growth of e-commerce across China echoes the expansion of internet use in China and embeds the transformation of the business pattern caused by the explosion of consumers and producers using the internet.

The growth of e-commerce in China has also witnessed competition between foreign and domestic players. An example is the case between the expansion of Amazon China and its Chinese rival, JD.com (Bloomberg 2018). Both companies are major business-to-consumer (B2C) shopping websites in China. In 1999, Amazon entered the Chinese market in a joint venture with a Chinese e-commerce company, Joyo China. Initially, Amazon provided technical support and conducted business as Joyo.com. In 2004, Amazon acquired the entire Chinese company and operated the business as Amazon China, which was the largest B2C company in the Chinese e-commerce market. In the same year, JD.com was established as a small B2C website selling computers

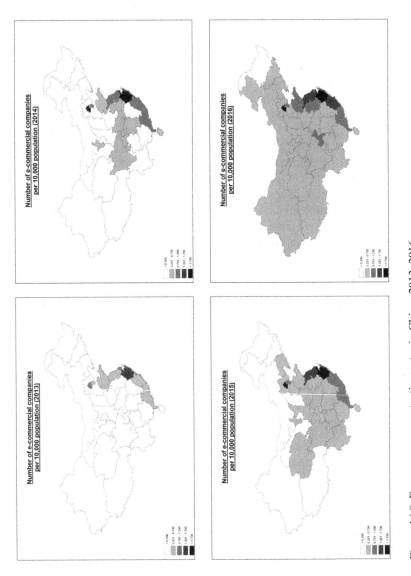

Figure 14.9 E-commerce penetration rates in China, 2013–2016.
Source: Author, using data from the National Bureau of Statistics.

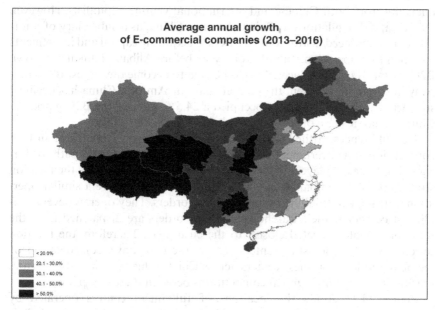

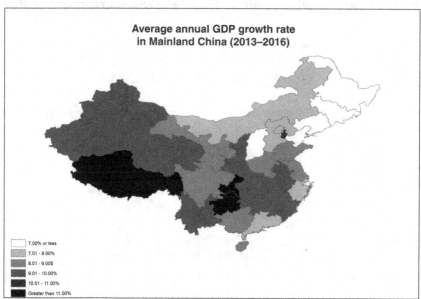

Figure 14.10 Average annual growth of economy and e-commerce in China, 2013–2016.

Source: Author, using data from the National Bureau of Statistics.

and other IT products. Since then, JD.com began to secure several rounds of financing from both Chinese and international venture capitalists. However, due to capital regulations in China, Amazon China, as a subsidiary of a foreign company, faced institutional barriers to acquiring capital and investments to expand its operation. In 2011, one year before Alibaba launched its own B2C portal, JD.com replaced Amazon China to become the largest B2C company in China. Since then, the market share of Amazon China has declined significantly. By 2017, JD.com occupied a 24.5% share of the B2C market in China, versus a share of less than 4% held by Amazon China.

The differences in the fates of JD.com and Amazon China explain how competition and government restrictions affect the development paths of foreign and domestic players in e-commerce growth. It also reveals the location strategy of B2C companies in China. Both companies share a similar operational strategy in fulfilling online purchase orders. They operate several fulfillment centers at their warehouses. Online orders are dispatched from the warehouse that is located closest to the customers. Therefore, the location selection of these logistics centers reflects the company's consideration of regional and local business centers across China. This selection also mirrors the market share and regional competitions between these companies.

Figure 14.11 shows the locations of fulfillment centers operated by Amazon China in 2016. Figure 14.12 shows these warehouses and their

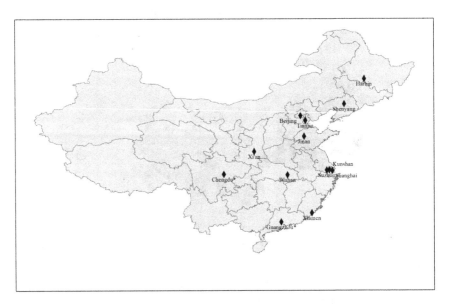

Figure 14.11 Locations of the fulfillment centers for Amazon China.
Source: Author, using data from Amazon.cn.

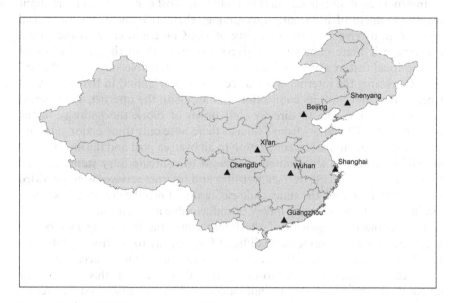

Figure 14.12 Locations of the logistics centers for JD.com.
Source: Author, using data from JD.com.

locations in relation to JD.com the same year. These two maps show that the site selections of warehouses from these two companies overlap in most regions. Amazon China operates more centers in the greater Beijing and Shanghai areas. JD.com operates from regional center cities, and it assigns a hinterland for each warehouse. The case of the site selection for regional centers shows that the centers are intentionally concentrated in big regional cities. This pattern questions the assumption that e-commerce tends to create a more even development of business across a region; it also suggests that e-commerce could become another vehicle to increase the centrality of big cities and contribute to more uneven development.

Controversies and challenges to the internet in China

Though growing rapidly with an excessive investment from both private and public sectors, domestic and international sources, the internet in China still needs to deal with several challenges due to the idiosyncratic Chinese economy, politics, culture, and society in comparison to the traditional Western society where the internet initially grew up. Over 30 years after the Chinese greeting to global cyberspace, it seems that China has begun to rethink the character and nature of the borderless information spread over the internet.

Internet censorship in China is probably the most controversial issue about the development of information technology there. It is also the most critical problem in determining the spatiality of the Chinese internet in the global cyberspace if there comes to a universal framework to theorize the social and political features of social relations in this intangible stage of globalization. Censoring the internet is not a recent event in China. In the early 1990s when China first built its connection and access to the internet, the Chinese government were able to directly shut down or block the pathway to the corresponding IP address from China as there were only three major gateways for Chinese users to build a connection to an oversea port and the information available online was still limited compared to contemporary standards. All these gateways to access oversea websites and internet services were provided by the state, therefore, the state could easily shut down the access gateway service in China to prevent any inflow of unfavorable information.

Due to the rapid growth of the internet and the increasing member of access gateways, it became more difficult for the state to control the information flows on the internet. Executive orders to shut down the access gateway to block unfavorable information became defected. For this reason, the Chinese government started to build a comprehensive censor system over all internet communications within its sovereign territory in 1998 (Goldman and Gu 2005). The Chinese legislative authority, the National People's Congress, passed the bill to officially monitor and censor all internet activities in 2000 and approved to invest and build a firewall to filter out unfavorable websites and information from oversea websites and online sources.

The Chinese censorship on the internet includes two major projections, a direct censoring scheme searching and blocking materials that contain sensitive/blacklisted words; and a firewall system blocking the access to a list of foreign websites and IP addresses. The censoring system filters both text and image sent to search engines, online portals, social media, and even personal emails, and block both receiving and sending process of materials containing these blacklisted words (King et al. 2014). This word filter system is capable of shutting down access to all content that contains blacklisted words. With this system, the state agency can simply control the information reaching terminals of Chinese internet users. That is, the Chinese internet users are given restricted access to the internet and provided selective materials by the order of the government. They are not capable to perform social and cultural functions on cyberspace freely and personally the same as the internet users in counties without such a heavily censored system. Therefore, the geography of the internet in China is different from that of democratic countries. And the study of netizen behaviors in China should never neglect such a fact as these users have only limited information to know and restricted space to move.

The second component of Chinese internet censorship is the Great Firewall (GFW), a comprehensive internet monitoring technology controlling the cross-border information flows between China and the rest of the world. It is a complex toolbox supported by state managed servers and supercomputers that are

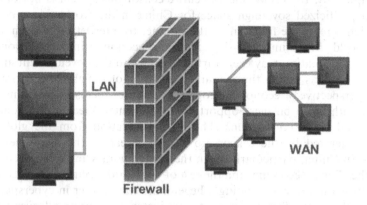

Figure 14.13 An illustration of the Great Firewall in the computer network.
Source: Wikimedia.

able to block foreign IP addresses, spoof the Domain Name System (DNS), reset internet protocol connections, and disrupt the usage of virtual private networks (VPN), etc. The GFW is a similar system to the traditional firewall used by local area network (LAN) as shown in Figure 14.13. That is, though connected to the global internet, the GFW serves as an artificial border between the internet terminals and ports in China and the rest of the world. All information in and out of the GFW, that is, sent and received internationally, is subjected to the sorting and filtering by this specific firewall. Only permitted and favorable information can enter and exit China, functioning like the immigration at ports. The firewall system creates a variable border between the Chinese cyberspace and the rest of the world, which ultimately challenges the borderless nature of this worldwide network. As the GFW is designed, maintained and enforced by the state agency, and performs like a border control system, the internet in China is now being heavily politicized and exhibiting some features of a sovereignty.

Yet China is not the only country in the world that censors and controls information on the internet. However, due to the size of the netizen group and its worldwide impact of the construction of the Chinese censorship system, the GFW and related maneuvers by the Chinese government have received global attentions from political entities and non-government organizations. MacKinnon (2011) calls this censorship "networked authoritarianism" to address its threat to the freedom of the internet in the world and the violation of the freedom of information and speech in China. Gunitsky (2015) criticizes this manipulation of the information on the internet as an autocratic regime. Han (2015) points out that there is a third element in the Chinese censorship system, a group of state-sponsored commentators that is active on the Chinese social media and news portals. Many social scientists have already noticed the significant social impact beyond the internet censorship.

Geographically, the GFW and the entire censorship system is not only creating a politicized sovereign space for China in the worldwide cyberspace, which threats the freedom of the internet to internet users from all over the world. More importantly, social and economic impacts beyond the cyberspace isolation may exert further damage to the development and growth of China, as well as the spatial organization of the global economy. From the perspective of economic geography, internet access is increasingly correlated with more business opportunities and intensified international economic and trade connections. The formal isolation from the global economy generated by the censorship system threatens the e-commerce operations in China, particularly given the contemporary interdependency between the Chinese economy and the rest of the world. Politically, the control of information flow is creating a hegemonic state power in cyberspace which may threat the free of access to the internet for users all over the world. It has recorded events that the DNS spoofing process by the GFW in China disrupted the web access of users in other countries (Anderson 2012). Therefore, the censorship may violate the law of some countries with constitutional protection of free of speech, create international dispute over the sovereignty issue in cyberspace, and ultimately hegemonize the globally shared cyberspace and create a less empowered group of internet users due to the lack of complete information. Culturally speaking, as the internet in China is providing selected information to majority internet users in China, it may further create a cultural and social gap between China and the rest of the world as two groups of internet users are reviewing two different sets of information online. The restricted information acquisition of Chinese netizens could easily be manipulated and indoctrinated to become a stereotyped group of individuals, which ultimately threats not only the cultural diversity in China, but also the diversity of ideas, thoughts, and creativities in the country. For these reasons, the censorship and the border construction on the Chinese cyberspace is not only an affair for Chinese netizens, but also a global affair related to a potentially new geography of economy, politics and culture in the world.

Conclusion

China has the largest group of internet users in the world. It also arguably has the most complicated social relations in cyberspace. Yet, due to a complicated context from both human and physical dimensions, the internet in China does have some problems that attract many social scientists for explanations and solutions. This chapter outlines the spatial pattern of the internet in the geography of the real world and argues the issue of the censorship in the context of spatiality of cyberspace. The internet penetrates different regions of China unequally due to various economic, social and geographical reasons. However, the underdeveloped area is still catching up. Specifically, the chapter uses the example of e-commerce in China to demonstrate the function and integrity of the internet to the economic development and social relations in the country. The internet brings together business opportunities across the country and all over the world. The internet changes the spatial organization of economic and social interactions. The internet redefines the location and position in a topological way which parallels the real world and the flattened cyberspace.

This chapter also discussed the potential challenge to the spatial organization of cyberspace brought by the internet censorship system in China. Such policies and manipulations restricting the public access to the international internet politicize cyberspace, disrupt social relations and network in cyberspace, and redistribute the power structure among the internet users. All these defects produced by the censorship system are challenging the globalizing geography of the internet and some basic human rights for internet users from all over the world.

References

Allen, F., Qian, J., and Qian, M. 2005. Law, finance, and economic growth in China. *Journal of Financial Economics* 77(1):57–116.

Anderson, D. 2012. Splinternet behind the great firewall of China. *Queue* 10(11):40–49.

Bloomberg. 2018. China's JD Prepares to Take on Amazon on Its Home Turf. www.bloomberg.com/news/articles/2018-01-26/jd-enlists-tencent-as-it-prepares-to-take-on-amazon-in-the-u-s.

Bureau of Statistics. 2015. *China Statistical Yearbook 2014*, Beijing: National Bureau of Statistics of China. www.stats.gov.cn/tjsj/ndsj/2014/indexeh.htm.

Castells, M. 2002. *The Internet Galaxy: Reflections on the Internet, Business, and Society*. Oxford: Oxford University Press.

Chadwick, A. 2006. *Internet Politics: States, Citizens, and New Communication Technologies*. New York: Oxford University Press.

Chen, B., and Feng, Y. 2000. Determinants of economic growth in China: Private enterprise, education, and openness. *China Economic Review* 11(1):1–15.

CNNIC (China Internet Network Information Center). 2003. *The Internet Timeline of China Part I*. www.cnnic.net.cn/html/Dir/2003/12/12/2000.htm (in Chinese).

CNNIC (China Internet Network Information Center). 2007. *The Internet Development Timeline in China (Zhongguo Hulianwang Fazhan Dashiji)*. Beijing: Beijing Weekly Press.

CNNIC (China Internet Network Information Center). 2016. *Statistical Report on Internet Development in China*. www.cnnic.net.cn/hlwfzyj/hlwxzbg/ (in Chinese).

CNNIC (China Internet Network Information Center). 2017. *Statistical Report on Internet Development in China*. www.cnnic.net.cn/hlwfzyj/hlwxzbg/ (in Chinese).

Curran, J., Fenton, N., and Freedman, D. 2016. *Misunderstanding the Internet*. London: Routledge.

Damm, J., and Thomas, S. (eds.) 2006. *Chinese Cyberspaces: Technological Changes and Political Effects*. London, UK: Routledge.

de los Santos, B. 2018. Consumer search on the internet. *International Journal of Industrial Organization* 58:66–105.

Ellison, N. 2007. Social network sites: Definition, history, and scholarship. *Journal of Computer-Mediated Communication* 13(1):210–230.

Freund, C., and Weinhold, D. 2002. The internet and international trade in services. *American Economic Review* 92(2):236–240.

Freund, C., and Weinhold, D. 2004. The effect of the internet on international trade. *Journal of International Economics* 62(1):171–189.

Goldman, M., and Gu, E. (eds.) 2005. *Chinese Intellectuals between State and Market.* London: Routledge.

Graham, M. 2010. Neogeography and the palimpsests of place: Web 2.0 and the construction of a virtual earth. *Tijdschrift voor Economische en Sociale Geografie* 101(4):422–436.

Gunitsky, S. 2015. Corrupting the cyber-commons: Social media as a tool of autocratic stability. *Perspectives on Politics* 13(1):42–54.

Han, R. 2015. Defending the authoritarian regime online: China's "voluntary fifty-cent army." *China Quarterly* 224:1006–1025.

Hausman, J., Sidak, J., and Singer, H. 2001. Cable modems and DSL: Broadband internet access for residential customers. *American Economic Review* 91(2):302–307.

Horrigan, J. 2010. *Broadband Adoption and Use in America.* Washington, DC: Federal Communications Commission.

Howard, P., and Jones, S. 2004. *Society Online: The Internet in Context.* Thousand Oaks, CA: Sage.

King, G., Pan, J., and Roberts, M. 2014. Reverse-engineering censorship in China: Randomized experimentation and participant observation. *Science* 345(6199):1251722.

Kraemer, K., Gibbs, J., and Dedrick, J. 2005. Impacts of globalization on e-commerce use and firm performance: A cross-country investigation. *The Information Society* 21(5):323–340.

Leamer, E., and Storper, M. 2014. The economic geography of the internet age. In *Location of International Business Activities.* pp. 63–93. London, UK: Palgrave Macmillan.

Li, N., and Zorn, W. 2006. An overview of China's early efforts to connect to the global Internet. *Xinhua Net* (Nov. 22). http://news.xinhuanet.com/newmedia/2006-11/21/content_5358804.htm (in Chinese).

Liu, X., Burridge, P., and Sinclair, P. 2002. Relationships between economic growth, foreign direct investment and trade: Evidence from China. *Applied Economics* 34(11):1433–1440.

Lum, T. 2006. *Internet Development and Information Control in the People's Republic of China. CRS Report for Congress.* Washington, DC: Library of Congress.

MacKinnon, R. 2011. China's "networked authoritarianism." *Journal of Democracy* 22(2):32–46.

Ministry of Commerce. 2017. *Report on E-commerce in China 2016* (in Chinese), http://data.mofcom.gov.cn/report/ECR(2016).pdf.

Mueller, M. 2010. *Networks and States: The Global Politics of Internet Governance.* Cambridge, MA: MIT Press.

National Development and Reform Commission of China. 2010. *The Outline of the Twelfth Five-Year Plan for National Economic & Social Development of the People's Republic of China.* Beijing: Renmin Press.

Papacharissi, Z. 2002. The virtual sphere: The internet as a public sphere. *New Media & Society* 4(1):9–27.

Pick, J., and Nishida, T. 2015. Digital divides in the world and its regions: A spatial and multivariate analysis of technological utilization. *Technological Forecasting and Social Change* 91:1–17.

Pick, J., and Sarkar, A. 2015. *The Global Digital Divide: Explaining Change.* Dordrecht: Springer.

Reed, T. 2014. *Digitized Lives: Culture, Power, and Social Change in the Internet Era.* London: Routledge.

Schuler, D., and Day, P. 2004. *Shaping the Network Society: The New Role of Civil Society in Cyberspace.* Cambridge, MA: MIT Press.

Sheehan, K. 2002. Toward a typology of Internet users and online privacy concerns. *The Information Society* 18(1):21–32.

Sila, I. 2013. Factors affecting the adoption of B2B e-commerce technologies. *Electronic Commerce Research* 13(2):199–236.

State Council. 2007. *State Council of the PRC Document 2007.* Beijing: State Council of the PRC.

Tai, Z. 2007. *The Internet in China: Cyberspace and Civil Society.* London: Routledge.

Tan, Z., and Ouyang, W. 2004. Diffusion and impacts of the internet and e-commerce in China. *Electronic Markets* 14(1):25–35.

Turban, E., Outland, J., King, D., Lee, J. K., Liang, T. P., and Turban, D. 2017. *Electronic Commerce 2018: A Managerial and Social Networks Perspective.* Dordrecht: Springer.

Warf, B. 2009. The rapidly evolving geographies of the Eurasian internet. *Eurasian Geography and Economics* 50(5):564–580.

Warf, B. 2011. Geographies of global internet censorship. *GeoJournal* 76(1):1–23.

Warf, B. 2013. Contemporary digital divides in the United States. *Tijdschrift voor Economische en Sociale Geografie* 104(1):1–17.

Wu, X. 2001. *Da Bai Ju (The Big Failure).* Hangzhou: Zhejiang People's Press (in Chinese).

Wymbs, C. 2000. How e-commerce is transforming and internationalizing service industries. *Journal of Services Marketing* 14(6):463–477.

Yang, G. 2003. The internet and civil society in China: A preliminary assessment. *Journal of Contemporary China* 12(36):453–475.

Zhang, Y., Bian, J., and Zhu, W. 2013. Trust fraud: A crucial challenge for China's e-commerce market. *Electronic Commerce Research and Applications* 12(5):299–308.

Zhou, L., Zhang, P., and Zimmermann, H. 2013. Social commerce research: An integrated view. *Electronic Commerce Research and Applications* 12(2):61–68.

Zhu, S., and Chen, J. 2013. The digital divide in individual e-commerce utilization in China results from a national survey. *Information Development* 29(1):69–80.

Zook, M. A. 2003. Underground globalization: Mapping the space of flows of the internet adult industry. *Environment and Planning A* 35(7):1261–1286.

Zook, M., and Graham, M. 2007. The creative reconstruction of the internet: Google and the privatization of cyberspace and DigiPlace. *Geoforum* 38(6):1322–1343.

Part III

The internet in everyday life

15 Google Earth

Todd Patterson

The introduction of computers brought about the potential for widespread innovations in computer mapping and understanding geography. Geographic information systems (GIS) improved the accuracy and precision of maps and opened up a new genre of educational opportunities employing maps for data visualization and broader geographic understanding. Remote sensing especially helped to change the way we perceive and comprehend the world by providing more objective details about the Earth.

Virtual globes, three-dimensional electronic representations of the Earth, became a natural extension of integrating geographic and remotely sensed data and have become widely employed in a variety of educational, personal, governmental/public service, and commercial applications. There were a variety of offline virtual globes introduced in the late 1990s but Google's Google Earth arguably became the most widely used online virtual globe since its initial release in 2001. Google Earth incorporated satellite imagery, aerial photographs, and various GIS data that allowed users to interact with the Earth in new ways and customize their own (and others') experiences. Two years later Google unveiled its ground-level view, Google Street View, which allowed users a panoramic view from various positions along streets throughout the world (Figure 15.1). Google Earth's data provision also includes a customizable eXtensible Markup Language (XML) text-based format that allows users to create and edit certain datasets.

Google Earth has touched most scientific disciplines, having been mentioned in 2,115 publications covering all of Scopus's 26 subject areas and was mentioned in an average of 229 publications per year since 2009 (Liang et al. 2018). Google Earth and Google Street View have stretched the concepts of space, tools of representation, and processes of reasoning related to spatial thinking (National Research Council 2006; Grossner 2012). The visualization, navigation, interoperability, accessibility and ease of use, and modeling and simulation make Google Earth a powerful tool for visualization, analysis, and education (*A Decade of Google Earth*).

Figure 15.1 Google Earth view of the Quadrant on Regent Street, London.
Source: Wikicommons.

Spatial thinking

Spatial thinking includes knowledge of space or the world around us – including concepts of "symmetry, orientation, scale, distance decay, and other" considerations (Schultz et al. 2008, 27). Spatial reasoning also includes ways of thinking and acting spatially, such as employing statistical data and GIS software and tools (Schultz et al. 2008; Gold et al. 2018). Spatial thinking and competencies are skills that can (and should be) taught and, accordingly, Google Earth can be an effective tool to both formally and informally improve spatial thinking. Formal means of employing Google Earth to learn spatial thinking skills include use in educational curricula and hands-on activities while informal means include navigating for travel plans and viewing various pictures and data of interest.

Google Earth in education

The presence of computers in educational and home settings has increased dramatically over the last several decades as the price of computers has continually declined while the capabilities have improved. While many schools have advanced beyond just adopting or providing technology for students, technology-rich curricular activities are still evolving (Cates et al. 2003; Davis 2017). Online resources have helped to improve student comprehension of major concepts and skills while also helping students to gain confidence in their knowledge of geographic issues (Solem and Gershmehl 2005). The recall of information also tends to be greater with visual images than text (Wager 2005), making Google Earth's visualization-based presentation a strong platform for learning.

Google Earth has become a formidable technological tool to help strengthen weaknesses in geography curricula and there have been numerous studies reflecting both how to implement Google Earth to support education as well as qualifying/quantifying Google Earth's (and other virtual globes') impact on educational outcomes. The most salient study examining the use of Google Earth and its positive affect on learning outcomes was Blank et al.'s (2016) development of curriculum to increase students' understanding of natural phenomena. They found that the Google Earth curriculum advanced students' science identity, earth science understanding, and science reasoning – but the curriculum was most transformative in scientific reasoning.

> Google Earth allows users to ask geographic questions, gather and organize geographic information, display geographic information, and answer geographically oriented questions. The intuitiveness of the interface allows educators to reduce the amount of time needed to teach the platform, thus providing a more immediate and responsive learning environment for students. Using Google Earth in the classroom also helps fulfill requirements of the National Geography Standards by "(1) being used in lessons that question how humans obtain and use earth materials as resources; (2) asking students to describe features of the earth; (3) inferring how human behavior changes the earth's surface; (4) measuring distances; and (5) using other thinking and analytical capacities."
>
> (Patterson 2007, 148–149)

A plethora of researchers have suggested curriculum components leveraging Google Earth as a platform, providing guidance and suggestions for using Google Earth in the classroom. Patterson (2007) developed a location-specific (South Carolina-based) lesson plan for seventh grade students to increase geographic awareness while building critical thinking, analysis, and inquiry skills. Kim (2018) had students leverage Google Earth's analytical and visualization tools to investigate the effects of a project-based community participation course, merging the notion of Google Earth in education and public service/governmental applications of the technology. Xiang and Liu (2017) added empirical evidence to the literature by having students evaluate spatio-temporal changes in Singapore. Williams and Davinroy (2015) went beyond the Google Earth focus and incorporated Street View in the curriculum by having students engage in distinguishing a suburban streetscape.

Spatial perception is an element of spatial thinking that can be overlooked in the classroom; Google Earth provides a strong platform for students to understand and challenge their own thinking about space and how they think space might be organized and related. Bodzin and Fu (2014) used Google Earth to help students examine their understanding and perceptions about climate change, and evaluated whether teacher and student level factors accounted for students' climate change knowledge in an urban-middle level educational setting. DePaor et al. (2017) expanded on that foundation and leveraged Google Earth to help students understand misconceptions about seasonality and better understand the role of the Sun's position and distance from the Earth in influencing seasonal changes. While environmentally oriented curriculum has been the strongest arena wherein Google Earth has been leveraged, those have not been the only areas for which Google Earth has had utility.

Google Earth in personal applications

Google Earth also has utility in "personal applications," or use outside of educational and other forums. Increasingly more consumers are leveraging technologies, including Google Earth, to buy homes because the visualization and data elements help users to understand the nature of the potential purchase and the context of the home being considered (McLaughlin 2017). More recently Google Earth has been considered for cultural studies, such as Laforest's (2016, 659) use of Google Earth to illustrate biographical narratives (his results reflected that

> Google Earth's interface has a lot more in common with the city than we might initially think … [and by using Google Earth we find that] the nature of the city involved is culturally-specific and is in some ways confining.

Similarly Martínez-Graña et al. (2013) created databases to develop a virtual tour for Las Quilamas Natural Park (Spain) for augmented reality and providing QR codes for visitors to scan to understand details at particular place marks. Essentially Google Earth can be used to enhance spatial thinking to better understand context of demography, culture, and other elements.

Although Google Maps tends to have more utility in personal applications, Google Earth's supporting data can be used to identify places of interest and understand context more broadly than Google Maps (by providing ancillary data and understanding more context as a result). For example, Giovenco et al. (2016) analyzed not only the location of vape shops in New Jersey but also considered tobacco outlet density (and correlated to demographic details to better understand locational variation of vape shops). Google Earth has also incorporated news feeds to provide context to users of events happening – underscoring the importance of where events occur and the geographic context of their occurrence.

Google Earth in public service/government

Google Earth has been used as a propaganda and information verification tool to monitor various governmental bodies' actions. In recent years governments have kept a watchful eye on activities claimed by governments for intelligence and national security purposes, such as China's crackdown on Muslim Uighurs (Cheng 2018) and North Korea's nuclear program (Fisher 2013).

Beyond security applications in the public service realm Google Earth has been widely used to understand urban development and public service provision. As reflected in education, urban sprawl and potential areas for development are a strong suite of using Google Earth's imagery. Rashid (2018) revisited a study of San Francisco urban sprawl completed in 1993 to evaluate Google Earth's applicability in change detection and validation of the Southworth and Owens study. Similarly Matamyo and Yuji (2017) developed a model for classifying urban land use using Google Earth imagery, reflecting that specialized image classification software is not required for simple model development and reasonably accurate land use assignment. Nadal et al. (2017 also engaged in the applicability of imagery in urban studies, leveraging Google Earth imagery and data and other remotely sensed imagery to evaluate building roof parameters to identify potential urban agriculture in housing for rooftop greenhouses.

As an extension of urban studies using Google Earth, Glaeser et al. (2018) used Street View images to predict incomes in New York City, developing a model for using imagery data to map wealth and poverty. Their model also reflected how technology-based survey techniques could better measure willingness of residents to fund urban amenities. Interestingly they also reflected how Internet data was used to improve the quality of urban services, including crowdsourcing-like feedback accumulation and response. Kang (2018) also used Street View to identify diverse urban types and like Rashid (2018) similarly identified the spread of urban areas across space. Gage et al. (2018) went beyond traditional uses of Google Earth by considering shade in public spaces; the assessment intended to identify health concerns (e.g., sunburn) as a means of improving public planning and usefulness of public spaces.

Google Earth has also been widely used to understand environmental concerns and analyze potential public responses to alleviate various issues. Hird et al. (2017) demonstrated a method of identifying wetland areas using the Google Earth engine along with R statistical software, reflecting that high-quality topographic variables could model wetland distribution at regional scales. Similarly Estoque et al. (2018) used Google Earth to compare image types of forested areas in the Philippines, using Google Earth points as reference for comparison across eight different imagery types. Importantly their results suggested the usefulness of various imagery types in classifying forest types and assisting public agencies evolve development and protection policies of environmental resources.

Google Earth's utility for land cover/land use classification has been demonstrated by Massey et al. (2018) by virtue of classifying a method of using Google Earth imagery to estimate total cropland for North America. Such models and uses reflect Google Earth's usefulness to understand space and improve spatial thinking skills (such as incorporating such results into educational curricula or being used for consideration by public agencies in decision making). Similarly Ragettli et al. (2018) implemented an automatic irrigation mapping algorithm to classify areas to distinguish productive irrigated fields from non-productive and non-irrigated areas (not employing ground truth data and achieving a classification accuracy between 77 and 96%) in Central Asia. Parente and Ferreira (2018) had similar accuracy success by classifying pasturelands in Brazil (at approximately 80%) to analyze the increase in pasture areas and herding activities between 2000 and 2016.

Even in the health sector Google Earth has been a useful platform. Masthi et al. (2015) leveraged Google Earth extensively in mapping cholera outbreak in Bengaluru, Karnataka in June 2013 by integrated house-to-house surveys and plotting the results to describe household causes of the outbreak and identify clustering around suspected water issues. It is important to also reflect that mashups, combining digital data or functionality, is also a useful capability that has expanded Google Earth's usefulness. Mashups have forged diverse ways of seeing with maps on the Web (Dalton 2013) and have been used in a variety of applications, such as the combination that Kamel Boulos et al. (2008) conducted to review public health geographies. (Their four mashups incorporated 2-D and 3-D virtual worlds, using Web 2.0 and GIS and creating avatar-inhabited lands for infectious disease surveillance.)

Google Earth in commercial applications

Earth classification has become a widespread use of Google Earth data to help users interpret and understand geographic phenomena. Many of the commercial applications for Google Earth have initiated in public sector uses and then evolved to commercial applications – such as leveraging land use/land cover classification studies to determine marketability of un/under-developed lands. For example, Cheng et al. (2018) leveraged Google Earth to classify palm oil areas to understand palm oil distribution, potential development areas, and socio-economic development and ecological and environmental problems as a result of such development.

Engineering and construction services have been employing Google Earth to understand changes in spatial patterns, including geomorphological processes (Moges et al. 2018), and conduct road survey and other engineering-related studies (Haibin et al. 2018; Markert et al. 2018; Elhakeem et al. 2017). Google Earth has even been transformational in archaeology, making it easier to identify potential sites as "within a few years tens of thousands of sites previously unrecorded and scarcely known to the ... world have been mapped" (Kennedy 2017, 153).

Google Earth looking forward

As contemporary students grow up in the digital age they have a different manner of learning and applying tools at their disposal because videos and images have been a primary means of information gathering for students.

The incorporation of technologies such as Google Earth further blurs the lines between educational, personal, government/public service, and commercial applications. Digital devices are a component of daily life and users are exposed more readily to maps and data and inherently understand how to interpret data (at least at a basic level). Games such as Pokemon Go have incorporated maps to help users understand location-based services and become more aware of their surroundings to learn the uniqueness of place (Chen et al. 2018). Pokemon Go, and other similar applications, has a natural evolution from Google Earth as a platform that incorporated the geographic visualization and data elements to help foster interaction with the real world (Gong et al. 2017). Google Earth itself not only fostered the evolution of entertainment it helped evolve learning platforms by paving the way for augmented reality games and learning environments (Lupton 2017).

As technology continues to evolve Google Earth is being used in innovative ways to support data analysis and visualization. Crowdsourcing has become a way to understand populations at risk or impacted by phenomena, such as earthquakes or floods (Choi et al. 2018), or even understand the "beauty ratings" of buildings based on number of postings (Saiz et al. 2018). Google Earth as a platform has integrated news sources to maintain details related to current events, such as providing up-to-date weather forecasts, mapping gas leaks in service for utilities, and providing real-time traffic details for travelers.

With broader uses of Google Earth issues of data quality, relevance, and "information overload" have become more relevant. Users inherently and unknowingly often rely on data with limitations and formal accuracy assessments available through metadata that are infrequently consulted (Jones et al. 2014). Volunteered geographic information inherently has its own limitations in reliability and accuracy (Comber et al. 2013), underscoring the need for users that rely on data presented in Google Earth to be cognizant of potential data quality and reliability concerns. Crowdsourcing could be a compromise between accuracy and timeliness but does present further issues related to data quality and reliability that will continue to evolve into the future.

Summary

As a virtual globe Google Earth has expanded in popularity and data availability since its inception. While Google Earth strives to improve users' awareness of the world around them, data quality and reliability has increasingly become a source of contention and consideration. As crowdsourcing becomes more popular the need for data validity and accuracy for reliability becomes increasingly more challenging and important.

Google Earth has utility, and certainly overlap, among educational, personal, governmental/public service, and commercial arenas. The lack of a clear consensus about spatial thinking or spatial literacy (Schultz et al. 2008), however, precludes absolute agreement and quantification of Google Earth's true value in any particular context. Given geography as an integrating discipline, though, it is clear that Google Earth has value in helping users understand the world around us.

References

Blank, L., H. Almquist, J., and J. Crews. 2016. Factors affecting student success with a Google Earth-based Earth science curriculum. *Journal of Science Education & Technology* 25(1):77–90.

Bodzin, A., and Fu, Q. 2014. The effectiveness of the geospatial curriculum approach on urban middle-level students' climate change understandings. *Journal of Science Education and Technology* 23(4):575–590.

Cates, W., B. Price, and A. Bodzin. 2003. Implementing technology-rich curricular materials: Findings from the exploring life project. In D. Lamont Johnson and Cleborne D. Maddux (eds.) *Technology in Education: A Twenty-Year Retrospective.* New York: Haworth Press.

Chen, C.-S., H.-P. Lu, and T. Luor. 2018. A new flow of location based service mobile games: Non-stickiness on Pokémon Go. *Computers in Human Behavior* 89:182–190.

Cheng, J. 2018. Razor wire evidence: Google Earth images show Uighur camps are not what China says they are. *World* (Nov.), p. 59.

Cheng, Y., L. Yu, Y. Xu, X. Liu, H. Lu, A. Cracknell, K. Kanniah, and P. Gong. 2018. Towards global oil palm plantation mapping using remote-sensing data. *International Journal of Remote Sensing* 39(18):5891–5906.

Choi, C., Y. Cui, and G. Zhou. 2018. Utilizing crowdsourcing to enhance the mitigation and management of landslides. *Landslides* 15(9):1889–1899.

Comber, A., L. See, S. Fritz, M. Van der Velde, C. Perger, and G. Foody. 2013. Using control data to determine the reliability of volunteered geographic information about land cover. *International Journal of Applied Earth Observation & Geoinformation* 23:37–48.

Dalton, C. 2013. Sovereigns, spooks, and hackers: An early history of Google geo services and map mashups. *Cartographica* 48(4):261–274.

Davis, M.. 2017. Social studies: Many of the digital tools teachers use are not tailored to the subject. *Education Week* 36(35):20.

De Paor, D., M. Dordevic, P. Karabinos, S. Burgin, F. Coba, and S. Whitmeyer. 2017. Exploring the reasons for the seasons using Google Earth, 3D models, and plots. *International Journal of Digital Earth* 10(6):582–603.

Elhakeem, M., A. Papanicolaou, and C. Wilson. 2017. Implementing streambank erosion control measures in meandering streams: Design procedure enhanced with numerical modelling. *International Journal of River Basin Management* 15(3):317–327.

Estoque, R., R. Pontius, Y. Murayama, H. Hou, R. Thapa, R. Lasco, and M. Villar. 2018. Simultaneous comparison and assessment of eight remotely sensed maps of Philippine forests. *International Journal of Applied Earth Observation & Geoinformation* 67:123–134.

Fisher, M. 2013. Google maps reveals exact site of North Korea's nuclear test, plus nearby test facility and gulag. *Washington Post* (Feb. 12). www.washingtonpost.com/news/worldviews/wp/2013/02/12/google-maps-reveals-exact-site-of-north-koreas-nuclear-test-plus-nearby-test-facility-and-gulag/.

Gage, R., N. Wilson, L. Signal, M. Barr, C. Mackay, A. Reeder, and G. Thomson. 2018. Using Google Earth to assess shade for sun protection in urban recreation spaces: Methods and results. *Journal of Community Health* 43(6):1061–1068.

Giovenco, D., D. Duncan, E. Coups, M. Jane Lewis, and C. Delnevo. 2016. Census tract correlates of vape shop locations in New Jersey. *Health & Place* 40:123–128.

Glaeser, E., S. Kominers, M. Luca, and N. Naik. 2018. Big data and big cities: The promises and limitations of improved measures of urban life. *Economic Inquiry* 56(1):114–137.

Gold, A., P. Pendergast, C. Ormand, D. Budd, and K. Mueller. 2018. Improving spatial thinking skills among undergraduate geology students through short online training exercises. *International Journal of Science Education* 40(18):2205–2225.

Gong, H., R. Hassink, and G. Maus. 2017. What does Pokémon Go teach us about geography? *Geographica Helvetica* 72(2): 227–230.

Grossner, K. 2012. *Finding the Spatial in Order to Teach It.* Geological Society of America Special Paper 486. Boulder, CO: Geological Society of America.

Haibin W., X. Luan, H. Li, J. Jia, Z. Chen, and L. Han. 2018. Elevation data fitting and precision analysis of Google Earth in road survey. *AIP Conference Proceedings* 1967(1):1–8.

Hird, J., E. DeLancey, G. McDermid, and J. Kariyeva. 2017. Google Earth engine, open-access satellite data, and machine learning in support of large-area probabilistic wetland mapping. *Remote Sensing* 9(12):1–27.

Jones, K., R. Devillers, Y. Bédard, and O. 2014. Visualizing perceived spatial data quality of 3D objects within virtual globes. *International Journal of Digital Earth* 7(10):771–788.

Kamel Boulos, M., M. Scotch, K.-H. Cheung, and D. Burden. 2008. Web GIS in practice VI: A demo playlist of geo-mashups for public health neogeographers. *International Journal of Health Geographics* 7:1–16.

Kang, J., M. Körner, Y. Wang, H. Taubenböck, and X. Zhu. 2018. Building instance classification using street view images. *ISPRS Journal of Photogrammetry & Remote Sensing* 145:44–59.

Kennedy, D. 2017. "Gates": A new archaeological site type in Saudi Arabia. *Arabian Archaeology & Epigraphy* 28(2):153–174.

Kim, M. 2018. Project-based community participatory action research using geographic information technologies. *Journal of Geography in Higher Education* 42(1):61–79.

Laforest, D. 2016. The satellite, the screen, and the city: On Google Earth and the life narrative. *International Journal of Cultural Studies* 19(6):659–672.

Liang, J., J. Gong, and W. Li. 2018. Applications and impacts of Google Earth: A decadal review (2006–2016). *ISPRS Journal of Photogrammetry & Remote Sensing* 146:91–107.

Lupton, Q. 2017. Pokémon in the midst: Collecting and using data from within the Pokémon GO ecosystem to facilitate ecology and wildlife biology education. *American Biology Teacher* 79(7):592–593.

Markert, K., C. Schmidt, R. Griffin, A. Flores, A. Poortinga, D. Saah, and R. Muench. 2018. Historical and operational monitoring of surface sediments in the lower Mekong Basin using landsat and Google Earth engine cloud computing. *Remote Sensing* 10(6):909.

Martínez-Graña, A., J. Goy, and C. Cimarra. 2013. A virtual tour of geological heritage: Valourising geodiversity using Google Earth and QR code. *Computers & Geosciences* 61:83–93.

Massey, R., T. Sankey, K. Yadav, R. Congalton, and J. Tilton. 2018. Integrating cloud-based workflows in continental-scale cropland extent classification. *Remote Sensing of Environment* 219:162–179.

Masthi, N., R. Ramesh, M. Madhusudan, and Y. Puthussery. 2015. Global positioning system & Google Earth in the investigation of an outbreak of cholera in a village of Bengaluru Urban District, Karnataka. *Indian Journal of Medical Research* 142(5):533–537.

Matamyo, S. and Y. Murayama. 2017. Integrating geospatial techniques for urban land use classification in the developing sub-Saharan African city of Lusaka, Zambia. *ISPRS International Journal of Geo-Information* 6(4):1–19.

McLaughlin, K. 2017. Buying a home sight unseen is easier than ever – and more common. *Wall Street Journal* (June 23). www.wsj.com/articles/buying-a-home-sight-unseen-is-easier-than-everand-more-common-1498140215.

Moges, M., C. Mulatu, A. Crosato, M. McClain, and E. Langendoen. 2018. Morphodynamic trends of the Ribb River, Ethiopia, prior to dam construction. *Geosciences* 8(7):255–276.

Nadal, A., R. Alamús, L. Pipia, A. Ruiz, J. Corbera, E. Cuerva, J. Rieradevall, and A. Josa. 2017. Urban planning and agriculture. Methodology for assessing rooftop greenhouse potential of non-residential areas using airborne sensors. *Science of the Total Environment* 601/602:493–507.

National Research Council. 2006. *Learning to Think Spatially*. Washington, DC: National Academies Press.

Parente, L., and L. Ferreira. 2018. Assessing the spatial and occupation dynamics of the Brazilian pasturelands based on the automated classification of MODIS images from 2000 to 2016. *Remote Sensing* 10(4):606–617.

Patterson, T. 2007. Google Earth as a (not just) geography education tool. *Journal of Geography* 106(4):145–152.

Ragettli, S., T. Herberz, and T. Siegfried. 2018. An unsupervised classification algorithm for multi-temporal irrigated area mapping in Central Asia. *Remote Sensing* 10(11):1823–1844.

Rashid, M. 2018. The evolving metropolis after three decades: A study of community, neighbourhood and street form at the urban edge. *Journal of Urban Design* 23(5):624–653.

Saiz, A., A. Salazar, and J. Bernard. 2018. Crowdsourcing architectural beauty: Online photo frequency predicts building aesthetic ratings. *PLoS ONE* 13(7):1–15.

Schultz, R., J. Kerski, and T. Patterson. 2008. The use of virtual globes as a spatial teaching tool with suggestions for metadata standards, *Journal of Geography* 107(1):27–34.

Shrestha, S., I. Miranda, A. Kumar, M. Escobar Pardo, S. Dahal, T. Rashid, C. Remillard, and D. Mishra. 2019. Identifying and forecasting potential biophysical risk areas within a tropical mangrove ecosystem using multi-sensor data. *International Journal of Applied Earth Observation & Geoinformation* 74:281–294.

Solem, M., and P. Gersmehl. 2005. Online global geography modules enhance undergraduate learning. *AAG Newsletter* 40(8):11.

Southworth, M., and P. Owens. 1993. The evolving metropolis. *Journal of the American Planning Association* 59(3):271.

Wager, W. 2005. *Integrating Technology into Instruction*. University of South Carolina Workshop.

Wikicommons. https://commons.wikimedia.org/wiki/File:Panorama_of_the_Quadrant_on_Regent_Street.jpg.

Williams, A., and T. Davinroy. 2015. Teachable moment: Google Earth takes us there. *Change* 47(1):62–65.

Xiang, X., and Y. Liu. 2017. Understanding "change" through spatial thinking using Google Earth in secondary geography. *Journal of Computer Assisted Learning* 33(1):65–78.

16 Augmented Reality

An overview

Mark Billinghurst

Since the early days of computing there has been a desire to remove the separation between the physical and digital domains. In a remarkable research paper published in 1965, Ivan Sutherland (Sutherland 1965) imagined a display in which a computer was able to generate reality and remove this separation entirely. He wrote "The ultimate display would, of course, be a room within which the computer can control the existence of matter. ... Such a display could literally be the Wonderland into which Alice walked." The ultimate display would be one in which the computer had become invisible and the real and digital domains were completely blended together.

While writing this, Sutherland was trying to make his vision of the ultimate display a reality. One of the projects that he was working on was a system that used small cathode ray tubes and transparent optical elements to overlay virtual images on the real world (Sutherland 1968). Completed in 1968, this became the first fully functioning Augmented Reality (AR) head-mounted display (HMD). Unlike other display technology at the time it overlaid digital content directly onto the real world, and so removed the separation between the physical and digital domains.

Over the next 50 years AR technology continued to be developed and the field of AR changed dramatically. This chapter provides an introduction to Augmented Reality and its applications, an overview of the early history, description of relevant technologies, and identifies some important areas for future research. However, first we provide a more rigorous definition of Augmented Reality.

Definition

Although the first AR HMD was developed in the 1960s, the term "Augmented Reality" was not coined until the 1990s, and so AR was not formally defined until well after Ivan Sutherland's work. In 1997 Ron Azuma (1997) wrote an influential survey of Augmented Reality that provided the first definition of AR. He said that AR systems had to have three key properties:

1. They combine real and virtual content
2. The virtual content is interactive in real time
3. The virtual content is registered in three dimensions

This means that the user should be able to see the real and virtual content at the same time, the virtual content should be able to be interacted with in real time, and the virtual content should appear fixed in three dimensional space while the user moves around it. For example, in the AR colouring book shown in Figure 16.1, a person can use their mobile phone to see virtual content popping out of the real book pages on the phone's live camera view. The live camera view combines real and virtual content, the virtual content can respond to the user's touch input and phone motion, and it is fixed to the real book pages. Thus this application satisfies the three requirements of Augmented Reality.

It is important to note that AR systems have to have all three of these properties. For example, many movies use special effects to combine real and virtual images, and the virtual content appears to be part of the real world, but the virtual images typically take many hours to render and are not shown

Figure 16.1 An AR colouring book.
Source: Author.

Figure 16.2 A VR experience, replacing the user's view of the real world.
Source: Author.

in real time, and so this is not an AR experience. Similarly, pointing a mobile phone camera at a QR code and triggering playback of a video clip is not AR because the virtual content is not registered in three dimensions.

There are many other ways to explain AR, but one of the most useful is to consider AR in the context of related technologies, and in particular Virtual Reality (VR) (Sherman and Craig 2018). VR typically uses a head mounted display to separate the user from the real world and completely fill their field of view with computer generated graphics (Figure 16.2). While AR attempts to enhance the user's real-world experience, VR seeks to replace it entirely and immerse the user into an artificial environment.

AR and VR are complementary in other ways as well. For example, VR displays typically have a wide field of view and show graphics that are as realistic as possible. In contrast, AR displays can be useful even if they have a smaller field of view with simple graphics. For example, an outdoor navigation application that simply showed a virtual arrow superimposed over the real world can still provide good directions. In AR the real and virtual content is aligned and so the user's viewpoint tracking should be as accurate as possible, however in VR the user can't see the real world at all, so the viewpoint position and orientation tracking can be less accurate. Table 16.1 shows the complimentary requirements for AR and VR.

Augmented Reality is also related to Mixed Reality (MR). This term was coined in 1994 by Paul Milgram and Fumio Kishino who defined it as "a

Table 16.1 The complimentary nature of AR and VR technologies

	Virtual Reality Replaces Reality	Augmented Reality Enhances Reality
Scene Generation	Requires realistic images	Minimal rendering okay
Display Device	Fully immersive, wide filed of view	Non-immersive, small field of view
Tracking	Low to medium accuracy is okay	The highest accuracy possible

Source: Author.

particular subclass of VR related technologies that involve the merging of real and virtual worlds" (Milgram and Kishino 1994). More specifically, they said that MR involves the blending of real and virtual worlds somewhere along the "Reality-Virtuality continuum" (RV) which connects completely real environments to completely virtual ones. The Reality-Virtuality continuum is a way of organizing computer interfaces in terms of how much of the user's visual perception they generate.

The RV continuum ranges from completely real to completely virtual environments and encompasses AR and Augmented Virtuality (AV). AV is a virtual world with elements of the real world introduced into it, in much the same way that AR is the real world with elements of virtual imagery introduced into it. Mixed Reality covers the portion of the continuum between the completely real environment, and the completely virtual environment. So Mixed Reality includes Augmented Reality experiences, but Virtual Reality does not.

These various ways of defining and viewing Augmented Reality help us to understand the technology and user experience more. In the next section I use Azuma's definition to give a deeper introduction to AR technology. This will help the reader understand the possibilities of AR further.

Technology

Azuma's three properties of Combination, Interaction and Registration are requirements for an Augmented Reality experience, and as such they define the technology required for AR. For example, to achieve combining real and virtual imagery there needs to be a display element which enables both real and virtual content to be seen at the same time. Interaction requires real time graphics and technology for capturing user input. Finally, registration of the virtual imagery with the real world is only possible if there is a tracking system that finds the user's viewpoint relative to the real world, and so draw the virtual graphics from the correct perspective. In this section I discuss each of these types of technology in more detail.

Display technology

The goal of AR display technology is to enable computer generated virtual imagery to be appear to be seamlessly blended with a view of the real world. Sutherland used as a see-through head mounted display to achieve this, but there are other methods for doing this as well. In their display taxonomy Bimber and Raskar (2005) identify three classes of AR displays (Figure 16.3); (1) Head-attached, (2) Hand-held, and (3) Spatial.

Head-attached displays include head mounted displays (HMDs) classified into optical see-through or video see-through. Optical see-through HMDs are those that allow the user to see the real world with their natural eyes, but use a see-through optical element to overlay graphics directly onto the real world. An example of one of these displays is the Microsoft HoloLens (Figure 16.4). Optical see-through HMDs have the advantage that they are generally simpler to produce, and that the user can still see the real world if the system fails, however, it is impossible for the virtual imagery to completely block out the real world, and it can be technically challenging to produce wide field of view optical see-through displays. Also, the view of the real world cannot be delayed, so any latency between the real and virtual imagery is immediately obvious, producing tracking and registration errors.

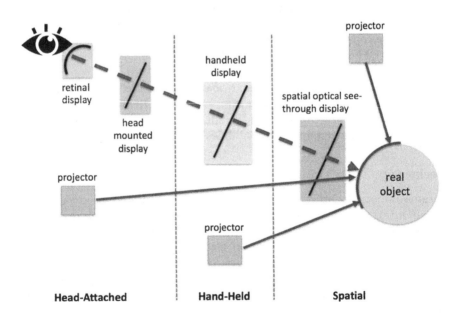

Figure 16.3 Types of AR displays.
Source: Author.

Figure 16.4 HoloLens optical see-through head mounted display.
Source: Author.

Video see-through HMDs are those that overlay virtual graphics onto a video of the real world. They typically have small video cameras mounted on them that provide a stereo video view of the real world, onto which the AR imagery can be shown. They have a number of advantages, such as enabling the virtual imagery to completely block the view of the real world, and making it easy to see a wide field of view by using wide angle cameras. Since the real world is seen through digital imagery this also provides a number of tracking and registration options, such as delaying the camera video to match the delays in the AR computer graphics and creating the illusion that the AR content is perfectly registered with the real world. However, the user only sees a video of the real world, so this may not have the same resolution and image quality as seeing the world with their natural eyes, and if the system fails they are not be able to see anything.

In addition to HMDs, the most popular way to experience AR is through a hand-held display, most commonly a mobile phone or handheld display. These have a camera with high resolution display, so virtual content can be overlaid on the live camera view of the real world. One of the advantages of a handheld display is that the displays typically have a touch screen and so output and input can be combined in one device. The user can view an AR scene on their mobile phone and then use touch input on the phone screen to interact with it.

Figure 16.5 Spatial AR – projecting AR content onto a real car.
Source: Author.

Spatial AR includes systems that use projectors to project imagery onto real object (Figure 16.5), such as projecting onto the sides of buildings or onto objects indoor. Until recently, projectors were large and expensive, but now they have become small enough to also be worn on the body or head. For example, Mistry's Sixth Sense interface uses a projector worn around the neck to project AR content onto nearly surfaces (Mistry and Maes 2009). Spatial AR interfaces allow several people to see the AR content at the same time, and don't require the user to wear anything. However, they typically need a dark environment and often complicated infrastructure.

Interaction technology

In order to interact with the AR content, AR applications need to be able to capture and respond to user input. At the most basic level, the application should respond to changes in the user's viewpoint. For example, if the user is viewing a virtual object on the table then as they move around the table the object should appear fixed in space. This is only possible if the user's viewpoint is tracked in real time, and the virtual content is continuously drawn from the user's viewpoint. Tracking is covered in more detail in the next section.

There are a wide range of methods to interact with AR applications, but the actual method used often depends on the type of AR display. For example, if the application is a mobile AR experience on a smart phone, then touch screen input and maybe device motion would be ideal. In head-mounted AR interfaces a range of different input options can be used, including hand-held stylus, head pointing, touch pad input on the HMD, handheld controller,

Figure 16.6 Using the tangible AR metaphor in an AR visualization application. The tangible user interface is on the left and AR view on the right.

Source: Author.

and speech, among others. Hand tracking technology has improved to the point where free handed gestures can be used for natural input (Billinghurst, Piumsomboon, and Bai 2014), and most recently researchers are exploring how gesture and speech can be combined together to produce multimodal systems (Piumsomboon et al. 2014). The ultimate aim is to make interaction with virtual content in an AR interface be as natural as interacting with objects in the real world.

One of the important aspects of AR interaction is deciding on the interaction metaphor to be used. This determines how the AR application responds to the user's input actions in the real world. One of the useful interaction metaphors for AR is the "Tangible AR" approach. This is where physical input objects are designed using a Tangible User Interface approach (Ishii 2008), and AR techniques are used to display the graphics output. For example, researchers at the HIT Lab NZ developed an AR visualization application where the user held a real ring mouse in their hand and looked at an AR tracking image (Looser, Billinghurst, and Cockburn 2004). Using the Tangible AR approach, in the AR view the ring mouse was transformed into a virtual magnifying glass, and the tracking image had the virtual content to be viewed overlaid on it. When the user looked through the virtual magnifying glasses, they could see different elements of the content data, such as different geospatial datasets on the Earth (Figure 16.6). The user could very naturally view different aspects of the dataset by simply moving the virtual magnifying glass to different positions.

Tracking and registration

The final requirement of Azuma's definition is that the AR content should appear fixed in the real world. In order to achieve this is it necessary to perform precise registration and tracking of the user's viewpoint. For example, in the AR colouring application shown in Figure 16.1, the virtual content should appear fixed to the real page, which requires the camera on the user's tablet to be tracked as it is move around.

Registration refers to adjustments made in the system while the user isn't changing their viewpoint. For example, when the user wears a see-through display and holds their head still then the virtual content they are seeing might need to be aligned with a real object, but inevitably there will be some misalignment due to how the AR display is positioned on the user's head, or image distortion due to the optics of the HMD. This can be overcome through a careful calibration process where the position is measured of the user's eye relative to the display elements.

Once registration is complete, then tracking is used to continually measure the position of the user's viewpoint as they move their head around. Tracking techniques can also be used to measure the user's hand position to support interaction. For example, Ivan Sutherland's original HMD used a mechanical tracker where the user's HMD was directly attached to the ceiling with a mechanical arm. This provided very accurate tracking, but only over a very limited range (Sutherland 1968). A second version used ultrasonic tracking for wireless position and orientation sensing. Other early AR systems used magnetic tracking, optical encoders, and inertial sensing.

For outdoor AR experiences, GPS and compass systems could be used to measure the user's position and orientation. However, GPS requires a satellite fix to multiple satellites and so is unusable indoors, or in some urban canyon outdoor environments between tall buildings. Modern mobile smartphones have GPS and compass sensors integrated into them, enabling many hand-held outdoor AR applications.

As processing power became more powerful, then computer vision has become the most popular tracking approach. There are many computer vision methods that can be applied. One of the simplest is to look for known visual features in the live camera view from the scene, and when found, calculate the camera position from those features. Figure 16.7 shows how the ARToolkit tracking library finds the camera position from square markers seen in the camera view. However, the limitation of this is that the user needs to place the tracking features in the environment in the first place. So this approach is useful for overlaying AR content onto known objects, but not useful for tracking in unknown environments.

Most recently, techniques from robotics have been used to enable AR tracking in places where there are not any known visual features. Known as Simultaneous Localization and Mapping (SLAM) (Thrun and Leonard 2008) this approach allows the system to recover both the camera pose and create a map of the environment while initially knowing neither. The first visual SLAM system using a single camera was developed in 2007 (Davison et al. 2007), and soon ported to mobile phones for AR tracking and extended

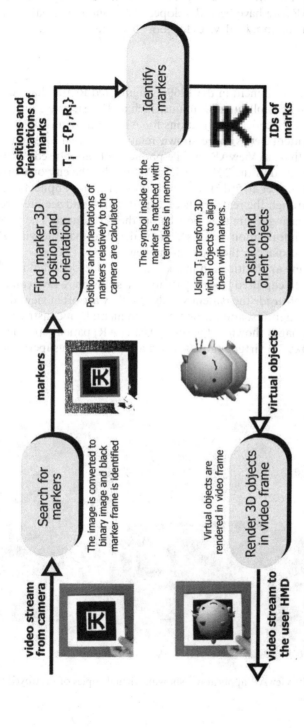

Figure 16.7 Marker-based AR tracking using ARToolKit.
Source: Author.

further (Klein and Murray 2007). Now, visual odometry libraries such as ARKit and ARcore have been developed that combine visual tracking with inertial sensors on mobile devices for more accurate tracking.

Applications

Since its first development in the 1960s, Augmented Reality has been applied in many different application domains, such as Engineering, Entertainment, and Education. The ideal applications for AR are those in which interactive 3D virtual content needs to be shown relative to a real location or object. For example, the CityViewAR mobile phone application used AR to replace buildings destroyed in the 2011 earthquake that hit the city of Christchurch, New Zealand (Lee et al. 2012) (Figure 16.8). Using this application, a person could walk through the real streets of Christchurch and see life-sized virtual buildings appearing where they were before the earthquake. The phone's GPS and compass information was used to make sure that the virtual content appears fixed in space in the live camera view.

AR has many applications in medicine, where doctors often need to interpret spatial data captured from a patient. CT scans or X-rays can be used to create images of inside the patient's body, but they are often viewed on a computer screen or lightbox and the doctor has to mentally reconstruct 3D shapes from the 2D images shown in the scans. Using AR, patient-specific scans can be used to make 3D virtual models which are then shown superimposed back

Figure 16.8 CityViewAR application, showing virtual copies of destroyed buildings.
Source: Author.

Figure 16.9 AR advertising. Virtual content appearing from a real newspaper.
Source: Author.

into the real patient's body (Blum et al. 2012). AR can also be used to show 3D anatomical models for medical education, allowing medical students to walk around larger than life content and clearly understand the structure of complicated organs.

Marketing is another important application domain for Augmented Reality. Advertising is about trying to capture a person's attention, and AR is a very effective way to have a person engage with a company's product. For example, in one of the first AR advertising campaigns, the Wellington Zoo advertised using a mobile AR application that had virtual animals popping out of real newspaper pages (Figure 16.9). AR can also make it easy for customers to understand difficult concepts, such as GEs very successful campaign using virtual windfarms to teach about Smart Grid technology. AR also enables people to try products before buying them, such as applications that enable people to virtually put on make-up or clothing.

AR can also be used in many different types of entertainment applications. The mobile game Pokemon Go uses the phone sensors to overlay virtual characters on the live camera view of the real world, and has become one of the most downloaded mobile games of all time. Companies such as Snap and Facebook provide camera effects applications that apply virtual enhancements to live video of people's faces (Figure 16.10). The MagicLeap AR headset can be used to play the popular game, Angry Birds, in the player's real living room, with virtual play pieces appearing on the real furniture. In all of these

Figure 16.10 AR special effects from Facebook Camera.
Source: Author.

examples, AR enables players to make their everyday world into something magical and very entertaining.

These are just a few of the possible types of applications for AR, made possible by the huge advances in technology over the last 50 years. In the next section we review some of the history of AR and show how it has evolved to the present day.

History

Since Sutherland's early experiments in the 1960s, the development of Augmented Reality has passed through four main phases:

(1) Pre-1980s: Early Experimentation
(2) 1980s – mid-1990s: Basic Research
(3) mid-1990s – early 2000s: Tools/Application Development
(4) early 2000s – present day: Commercialization

In this section we quickly discuss each of these in turn to give a historical perspective on Augmented Reality.

Pre-1980s early experimentation

The first two decades of AR was in spent in early experimentation by a number of different pioneers. In 1968 Ivan Sutherland developed the first complete AR system, but he soon left Harvard University and moved to the University of Utah where his attention turned to computer graphics. In the 1960s and 1970s Tom Furness III in the US Air Force was developing a SuperCockpit where an airplane pilot could see virtual flight information superimposed over the real world (Furness 1986). In 1982 a working version of this was developed as part of the Visually Coupled Airborne Simulator System (VCASS), and companies such as Honeywell and Hughes produced a wide variety of designs. For example, in 1985 the Honeywell Integrated Helmet and Display Sighting System (IHADSS) was deployed on the AH-64 Apache attack helicopter. This allowed the pilot to see overlaid computer graphic symbols on a live video view from an IR camera mounted on the nose of the aircraft. The pilot's head orientation was tracked within the helicopter and used to steer the camera.

Furness conducted much of his research in secret, but by the early 1980s information about the Airforce's AR and VR displays was becoming public. NASA decided that the same technology could be used to develop training experiences, and so created their own head worn VR system from cheap LCD panels from Sony Watchman mobile TV sets. They were assisted in this by VPL research, the first VR company, which was founded in 1985 by Jaron Lanier. VPL created the first complete turnkey VR system with head mounted displays, head and hand tracking, graphics hardware, a programming language and glove input devices. So by the end of the 1980s technology was available to enable universities and industry to begin conducting research in VR and AR. In 1989 Tom Furness left the Air Force to found the Human Interface Technology Laboratory (HIT Lab) at the University of Washington.

1980s–1990s basic research

By the late 1980s AR technology was available outside the military laboratories and a number of academic institutions began to conduct basic research to solve key problems in AR tracking, display, and interaction. The University of Washington was a leader in the field and developed such innovations as the first AR retinal display (Pryor, Furness, and Viirre 1998), collaborative AR experiences (Billinghurst, Weghorst, and Furness 1998) and ARToolKit tracking technology. Other significant AR research was conducted at the University of North Carolina where Fred Brooks and Henry Fuchs led a team of researchers who developed see-through head mounted displays (Rolland, Holloway, and Fuchs 1995), and novel tracking solutions (Billinghurst, Weghorst, and Furness 1998). At Columbia University in the early 1990s, Steve Feiner (Feiner et al. 1997) explored early AR interaction methods and developed the first wearable AR system. At the same time Paul Milgram at the University of Toronto was exploring the human factors of AR, using AR for tele-operation, and developing the Mixed Reality taxonomy (Milgram and Kishino 1994).

Apart from these efforts in academia, a number of companies were also conducting basic research in AR. In the late 1960s the Polhemus company spun out of the US Air Force's SuperCockpit Program and became the leading supplier of magnetic tracking technology for AR and VR systems. In 1989 Reflection Technologies invented the Private Eye, a new type of AR display based on scanning a column of LEDs. This was used by many AR and wearable computer researchers, and also commercialized in the Nintendo Virtual Boy. Boeing was exploring how AR could be used to simplify wire harness bundling, and in a 1992 paper describing the system, Boeing engineer Tom Caudell (Caudell and Mizell 1992) coined the term "Augmented Reality."

1990s–2000s tools/application development

Once AR research had left the military and was becoming available in academia, researchers began exploring the application space and creating tools for developers to use. Some of the first applications where in the medical space, with the team at the University of North Carolina showing how AR could be used to create the illusion that laparoscopic imagery could be viewed back inside the patient (Fuchs et al. 1998). Work was also done on providing virtual cues for image guided surgery (Leventon 1997), and viewing bones inside the patient's body (Navab, Feuerstein, and Bichlmeier 2007).

Around this time researchers also developed the first shared AR experiences, exploring how people in a face to face settings could see virtual content in their midst. Jun Rekimoto's Transvision application allowed people to sit around a table and see virtual content shown in the space between them, viewed on handheld displays (Rekimoto 1996). Billinghurst, Weghorst, and Furness (1998) and Schmalstieg et al. (1996) were the first to explore collaborative AR experiences delivered on head mounted displays. They found that one of the advantages of this approach was that it allowed the virtual content to be seen at the same time as face to face communication cues, seamlessly blending the task space and communication space.

As computing became more portable and wearable, AR became mobile and could be used outdoors. One of the first of these systems was the Touring Machine, developed by Steve Feiner at Columbia University in 1997 (Feiner et al. 1997). This was a large backpack computer combined with GPS tracking and a see-through head mounted display which enabled people to walk around the university campus and see virtual annotations attached to real buildings. Researchers at other universities expanded on this work, such as Bruce Thomas from the University of South Australia who developed the first outdoor AR game, porting the popular game of Quake to a backpack

AR system (Piekarski and Thomas 2002). These wearable systems were typically very bulky due to the custom computer hardware, batteries, tracking sensors and display hardware.

Head-mounted displays are not the only way to have a mobile AR experience, and by the mid-1990s people were exploring AR on handheld devices. The first group of these were tethered to desktop computers. For example, Jun Rekimoto's Navicam allowed people to use a handheld display to show virtual tags on a live camera view of the user's surroundings. Similarly, the ARpad application allowed two people to view and interact with virtual content on handheld displays at the same time Mogilev et al. (2002). As the handhelds got more powerful the AR application could be run entirely on the device, as first shown by Wagner, Langlotz, and Schmalstieg in 2008. Shortly after mobile phones had enough graphics and CPU power to also show AR content, and by 2007 phones had integrated GPS and compass tracking, enabling an outdoor AR experience similar to the Touring Machine of ten years before.

Developing early AR systems was very complicated. For example, in 1992 Ron Azuma was awarded his Ph.D. thesis for the years of work it took to develop a tracking and calibration system that enabled a person to see a virtual axes positioned over a real box. One of the things that led to significant advances in the field was the development of tools that solved some of the problems of tracking and interaction. One of the first of these was a computer vision based matrix tracking software developed by Jun Rekimoto (1998). This enabled people to point a camera at a square tracking marker, track the camera position and place virtual content at a fixed position relative to the marker. Soon after, the ARToolKit library was developed that supported tracking from user-defined square markers (Figure 16.11). ARToolKit was released as an open-source tool enabling tens of thousands of developers everywhere access to a fundamental tool for AR. Now they could start focusing on application development, not solving basic research problems. Using a single ARToolKit marker enabled a developer to track the camera position, while placing markers on several real objects enabled them to be used for interaction with the virtual content.

Important AR technology developments were not just restricted to academic researchers. A number of countries realized the importance of AR and launched national research projects. The first of these was the Mixed Reality Systems Laboratory in Japan from 1997 to 2001. This was a joint venture between Canon and the Japanese Government, which together invested close to $50 million USD on research into various Mixed Reality topics. One of the outcomes of this was Canon's Mixed Reality platform, and a range of innovative applications in entertainment, broadcasting and industry. Around the same time (1999–2003) the German government invested 23 million Euros into the ARVIKA project. This project involved 23 partners from universities and the automotive and airplane industry, with a focus on how AR could be used to improve manufacturing. For example, a person could wear an AR

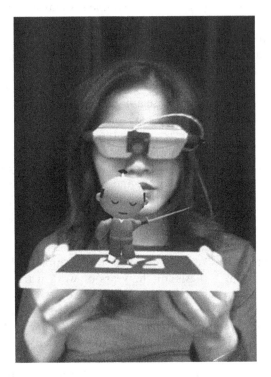

Figure 16.11 ARToolKit tracking example.
Source: Author.

HMD to see virtual information superimposed over real machine parts to help with the assembly process. Weidenhausen, Knoepfle, and Stricker (2003) provide an excellent summary of the lessons learned from ARVIKA and the key research outputs.

By the early 2000s researchers from academia and industry had begun to explore a wide variety of AR applications, from medical, to teleconferencing, gaming and industrial assembly. Perhaps more importantly, the tools were beginning to emerge to make it easier to people outside of research to create their own applications. This led to the widespread commercialization of Augmented Reality, which we review in the next section.

Commercialization

Although AR was developed rapidly in academic and industrial research laboratories, it was not until 1998 when the first dedicated AR company was formed. This was Total Immersion, a company using AR for marketing and location-based experiences. Shortly after, in 2001, ARToolworks was founded

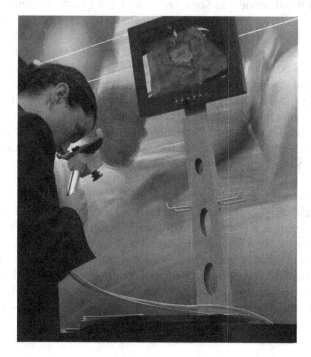

Figure 16.12 AR Kiosk for the America's Cup.
Source: Author.

to commercialize the ARToolKit software and provide AR consulting services. Then, from the ARVIKA project, Metaio was founded in 2003, focusing on industrial AR. In the years following many other companies started, mostly focused on niche application areas for AR.

At the same time, AR was becoming more commonplace in the consumer space. A number of museums and theme parks started having AR based experiences. For example, during the America's Cup competition in 2005 in New Zealand, over 250,000 visitors saw an AR-enhanced book explaining the history of the America's Cup (Figure 16.12) (Woods et al. 2004). At the same time people were seeing live AR enhanced images on their televisions. During live sports broadcasts, virtual lines were shown on the field during American Football games where the first down line was, country flags were shown in the swimming lanes during the Olympics swimming competition, and distance and heading information shown during the America's Cup races. Similarly, virtual set technology was used to bring live graphics into the newsfeeds. For example, in the 2008 US election campaign, CNN used AR technology to bring a video avatar of a remote journalist into their live coverage of Barak Obama's victory celebrations. The TV viewing public grew used to seeing AR enhanced broadcasts on a regular basis.

Another important milestone was Sony's release of the Eye of Judgement Playstation game in 2007. They released a camera accessory for the Playstation 3, and the Eye of Judgement used this to track physical playing cards and overlay game characters on them. The AR tracking was based on Jun Rekimoto's Cybercode matrix tracking library developed almost a decade earlier (Rekimoto and Ayatsuka 2000). Two players could play together in the real world, while watching the characters interact with each other in an AR view on their TV screens. The game was reasonably popular, selling over 330,000 copies and becoming the most widely used AR application at the time. This showed that there was a market for AR in the consumer console space, and Sony's product was followed later by X-box, Nintendo 3DS, and PS Vita AR games.

In March 2007, the MIT Technology Review placed AR on its list of the ten most exciting technologies, and at the end of the year *The Economist* did a feature on AR, saying that it was like "Reality, only Better." However it was 2009 that was a key year for AR commercialization. Figure 16.13 shows the Google Trends graph of relative number of Google searches that contain the term "Augmented Reality" compared to "Virtual Reality." Since 2004 there was a steady decrease in searches on VR, while activity in AR remained low. Then interest in AR quickly rose sharply in 2009, passing VR searches by June and being twice as popular by the end of the year.

There were three factors that contributed to this situation:

(1) mobile phone-based AR
(2) Adobe Flash support for AR
(3) high profile AR advertising campaigns

The first iPhone was launched in 2007 and Android phone in 2008. Both of these smart phones were powerful enough to run AR applications on the device,

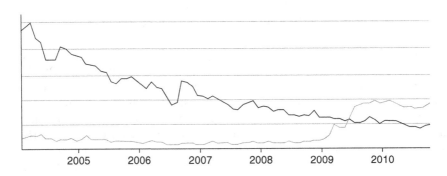

Figure 16.13 Relative Google searches for "VR" (in dark line) and "AR" (in light line). Source: Author.

Figure 16.14 Wikitude AR browser.
Source: Author.

using highly optimized computer vision and simple graphics. The ARToolKit library was ported to iOS in 2007 and by 2008 Daniel Wagner and others from the Technical University of Graz developed markerless natural feature tracking software for mobile phones (Wagner, Langlotz, and Schmalstieg 2008). The first generation of Android phones had GPS and compass sensors in them which meant they could provide outdoor AR experiences without the need for bulky backpack computing. Wikitude used this feature to develop the first outdoor AR browser in 2008, which allows users to see virtual tags in the real world around them (Figure 16.14). This became a popular type of application with Metaio, Layer, Sekai Camera, and others also releasing AR browser products. So by 2009, anyone with an iPhone or Android device in their pocket could have an AR experience.

In 2008 Adobe added camera support to its popular Flash plug-in for the web, meaning that people could view live camera input in their browser. Soon after ARToolKit was ported to Flash in the FLARToolKit library and for the first time people could have an AR experience on the web. Flash was far easier to develop programs for than was C/C++, and so the number of developers able to create AR experiences significantly increased. Flash was also installed on over 1.6 billion computers and so the potential user base was huge. This led to a large number of AR web experiences, such as the Jack Link's Beef Jerky website which allowed people to pose with a virtual sasquatch in the real world. By the end of 2009 anyone with a web browser could have an AR experience on their computer.

The final factor driving awareness of Augmented Reality was a number of high-profile AR advertising campaigns. Advertising agencies are always looking for the next new thing to grab consumer attention and mobile and web-based AR platforms were perfect for this. The first mobile AR advertising campaign was developed in 2007 by Saatchi and Saatchi and the HIT Lab NZ. In December 2009 *Esquire* magazine was the first major print magazine with AR enhanced pages, including a cover showing a virtual video of Robert Downey Jr. Since that time there have been thousands of campaigns run on mobile, print, and the web that used AR, enabling hundreds of millions of people to experience AR for the first time.

Taken together, 2009 was a real milestone in the public exposure to AR. In the ten years since then commercialization and technology has developed at an even faster pace. Following on from Total Immersion, ARToolWorks, and Metaio, hundreds of start-ups have appeared at all parts of the AR eco-system, attracting considerable venture investment. Most notably MagicLeap, founded in 2010, has raised over $2.3 billion USD in venture investment to date. More importantly all of the top five technology companies (Amazon, Apple, Facebook, Google, and Microsoft) have significant internal projects or products in the AR space. These companies are attracted to the AR space by the rapidly growing market size, projected to grow to between $85–120 billion USD by 2025.

On the technical front, notable milestones include the release of the Vuforia markerless tracking library in 2011, which quickly become the dominant platform for computer vision based mobile AR. Most recently, in 2017 Apple with ARkit and Google with ARcore both released visual-inertial odometry software that enables mobile phone users to track their phones indoors and outdoors without any visual markers. On the display side, in 2013 Google released Google Glass, the first truly wearable smart glasses that could be worn all day long, and which were also capable of simple AR experiences. The Microsoft HoloLens came to market in 2016, the most advanced self-contained see-through AR display which has sophisticated visual tracking, gesture, and voice interaction and thin holographic display elements.

However there is still consider research and development needed before Augmented Reality reaches its full potential. In the next section we review research that might change the AR experience even more in the future.

Research directions

Despite over 50 years of development, Ivan Sutherland's vision of the Ultimate Display still hasn't been achieved. There are many areas of research in AR that still need to be explored. Klein and Murray (2007) reviewed ten years of ISMAR conference papers, and this was a continuation of an earlier work that provided an overview of the first ten years (Zhou, Duh, and Billinghurst 2008). These two papers summarized the research trends of the past, and also identified the important directions going forward.

In particular, they identify six key areas for research: Tracking, Interaction techniques and user interfaces, Displays, Applications, Evaluation, Rendering and Visualization. Within each of these areas there are a number of research topics that could be addressed. For example, in Tracking, there has been relatively little research on how to provide pervasive/ubiquitous AR tracking services, such as wide area tracking from combining multiple sensors (Pustka and Klinker 2008) and information about the physical environment (Chatzopoulos et al. 2017). There is also an opportunity to explore more advanced computer vision techniques, such as basing tracking on a semantic understanding of the user's surroundings (Grundhöfer et al. 2007). This extended tracking beyond traditional camera and object tracking, and into more context awareness and space understanding. Finally, the use of Deep Learning to improve AR tracking is another area that could be very impactful, which is just beginning to be explored (Garon and Lalonde 2017).

Display technology is another area with many possible research directions. For example, with optical see-through displays there is a need for better field of view and higher resolution. Another interesting area is multiple focal plane displays. Most existing head mounted displays have a single focal plane, which makes it difficult to see the AR content always in focus which the real world background. However some research has begun on multi-focal plane displays, and adaptive displays (Liu, Cheng, and Hua 2008). One particularly interesting area is Light Field Displays, and using them to deliver photorealistic content to the user (Schmalstieg et al. 1996), and being able to capture Light Field content.

On the interaction side, one emerging area of research is for Multimodal interfaces. Low cost gesture hardware is beginning to appear alongside speech input software, but until now most AR devices just use one input modality. There is a research opportunity to explore how gesture and speech could be combined to provide intuitive input. Early research in this area has found that AR multimodal interfaces perform better than those using a single modality. Improvements in gaze tracking technology have also meant that gaze could be applied as another input modality in the multimodal interface. It could be very natural to be able to gaze at an AR object and the issue speech commands such as "delete that" to delete an object.

Conclusions

This chapter provided an overview of Augmented Reality, beginning with Sutherland's vision of the ultimate computer and tracking the growth to a multi-billion dollar industry 50 years later. The overarching goal of AR has always been to remove the boundary between the digital and physical domains and we have also provided an overview of some of the key technologies to do that. Developments in display, interaction and tracking technologies have made AR experiences more realistic than ever before. However, there is still potential for continuing developments. The research section provides an idea of current areas of active work and directions for future opportunities. It is clear from this overview that there is still more work to be done to achieve Sutherland's vision, but overall Augmented Reality is an incredibly exciting field advancing at a rapid pace in the right direction.

References

Azuma, R. 1997. A survey of augmented reality. *Presence: Teleoperators & Virtual Environments 6*(4):355–385.

Bajura, M., and Neumann, U. 1995. Dynamic registration correction in video-based augmented reality systems. *IEEE Computer Graphics and Applications 15*(5):52–60.

Billinghurst, M., Piumsomboon, T., and Bai, H. (2014). Hands in space: Gesture interaction with augmented-reality interfaces. *IEEE Computer Graphics and Applications 34*(1):77–80.

Billinghurst, M., Weghorst, S., and Furness, T. 1998. Shared space: An augmented reality approach for computer supported collaborative work. *Virtual Reality 3*(1):25–36.

Bimber, O., and Raskar, R. 2005. *Spatial Augmented Reality: Merging Real and Virtual Worlds*. AK Peters/CRC Press.

Blum, T., Kleeberger, V., Bichlmeier, C., and Navab, N. 2012. Miracle: Augmented reality in-situ visualization of human anatomy using a magic mirror. In *2012 IEEE Virtual Reality Workshops (VRW)* pp. 169–170. IEEE.

Caudell, T., and Mizell, D. 1992. Augmented reality: An application of heads-up display technology to manual manufacturing processes. In *Proceedings of the Twenty-fifth Hawaii International Conference on System Sciences* vol. 2, pp. 659–669. IEEE.

Chatzopoulos, D., Bermejo, C., Huang, Z., and Hui, P. 2017. Mobile augmented reality survey: From where we are to where we go. *IEEE Access 5:*6917–6950.

Davison, A., Reid, I., Molton, N., and Stasse, O. 2007. MonoSLAM: Real-time single camera SLAM. *IEEE Transactions on Pattern Analysis & Machine Intelligence* (6):1052–1067.

Feiner, S., MacIntyre, B., Höllerer, T., and Webster, A. 1997. A touring machine: Prototyping 3D mobile augmented reality systems for exploring the urban environment. *Personal Technologies 1*(4):208–217.

Fuchs, H., Livingston, M., Raskar, R., Keller, K., Crawford, J., Rademacher, P., and Meyer, A. 1998. Augmented reality visualization for laparoscopic surgery. In W. Wells, A. Colchester, and S. Delp (eds.) *International Conference on Medical Image Computing and Computer-Assisted Intervention.* pp. 934–943. Heidelberg: Springer.

Furness III, T. 1986. The super cockpit and its human factors challenges. In *Proceedings of the Human Factors Society Annual Meeting 30*(1):48–52. Thousand Oaks, CA: Sage.

Garon, M., and Lalonde, J. 2017. Deep 6-DOF tracking. *IEEE Transactions on Visualization and Computer Graphics 23*(11):2410–2418.

Grundhöfer, A., Seeger, M., Hantsch, F., and Bimber, O. 2007. Dynamic adaptation of projected imperceptible codes. In *Proceedings of the 2007 6th IEEE and ACM International Symposium on Mixed and Augmented Reality.* pp. 1–10. IEEE Computer Society.

Ishii, H. 2008. The tangible user interface and its evolution. *Communications of the ACM 51*(6):32.

Kim, K., Billinghurst, M., Bruder, G., Duh, H., and Welch, G. 2018. Revisiting trends in augmented reality research: A review of the 2nd decade of ISMAR (2008–2017). *IEEE Transactions on Visualization and Computer Graphics 24*(11):2947–2962.

Klein, G., and Murray, D. 2007. Parallel tracking and mapping for small AR workspaces. In *Proceedings of the 2007 6th IEEE and ACM International Symposium on Mixed and Augmented Reality*. pp. 1–10. IEEE Computer Society.

Lee, G., Dünser, A., Kim, S., and Billinghurst, M. 2012. CityViewAR: A mobile outdoor AR application for city visualization. In *2012 IEEE International Symposium on Mixed and Augmented Reality-Arts, Media, and Humanities (ISMAR-AMH)*. pp. 57–64. IEEE.

Leventon, M. 1997. *A Registration, Tracking, and Visualization System for Image-guided Surgery*. Ph.D. dissertation, Massachusetts Institute of Technology.

Liu, S., Cheng, D., and Hua, H. 2008. An optical see-through head mounted display with addressable focal planes. In *2008 7th IEEE/ACM International Symposium on Mixed and Augmented Reality*. pp. 33–42. IEEE.

Looser, J., Billinghurst, M., and Cockburn, A. 2004. Through the looking glass: the use of lenses as an interface tool for augmented reality interfaces. In *Proceedings of the 2nd International Conference on Computer Graphics and Interactive Techniques in Australasia and South East Asia*. pp. 204–211. ACM.

Milgram, P., and Kishino, F. 1994. A taxonomy of mixed reality visual displays. *IEICE TRANSACTIONS on Information and Systems 77*(12):1321–1329.

Mistry, P., and Maes, P. 2009. SixthSense: A wearable gestural interface. In *ACM SIGGRAPH ASIA 2009 Sketches*. p. 11. ACM.

Mogilev, D., Kiyokawa, K., Billinghurst, M., and Pair, J. 2002. AR pad: An interface for face-to-face AR collaboration. In *Conference on Human Factors in Computing Systems: CHI'02 Extended Abstracts on Human Factors in Computing Systems* 20(25):654–655.

Navab, N., Feuerstein, M., and Bichlmeier, C. 2007. Laparoscopic virtual mirror new interaction paradigm for monitor based augmented reality. In *2007 IEEE Virtual Reality Conference*. pp. 43–50. IEEE.

Piekarski, W., and Thomas, B. 2002. ARQuake: the outdoor augmented reality gaming system. *Communications of the ACM 45*(1):36–38.

Piumsomboon, T., Altimira, D., Kim, H., Clark, A., Lee, G., and Billinghurst, M. 2014. Grasp-Shell vs gesture-speech: A comparison of direct and indirect natural interaction techniques in augmented reality. In *2014 IEEE International Symposium on Mixed and Augmented Reality (ISMAR)* pp. 73–82. IEEE.

Pryor, H., Furness III, T., and Viirre III, E. 1998. The virtual retinal display: A new display technology using scanned laser light. In *Proceedings of the Human Factors and Ergonomics Society Annual Meeting* 42(22):1570–1574. Thousand Oaks, CA: Sage.

Pustka, D., and Klinker, G. 2008. Dynamic gyroscope fusion in ubiquitous tracking environments. In *Mixed and Augmented Reality, 2008. ISMAR 2008. 7th IEEE/ACM International Symposium*. pp. 13–20. IEEE.

Rekimoto, J. 1996. Transvision: A hand-held augmented reality system for collaborative design. In *Proceeding of Virtual Systems and Multimedia* 96:18–20.

Rekimoto, J. 1998. Matrix: A realtime object identification and registration method for augmented reality. In *Proceedings. 3rd Asia Pacific Computer Human Interaction*. pp. 63–68. IEEE.

Rekimoto, J., and Ayatsuka, Y. 2000. CyberCode: Designing augmented reality environments with visual tags. In *Proceedings of DARE 2000 on Designing Augmented Reality Environments*. pp. 1–10. ACM.

Rolland, J. P., Holloway, R. L., and Fuchs, H. 1995. Comparison of optical and video see-through, head-mounted displays. In *Telemanipulator and Telepresence Technologies* 2351:293–308. International Society for Optics and Photonics.

Sandor, C., Fuchs, M., Cassinelli, A., Li, H., Newcombe, R., Yamamoto, G., and Feiner, S. 2015. Breaking the barriers to true augmented reality. *arXiv preprint arXiv:1512.05471*. Cornell University. https://arxiv.org/abs/1512.05471.

Schmalstieg, D., Fuhrmann, A., Szalavari, Z., and Gervautz, M. 1996. Studierstube – An environment for collaboration in augmented reality. In *CVE'96 Workshop Proceedings* vol. 19.

Sherman, W., and Craig, A. 2018. *Understanding Virtual Reality: Interface, Application, and Design*. Morgan Kaufmann.

Sutherland, I. 1965. The ultimate display. *Proceedings of IFIPS Congress* 2:506–508. New York.

Sutherland, I. 1968. A head-mounted three dimensional display. In *Proceedings of the December 9–11, 1968, Fall Joint Computer Conference, Part I*. pp. 757–764. ACM.

Thrun, S., and Leonard, J. 2008. Simultaneous localization and mapping. Springer *Handbook of Robotics*. pp. 871–889. Dordrecht: Springer.

Wagner, D., and Schmalstieg, D. 2003. *First Steps towards Handheld Augmented Reality*. p. 127. IEEE.

Wagner, D., Langlotz, T., and Schmalstieg, D. 2008. Robust and unobtrusive marker tracking on mobile phones. In *2008 7th IEEE/ACM International Symposium on Mixed and Augmented Reality*. pp. 121–124. IEEE.

Weidenhausen, J., Knoepfle, C., and Stricker, D. 2003. Lessons learned on the way to industrial augmented reality applications, a retrospective on ARVIKA. *Computers & Graphics* 27(6):887–891.

Woods, E., Billinghurst, M., Looser, J., Aldridge, G., Brown, D., Garrie, B., and Nelles, C. 2004. Augmenting the science centre and museum experience. In *Proceedings of the 2nd International Conference on Computer Graphics and Interactive Techniques in Australasia and South East Asia*. pp. 230–236. ACM.

Zhou, F., Duh, H., and Billinghurst, M. 2008. Trends in augmented reality tracking, interaction and display: A review of ten years of ISMAR. In *Proceedings of the 7th IEEE/ACM International Symposium on Mixed and Augmented Reality*. pp. 193–202. IEEE Computer Society.

17 Twitter

Matthew Haffner

Twitter is a unique social media platform. As of 2018, it has around 335 million active users (Twitter 2018a), but volume alone is not what makes it noteworthy. Facebook and Instagram have more users, but Twitter's relative openness and accessibility have made it of special interest, particularly within the academy. Twitter's public application programming interfaces (APIs) make it easy to extract data, and users' profiles are less frequently "locked," or made private, than other platforms (Haffner et al. 2018). These factors, combined with the mobile nature of users and their ability to tag their location, have made Twitter the target of much geographic research. A recent meta-analysis of articles that utilize location information in social media posts reveals that Twitter dominates the academic landscape – over 54% of studies make use of the platform compared to the second most used platform, Flickr, at only 20% (Stock 2018).

At the same time, Twitter is utilized in geography more than any other domain (Williams, Terras, and Warwick 2013). Applications are wide ranging and have included the study of content variation over space, linguistic variability, natural disaster reporting, and socio-spatial inequality. Yet, appropriate and ethical use of geolocated Twitter data remains a challenge. Many questions surround its applicability, with concerns over demographic and spatial biases, representativeness, and issues of privacy. In this chapter, I first discuss the history of location tagging on Twitter and geographic web contributions. Then, I discuss how geolocated Twitter data has been conceptualized, ways in which has been used, and challenges of using Twitter as a data source.

History of location tagging

In August 2009, Twitter developed a feature allowing users to attach their location to a post, often termed "geotagging" (Twitter 2009). The feature augmented data streamed to Twitter APIs and the way tweets appeared on users' profiles, allowing developers to filter tweets by location. Users for the first time could view content relevant to a particular place. While this was novel at the time, initial location tagging from the user's perspective was cumbersome. They could only enable location by navigating through multiple tiers of menus within their profile settings. Users were restricted to geotagging through their device's global positioning system (GPS) system, broadcasting their precise location as a latitude–longitude coordinate pair. Location tagged posts remained enabled until manually disabled, making it difficult to geotag on a tweet-by-tweet basis. In turn, this configuration also made users vulnerable to forgetting that the feature was enabled, causing them to unintentionally broadcast their location with every post.

The problematic nature of Twitter's location sharing feature was notably exposed when Adam Savage, host of the popular TV show *Mythbusters*, posted a geotagged picture of his car on Twitter. The post's caption contained "Now it's off to work...," thus revealing his precise home location to thousands of followers at the exact moment he was leaving (Murphy 2010). While ethical and privacy concerns still abound, Twitter has significantly modified its location-sharing options, among other features, since then. Notable changes came in 2015 when Twitter announced they would be partnering with the location tagging platform Foursquare, allowing users to tag a more ambiguous "general" location such as a city, state, or neighborhood (Twitter 2018b). Alternatively, users can now tag a more specific location like a restaurant, business, or park without using precise latitude–longitude coordinates. Further, tweets now have a location icon on the prompt of every tweet, making it possible to geotag individual tweets and thus render it unnecessary to alter profile settings first.

Crowdsourced geographic information

Frequent changes to its location sharing options have made Twitter a difficult platform to conceptualize. Regardless, at a basic level it is part of a larger trend in web contributions that have dismantled the traditional hierarchical structure of the Internet. Whereas "Web 1.0" was characterized by Netscape, web encyclopedias, and top-down production, "Web 2.0" is characterized by Google, Wikipedia, and user-generated content, which now dominate (O'Reilly 2005). Web 2.0 paved the way for geographic contributions on the Internet, which Goodchild (2007) termed volunteered geographic information (VGI) in his seminal paper "Citizens as sensors: The world of volunteered geography." This phenomenon sparked a wave of research in the contributions of geographic information by non-experts, including platforms such as OpenStreetMap (OSM), Wikimapia, eBird, and Twitter.

The challenge of classification: VGI or something else?

A number of different labels have been put forth to classify Twitter, among other social media platforms, within the geospatial web. Terms include location-based social network (LBSN) (Evans 2015a; Evans 2015b), georeferenced social media (Shelton et al. 2014), geosocial media (Shelton,

Poorthius, and Zook 2015); geosocial services (Zickuhr 2013), and location-based social media (LBSM) (Wilken 2014; Evans 2015a; Schwartz and Halegoua 2015; Haffner et al. 2018; Yuan, Wei, and Lu 2018; Zhao, Sui, and Li 2017). Though this last term is seemingly dominant, situating Twitter's role within the broader network of crowdsourced geographic content is difficult. Early studies using geolocated Twitter data considered it a type of VGI, but recently scholars have questioned the amount of "volunteerism" in geotagged tweets.

Harvey (2013) particularly makes a careful distinction between VGI and what he calls "contributed" geographic information (CGI), that is, locational information which is not necessarily provided willingly. Harvey (2013) also differentiates between opt-out services – those that enable location features by default – and opt-in services, which disable location features by default. While Twitter does not enable location tagging automatically, it still does not cleanly fit into the opt-in category. Once a user geotags a post, their tweets remain location-enabled until the location icon is deselected. This typically involves only the click of a button, but users can still unintentionally broadcast their location, as mentioned in the earlier example of Adam Savage.

Beyond this, Twitter's internal use of location complicates categorization. The company uses location-based advertisements based on a user's web IP address and GPS signal, regardless of whether or not a user geotags their posts (Twitter 2018c). Additionally, advertisements can be gender-, age-, and language- targeted. Though users can disable targeted advertisements, they are *enabled* by default and therefore should be considered opt-out. Even if other Twitter *users* cannot view an individual's location from a tweet, the company and those with whom they wish to share data still have access to locational information of the users. Twitter is forthright about this practice – even positive – as they emphasize that targeted advertisements enhance the user's experience (Twitter 2018d). From Twitter's perspective, effective advertising is desirable as it engenders lucrative financial partnerships with other organizations. Such subtleties blur the lines between opt-in and opt-out services, further complicating classification of geolocated Twitter data.

Another salient challenge lies in conceptualizing the nature of spatial information in tweets since an abundance of locational information is available – far more than a singular coordinate pair determined by the user's device. The JavaScript Object notation (JSON) object created when a tweet is sent contains numerous fields, with many directly or indirectly referring to location. Two user-defined (i.e., manually selected) locational references are attached to the user's account: (a) a location field which provides a link to content from other users claiming that place and (b) a time zone, selected when user creates an account and only exposed through APIs. Additionally, each user is assigned a profile language based on the language they used to access Twitter with during account creation. Though not explicitly location-based, language is indicative of a user's characteristics and cultural preferences.

Implicit vs. explicit spatial information

On an individual level, tweets can reference location in numerous ways with varying degrees of spatiality, from explicit to implicit (Elwood, Goodchild, and Sui 2012; Graham and Shelton 2013). The contribution of a building footprint on OSM, for instance, is clearly an explicitly geographic contribution, but Twitter contributions are more fluid. The underlying spatial information produced by a user who formally geotags a tweet is explicit, but other references to location, such as mentioning a city or restaurant or attaching a photograph, tend towards implicit spatial information. These more benign references to location, which are common in social media, are termed "ambient geospatial information" (Stefanidis, Crooks, and Radzikowski 2013). Indeed, Humphreys (2012) notes that roughly 20% of tweets reference an individual's location in some way, though the number of tweets formally geotagged is very small (Leetaru et al. 2013).

How Twitter is used

Geolocated Twitter data has been used in a wide array of applications, and motivations for its use have been multifaceted. While early applications were simple, analyses have become more complex, coinciding with Crampton et al.'s (2013) call to move "beyond the geotag" – a petition to study connections, networks, and relationships; supplement Twitter with other data sources; and incorporate time into analyses, rather than mapping a static snapshot of points.

Studying differential content production over space is one principal way geolocated Twitter data has been used (e.g., Zook and Poorthuis 2014; Shelton, Poorthuis, and Zook 2015; Lansley and Longley 2016; Haffner 2018a), whereby social media discourse often complicates traditional or assumed boundaries. The relationship between country and language in Europe, for instance, is blurred by overlapping and disjunct language clusters of geolocated tweets throughout the continent (Figure 17.1). Similarly, the commonly held notion of the "9th Street Divide" in Louisville, Kentucky is rendered more fluid by noteworthy references to the term "ghetto" on both sides of the dividing line (Shelton, Poorthuis, and Zook 2015). The use of "#AllLivesMatter" as a counter-protest narrative on Twitter is challenged by the unexpected correlation coefficients of municipalities' racial demographics, in which percent Black has a positive relationship with the number of tweets containing this phrase (Haffner 2018a).

Social media data are continually streaming, as opposed to conventional data sources like those from the U.S. Census, which are collected only periodically and expensive when done so. Since users are mobile and can tag locations from a location-enabled device, a logical application is the study of day-to-day mobility patterns and urban dynamics, as a potential

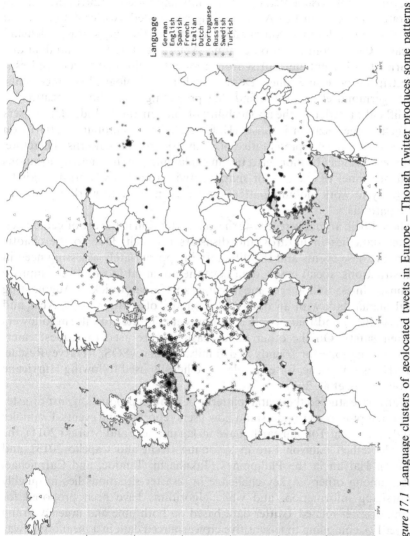

Figure 17.1 Language clusters of geolocated tweets in Europe – Though Twitter produces some patterns corresponding to each language's country of origin, significant blending of languages is present, especially near country borders (cartographic work by L. J. Bergevin).

replacement to sources like the longitudinal origin-destination employment statistics (LODES) or other datasets collected using travel diaries. Observing the locations of geolocated tweets over different daily time periods (e.g., day vs. night), for instance, produces noticeably distinct spatial patterns (Figure 17.2).

Beyond simple visual methods, Lee, Gao, and Goulias (2016) predict trip generation patterns in Los Angeles using geolocated Twitter data collected over two days, finding comparable results to the city's travel demand forecasts. Over a longer period of time, Abbasi et al. (2015) build data dictionaries to differentiate tourist from resident geolocated tweets and estimate trip purpose in Sydney. Despite such methodological advancements, van Eggermond et al. (2015) find that predicting users' home locations is still difficult, restricting the applicability of such methods. Indeed, few users tag their home location in posts (Haffner 2018b), so mobility patterns on Twitter may not be representative of daily commuting patterns. Therefore, such methods may be more effective in explaining popular activity locations at various times of the day, or simply activity-to-activity, work-to-activity, or activity-to-work mobility patterns rather than home-to-work or work-to-home patterns.

The subfield arguably receiving the most attention in the use of geolocated Twitter data lies in hazards and disasters research. Due to the chaotic nature of these events and the difficulty in appropriately assessing need in such situations, social media data have the potential to save lives, improve response times, and target most vulnerable areas quickly. Augmenting social media data with an open-ended survey of Twitter users, Acar and Muraki (2011) find that people commonly use Twitter to inform followers of their safety. On the other hand, hashtags are used to request emergency evacuations for friends and family. #HarveySOS, #HarveyRescue, and #HoustonRescue, for example, were heavily used following Hurricane Harvey (Yang et al. 2017).

Many have studied the spatial patterns of content production *post*-disaster in an attempt to better understand how users report during events. Examples include the Great Tohoku earthquake in Japan (Acar and Muraki 2011), the 2012 Horsethief Canyon Fire in Wyoming (Kent and Capello 2013), and Typhoon Haiyan in the Philippines (Takahashi, Tandoc, and Carmichael 2015), among others. A key challenge in disaster situations lies in quickly synthesizing information, and while algorithms have been proposed for sorting disaster-related Twitter data based on both time and urgency (Yang et al. 2017), compiling time-sensitive crowdsourced data in a meaningful way remains an arduous task. Rather than used on their own, a more effective approach may be combining social media data with conventional data (de Albuquerque et al. 2015) or other crowdsourced datasets to obtain a more complete picture.

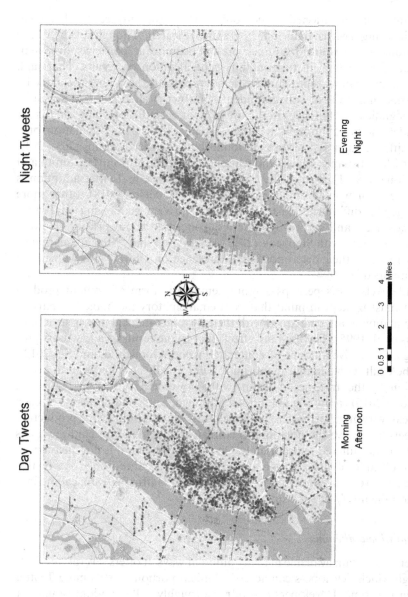

Figure 17.2 Differences in New York City tweets by time of day over a five-day period. Bifurcating data by time of day produces subtle but apparent differences in content production patterns which could be useful in studying urban mobility (cartographic work by Kristina Kay Emery).

Challenges of using Twitter data

Demographic and spatial bias

The limitations mentioned earlier only scratch the surface of challenges faced in using Twitter data for geographic research. Though all data have limitations, crowdsourced sources present significant concerns over representativeness, which in turn inhibits generalization of results. That said, Twitter appears more representative demographically than other social media platforms. In a study of college students, Haffner et al. (2018) find no statistically significant differences in the use of Twitter and geotagging on Twitter by gender and race. Similarly, in a larger study of the general population, only small differences in Twitter use are found between racial groups: 20%, 26%, and 20% for Hispanic, Black, and white respondents, respectively (Smith and Anderson 2018). Further, while only a 1% difference in Twitter adoption is found between women and men (Smith and Anderson 2018), other more subtle gender differences exist. Women tend to view geotagging on Twitter more positively and surprisingly are less concerned about privacy (Haffner 2018). This aligns with Stephens' (2013) observation that "Men are mapping" (in reference to their disproportionate presence on OSM) "and women are being mapped" (p. 994).

While Web 2.0 is perhaps a more democratic form of content production, it must be kept in mind that even emancipatory methods, like participatory mapping, are embedded with their own hierarchical power structures (Harris et al. 1995). The long tail effect, that is, the result of a few users producing an excessive amount of content (Elwood, Goodchild, and Sui 2012), must be dealt with, though there is not a universally accepted way to do so. Additionally, the strategies of dealing with "power users" (Shelton, Poorthuis, and Zook 2015) cannot be ubiquitously applied, as the requirements of each study can vary greatly. Twitter "bot" accounts, which automate the production of tweets, also have the potential to conflate datasets and obscure processes if they are not filtered appropriately. On the other hand, it is not uncommon for a multitude of users within a study to have produced only one or two geolocated tweets over long periods of time (e.g., Haffner 2018b), calling into question the utility of such sparse data.

Computational dilemmas

Twitter makes data available in many different ways. It offers several free APIs through which developers can access a limited portion of streaming Twitter data in real time. Developers can retrieve roughly a 1% random sample of all tweets, or they can filter tweets by term or location. By using four coordinate pairs as a bounding box around an area of interest, all geolocated tweets in the area can be gathered as long as relatively generous rate limits are not violated. Though only a limited sample of past tweets are available through the standard (i.e., free) API, premium and enterprise APIs, offer access to all past tweets plus fewer (or no) restrictions on the amount of streaming data that can be retrieved.

Yet even if data can be accessed, it cannot always be processed, queried, or analysed effectively. The US Library of Congress collected every tweet from Twitter's inception in 2006 until 2017, with an estimated total of 170 billion unique messages (Library of Congress 2017). In late 2017, however, the Library of Congress announced that they would no longer continue to harvest these data. Increasingly large data volumes, along with an expanding range of content (i.e., pictures and videos), inhibited data collection procedures. Further, the Library has never made the data available due to an inability to develop a suitable distributional procedure. Such data handling challenges are common today in Web 2.0's world of "big data," and the computational expense incurred by spatial data queries makes handling geolocated social media data particularly difficult.

Twitter data does not follow a traditional tabular structure. Each tweet can return a variable number of keys and values representing the hashtags used, a tweet that was quoted, users referenced, and links. Twitter objects produced through the APIs can possess hundreds of different keys that may be present in some records and absent in others. This data format, termed "semi-structured," has challenged conventional forms of computing and storage, particularly relational database management system (RDBMS) model which often makes use of Structured Query Language (SQL). To tackle the challenges created by social media data and other semi-structured formats, many different "Not only SQL" (NoSQL) data models have emerged. These include a document store (as with MongoDB and CouchDB), a wide column store (as with Cassandra), and a key-value store (as with Riak and Redis) (Cattell 2011). This lack of standardization means that NoSQL principles cannot be universally applied across systems. Additionally, the sheer volume of social media data often requires distributed computing to efficiently extract results. Using distributed resources with geolocated social media data simultaneously requires knowledge of GIS and cyberinfrastructure, both of which can change at fast rates (Wang et al. 2009).

Despite these challenges, several successful applications integrating geolocated Twitter into a NoSQL framework have been documented in detail. Using MongoDB, Cao et al. (2015) introduce a framework for analyzing social media data across time and space. Developing a system called FluMapper, they demonstrate a novel application that synthesizes the travel patterns of potentially flu-infected Twitter users. Similarly, Soltani et al. (2016) have developed a system called UrbanFlow, which combines geotagged tweets with land use parcels in the Chicago area. Such case studies can provide a useful reference point for others, but subtle differences in requirements can greatly change cyberinfrastructure needs.

Privacy

Privacy issues associated with location tagging are often only mentioned in vague passing, barring a few, albeit notable, exceptions. In their *Geoprivacy*

Manifesto, Keßler and McKenzie (2018) notably point out that both geoprivacy and its advancements are difficult to assess. Zook et al. (2017), in developing *Ten simple rules for responsible big data research*, importantly acknowledge that "data are people" (p. 2) and that privacy is more complex than a simple binary state. While privacy issues most often surface in *social* applications of Twitter data, Crawford and Finn (2015) highlight the lack of privacy discussions in the context of disaster situations. Some scenarios are time sensitive and represent life or death situations, yet the data used are not immune to abuse, shedding light on the need for greater ethical scrutiny and critical perspectives on big data regardless of its application.

Yet, corporations such as Google, Apple, Facebook, and of course, Twitter, have far greater detail about individuals and their locational practices than the academy, which often relies on publicly available data. Recent privacy breaches, brought to light through the Facebook – Cambridge Analytica Data Scandal (Granville 2018), have exposed the extent of knowledge that corporations possess and how they will use that information to manipulate the public opinion. This seemingly confirms the fears of Miller and Goodchild (2015), who wrote

> We must be cognizant about where this research is occurring – in the open light of scholarly research where peer review and reproducibility is possible, or behind closed doors of private-sector companies and government agencies, as proprietary products without peer review and without full reproducibility.
>
> (p. 260)

Knowing that much research is indeed occurring outside of the light of peer review, a question logically follows: what breaches of privacy have occurred but will never become public?

Before the advent of the "data avalanche" (Miller 2010), Dobson and Fisher (2007) warned of the "panopticonic" capabilities of human location tracking. Such technologies are particularly dangerous when they force the user to relinquish something (i.e., their privacy) while simultaneously providing some benefit. Given that social media is being adopted by an ever-increasing number of the people (Smith and Anderson 2018) and that many users view geotagging as an effective way to gain social capital (Haffner et al. 2018), scrutiny of the way these data collected and used is more important than ever.

Twitter data and social disparity

Despite these clearly concerning drawbacks, analyses of geolocated Twitter data can be powerful if the limitations are properly understood, particularly in their ability to elicit patterns of social disparity. Uncovering differences – or simply absences – in spatial patterns of production has the capability of evoking geographic "digital divides" (Warf 2001). While lack of content in itself is not positive evidence of inequality, online landscapes are reflective of material, offline processes (Shelton et al. 2014). Going a step further, in the information age – where decisions about places are made via online and social media presence (or a lack thereof) – online content has increasingly greater potential to affect the perception and configuration of physical space.

Evaluating the Twitter data associated with Hurricane Sandy, Shelton et al. (Shelton et al. 2014) find an expected association between areas most affected by the Hurricane and the Twitter content referencing Sandy, yet the correspondence is not one-to-one. They importantly acknowledge that some locations might be neglected due to a lack in material capabilities in producing content and that analyses *after* a disaster could be effective in providing insight on social inequality. Extending on this approach, Shelton et al. (2015) find stark segregation between Louisville, Kentucky's West End, and East End Twitter users, not just within the content production patterns in these neighborhoods but also in the places throughout the metropolitan area visited by both groups. Those who mostly produce tweets from the more affluent East End rarely navigate the spaces visited by West End users.

Taking a broader, place-based approach, Haffner (2018a) examines the spatial patterns and correlations between municipalities' census data and their number of references to #BlackLivesMatter and #AllLivesMatter in cities across Texas and Louisiana. Contrary to commonly held notions that #AllLivesMatter is used as counter-narrative to #BlackLivesMatter, there is a considerable association between a city's number of #AllLivesMatter references and its percent black population. On the other hand, there is a strong negative association between the number of references to either phrase and a city's percent white population, and the same holds true for percent Hispanic. This aligns with the notion that whites are mostly unwilling to engage in thoughtful racial dialogue, albeit in an unexpected way.

Conclusion

This chapter has highlighted the history, applications, and challenges of working with geolocated Twitter data. This history is marked with tension in ways of conceptualizing the data, Twitter's value in light of its biases, and how to handle such sources from a computational perspective. Twitter's location tagging features are disabled by default – from the perspective of allowing other users to view the data – but the company's use of IP addresses for geo-targeted advertisements, complicate its "opt-in" verses "opt-out" status. Further, even if a tweet is not formally geotagged, locational information can still be extracted or inferred. Compared to other social media platforms, Twitter is more representative of the general population in terms of its user base, but evidence suggests that various subgroups use Twitter in quite different ways.

Throughout Twitter's history, privacy has received notable attention, but many questions in this realm remain unanswered. Since Twitter can – and has – changed its policies and features over time, the way that users participate in the social network is also continually evolving. A significant challenge moving forward lies in keeping up with these shifts in order to effectively explain user behavior, in addition to addressing existing issues mentioned in this piece. Despite its many drawbacks, geolocated Twitter data possesses power to reflect material, offline inequalities in ways not possible before its inception. Social media data can complicate traditionally held assumptions and provide a voice to historically marginalized groups. For these reasons – if nothing else – its use is warranted by geographers and others within the academy.

References

Abbasi, A., T. Rashidi, M. Maghrebi, and S. Waller. 2015. Utilising location-based social media in travel survey methods: Bringing Twitter data into the play. *Proceedings of the 8th ACM SIGSPATIAL International Workshop on Location-based Social Networks.* pp. 1–9.

Acar, A., and Y. Muraki. 2011. Twitter for crisis communication: Lessons learned from Japan's tsunami disaster. *International Journal of Web Based Communities* 7(3):392–402.

Cao, G., S. Wang, M. Hwang, A. Padmanabhan, Z. Zhang, and K. Soltani. 2015. A scalable framework for spatiotemporal analysis of location-based social media data. *Computers, Environment and Urban Systems* 51:70–82.

Cattell, R. 2011. Scalable SQL and NoSQL data stores. *ACM SIGMOD Record* 39 (4):12–27.

Crampton, J., M. Graham, A. Poorthuis, T. Shelton, M. Stephens, M. Wilson, and M. Zook. 2013. Beyond the geotag: Situating big data and leveraging the potential of the geoweb. *Cartography and Geographic Information Science* 40(2):130–139.

Crawford, K., and M. Finn. 2015. The limits of crisis data: Analytical and ethical challenges of using social and mobile data to understand disasters. *GeoJournal* 80(4):491–502.

de Albuquerque, J., B. Herfort, A. Brenning, and A. Zipf. 2015. A geographic approach for combining social media and authoritative data: Towards identifying useful information for disaster management. *International Journal of Geographical Information Science* 29(4):667–689.

Dobson, J., and P. Fisher. 2007. The panopticon's changing geography. *Geographical Review* 97(3):307–323.

Elwood, S., M. Goodchild, and D. Sui. 2012. Researching volunteered geographic information: Spatial data, geographic research, and new social practice. *Annals of the Association of American Geographers* 102(3):571–590.

Evans, L. 2015a. Being-towards the social: Mood and orientation to location-based social media, computational things, and applications. *New Media Society* 17(6):845–860.

Evans, L. 2015b. *Locative Social Media: Place in the Digital Age.* New York: Palgrave Macmillan.

Goodchild, M. 2007. Citizens as sensors: The world of volunteered geography. *Geojournal* 69(4):211–221.

Graham, M., and T. Shelton. 2013. Geography and the future of big data, big data and the future of geography. *Dialogues in Human Geography* 3(3):255–361.

Granville, K. 2018. Facebook and Cambridge Analytica: What you need to know as the fallout widens. *New York Times* (March 19). www.nytimes.com/2018/03/19/technology/facebook-cambridge-analytica-explained.html.

Haffner, M. 2018a. A place-based analysis of #BlackLivesMatter and counter-protest content on Twitter. *Geojournal* 1–24.

Haffner, M. 2018b. A spatial analysis of non-English Twitter activity in Houston, Texas. *Transactions in GIS* 22(4):913–929.

Haffner, M., A. Mathews, E. Fekete, and G. Finchum. 2018. Location-based social media behavior and perception: Views of university students. *Geographical Review* 108(2):203–224.

Harris, T., D. Weiner, T. Warner, and R. Levin. 1995. Pursuing social goals through participatory GIS: Redressing South Africa's historical political ecology. In J. Pickles (ed.) *Ground Truth: The Social Implications of Geographic Information Systems.* pp. 196–222. New York: Guilford Press.

Harvey, F. 2013. To volunteer or to contribute locational information? Towards truth in labeling for crowdsourced geographic information. In D. Sui, S. Elwood, and M. Goodchild (eds.) *Crowdsourcing Geographic Knowledge: Volunteered Geographic Information (VGI) in Theory and Practice.* pp. 31–42. Dordrecht: Springer.

Humphreys, L. 2012. Connecting, coordinating, cataloguing: Communicative practices on mobile social networks. *Journal of Broadcasting & Electronic Media* 56(4):494–510.

Keßler, C., and G. McKenzie. 2018. A geoprivacy manifesto. *Transactions in GIS* 22(1):3–19.

Kent, J., and H. Capello Jr. 2013. Spatial patterns and demographic indicators of effective social media content during the Horsethief Canyon Fire of 2012. *Cartography and Geographic Information Science* 40(2):78–89.

Lansley, G., and P. Longley. 2016. The geography of Twitter topics in London. *Computers, Environment and Urban Systems* 58:85–96.

Lee, J.H., S. Gao, and K. Goulias. 2016. Can Twitter data be used to validate travel demand models? *95th Annual Transportation Research Board Meeting.* pp. 1–27.

Leetaru, K., S. Wang, A. Padmanabhan, and E. Shook. 2013. Mapping the global Twitter heartbeat: The geography of Twitter. *First Monday* 18(5).

Library of Congress. 2017. Update on the Twitter archive at the Library of Congress. Washington, DC: Library of Congress. https://blogs.loc.gov/loc/files/2017/12/2017dec_twitter_white-paper.pdf.

Miller, H. 2010. The data avalanche is here. Shouldn't we be digging? *Journal of Regional Science* 50(1):181–201.

Miller, H., and M. Goodchild. 2015. Data-driven geography. *Geojournal* 80(4):449–461.

Murphy, K. 2010. Web photos that reveal secrets, like where you live. *New York Times* (April 12). www.nytimes.com/2010/08/12/technology/personaltech/12basics.html.

O'Reilly, T. 2005. What is Web 2.0: Design Patterns and Business Models for the Next Generation of Software. www.oreilly.com/pub/a/web2/archive/what-is-web-20.html.

Schwartz, R., and G. Halegoua. 2015. The spatial self: Location-based identity performance on social media. *New Media & Society* 17(10):1643–1660.

Shelton, T., A. Poorthuis, M. Graham, and M. Zook. 2014. Mapping the data shadows of Hurricane Sandy: Uncovering the sociospatial dimensions of Big Data. *Geoforum* 52:167–179.

Shelton, T., A. Poorthuis, and M. Zook. 2015. Social media and the city: Rethinking urban socio-spatial inequality using user-generated geographic information. *Landscape and Urban Planning* 142:198–211.

Smith, A., and M. Anderson. 2018. Social Media Use in 2018. Washington, DC: Pew Research Center. www.pewInternet.org/2018/03/01/social-media-use-in-2018/.

Soltani, K., A. Soliman, A. Padmanabhan, and S. Wang. 2016. UrbanFlow: Large-scale framework to integrate social media and authoritative landuse maps. *Proceedings of the XSEDE16 on Diversity, Big Data, and Science at Scale*. pp. 1–8. New York: ACM.

Stefanidis, A., A. Crooks, and J. Radzikowski. 2013. Harvesting ambient geospatial information from social media feeds. *Geojournal* 78(2):319–338.

Stephens, M. 2013. Gender and the GeoWeb: Divisions in the production of user-generated cartographic information. *Geojournal* 78(6):981–996.

Stock, K. 2018. Mining location from social media: A systematic review. *Computers, Environment, and Urban Systems* 71:209–240.

Takahashi, B., E. Tandoc Jr., and C. Carmichael. 2015. Communicating on Twitter during a disaster: An analysis of Tweets during Typhoon Hayan in the Philippines. *Computers in Human Behavior* 50:392–398.

Twitter. 2009. Think globally, Tweet locally. *twitter.com*. https://blog.twitter.com/official/en_us/a/2009/think-globally-tweet-locally.html.

Twitter. 2018a. Q2 2018 earnings report. *investor.twitterinc.com*. https://investor.twitterinc.com/static-files/246ced32-c085-4370-8308-2abea420d11a.

Twitter. 2018b. Tweet location FAQs. *help.twitter.com*. https://help.twitter.com/en/safety-and-security/tweet-location-settings.

Twitter. 2018c. Geo, Gender, Language, and Age Targeting. *business.twitter.com*. https://business.twitter.com/en/help/campaign-setup/campaign-targeting/geo-gender-and-language-targeting.html.

Twitter. 2018d. Your Privacy Controls for Personalized Ads. *help.twitter.com*. https://help.twitter.com/en/safety-and-security/privacy-controls-for-tailored-ads.

van Eggermond, M., H. Chen, A. Erath, and M. Cebrian. 2015. Investigating the potential of social network data for transport demand models. *Transportation Research Board 95 Annual Meeting*. https://pdfs.semanticscholar.org/c18d/aa9f26 5e31721c5432c3e1e51240f85a8466.pdf.

Wang, S., Y. Liu, N. Wilkins-Diehr, and S. Martin. 2009. SimpleGrid Toolkit: Enabling geosciences gateways to cyberinfrastructure. *Computers & Geosciences* 35(12):2283–2294.

Warf, B. 2001. Segueways into cyberspace: Multiple geographies of the digital divide. *Environment and Planning B* 28(1):3–19.

Wilken, R. 2014. Places nearby: Facebook as a location-based social media platform. *New Media and Society* 16(7):1087–1103.

Williams, S. A., M. Terras, and C. Warwick. 2013. What people study when they study Twitter. *Journal of Documentation* 69(3):384–410.

Yang, Z., L. Nguyen, J. Stuve, G. Cao, and F. Jin. 2017. Harvey flooding rescue in social media. *2017 IEEE International Conference on Big Data*. pp. 2177–2185. Boston.

Yuan, Y., G. Wei, and Y. Lu. 2018. Evaluating gender representativeness of location-based social media: A case study of Weibo. *Annals of GIS* 24(3):163–176.

Zhao, B., D. Sui, and Z. Li. 2017. Visualizing the gay community in Beijing with location-based social media. *Environment and Planning A: Economy and Space* 49(5):977–979.

Zickuhr, K. 2013. Location-based services. *Pew Research Center: Internet & Technology.* www.pewinternet.org/2013/09/12/location-based-services/.

Zook, M., and A. Poorthuis. 2014. Offline brews and online views: Exploring the geography of beer Tweets. In M. Patterson and N. Hoalst-Pullen (eds.) *The Geography of Beer.* pp. 201–209. Dordrecht, Netherlands: Springer.

Zook, M., S. Barocas, d. boyd, K. Crawford, E. Keller, S. Gangadharan, A. Goodman, R. Hollander, B. Koenig, J. Metcalf, A. Narayanan, A. Nelson, and F. Pasquale. 2017. Ten simple rules for responsible Big Data research. *PLoS Computational Biology* 13(3):1–10.

18 Neogeography

Wen Lin

In 2006, Di-Ann Eisnor, co-founder of Platial.com, wrote a post on Platial entitled "What is Neogeography Anyway?" Eisnor noted neogeography means "new geography." But the prefix "neo" means more than just new. Eisnor then continued

> Neogeography, as we see it, is a diverse set of practices that operate outside, or alongside, or *in the manner of*, the practices of professional geographers. Rather than making claims on scientific standards, methodologies of neogeography tend toward the intuitive, expressive, personal, absurd, and/or artistic, but may just be idiosyncratic applications of "real" geographic techniques.
>
> (Eisnor 2006, para. 3)

Platial was an online platform launched in 2005, a website for personal mapping that utilized Google Maps and "one of a new breed of map mashups – web applications created by mixing an already-existing open mapping platform with original software" (Wired 2006, para. 6). Users used mashups to record their personal stories. Platial closed down in 2010, with over 500,000 sites being mapped by the Platial members (Eisnor 2010).

Platial.com is only one example of the explosive growth of user-generated geographic contents production since the mid-2000s, marked by the emergence of an array of interactive Web 2.0 technologies and location aware mobile devices (e.g., Haklay et al. 2008). Examples of mapping platforms that allow user contributions include Google Earth, Google Maps, Wikimapia, OpenStreetMap, Microsoft Virtual Earth, and Ushahidi. Neogeography is one of the terms used to describe such a phenomenon, a term further popularized by Andrew Turner's (2006) book *Introduction to Neogeography*. Other terms used to address this phenomenon include volunteered geographic information (VGI) (Goodchild 2007), the geoweb (Scharl and Tochtermann 2007), wikification of GIS (Sui 2008), maps 2.0 (Crampton 2009), and spatial media (Kitchin et al. 2017).

Such a phenomenon has brought profound changes to the ways geographic information is collected and shared. It has stimulated tremendous efforts from researchers within geography studying the impacts and implications of this phenomenon. The main goal of this chapter is to provide a brief account of its emergence and relevant debates associated with the phenomenon (see also Lin 2018a). It begins with a short discussion of the origin of the term and associated practices, which does not aim to provide a comprehensive history of the emergence of neogeography, but to foreground key timelines and events as a way of situating the highly diverse realm of user-generated geographic information production. The chapter subsequently focuses on addressing debates revolved around neogeography.

Tracing the birth of neogeography

Di-Ann Eisnor's post on Platial discussing neogeography has been recognized as the initial account of defining neogeography. In these early discussions, neogeography practiced by those who are not professional geographers was emphasized. The story of Platial was reported by several media outlets, including *Wired* magazine and National Public Radio. These reports highlight how neogeography enables personalized mapmaking by people without conventional training in GIS or cartography. Common practices include adding narratives with information extracted from other websites to specific locations on the online map.

Andrew Turner's book (2006) has been widely cited in the research literature, recognizing its role of further popularizing the term neogeography. Turner (2006, 2) notes "neogeography means 'new geography' and consists of a set of techniques and tools that fall outside the realm of traditional GIS, Geographic Information Systems." Similarly, Turner's definition of neogeography emphasizes its distinction from techniques and tools used in the conventional realm of GIS. For example, a neogeographer might engage with mapping application programing interfaces (API) such as Google Maps as opposed to GIS software such as ArcGIS. Turner suggests that neogeography "is about people using and creating their own maps, on their own terms and by combining elements of an existing toolset," and that neogeography "is about sharing location information with friends and visitors, helping shape context, and conveying understanding through knowledge of place" (p. 3). Moreover, neogeography is fun. Turner then moves on to introduce a range of terms, technologies, and techniques concerning neogeography. As such, personalized mapmaking, interest in sharing knowledge of places and locations with others, and a sense of playfulness are the major characteristics identified by Turner (2006).

Acknowledging the growing efforts outside of the academy to engage with neogeography, Wilson and Graham (2013) chronicle a number of events within geography that mark the developments of neogeography, starting with a VGI specialist meeting held in Santa Barbara, USA, in 2007, which led to a special issue of *GeoJournal* in 2008. Following this event are several specialist meetings, designated conference sessions, and special issues published in geography journals. These developments of neogeography, Wilson and Graham

(2013) note, focus on issues related to data, including issues of "proliferation, standardization, interoperability, quality/accuracy, and visualization" (p. 3).

With these efforts, it is widely recognized that neogeographic mapping practices are highly diverse. One broad category of these interactive mapping practices involves engaging with personalized mapping on issues related to everyday experience (e.g., Kingsbury and Jones 2009; Caquard 2011). Another category may involve mappings underpinned by efforts of civic engagement, such as crisis mapping and relief needs mapping (e.g., Crutcher and Zook 2009; Elwood and Mitchell 2013). Neogeography marks the increasing role for amateurs and non-geographers to participate in spatial data generation and mapmaking. The next section reviews debates on neogeography from geography and related fields.

Debates on neogeography

The emergence of neogeography has stimulated much discussion among geographers, not only with respect to examining ways of conducting geographic research, but also regarding possible impacts on the public notion of geography (e.g., Sui 2008; Hudson-Smith and Crooks 2008; see also Lin 2018a). Later discussions have engaged with a number of areas of research, resulting in a rich body of work on neogeography and related practices. Wilson and Graham (2013) view neogeography as practices that enact new relationships. These practices are contingent, partial, and collaborative. They further suggest five areas of researching neogeography:

> (1) explore the conditions that enable the emergence of neogeography, (2) unpack the implications for digital representations that are produced by and through attention and bias, (3) trace the subject formations necessary for such developments, (4) reflect on the changing role of geography and geographers, and (5) constitute the possibility for responsive interventions and interruptions.
>
> (Wilson and Graham 2013, 4)

The following three subsections review debates on neogeography through several aspects, which, while might not mirror the above five areas identified by Wilson and Graham (2013), have addressed these areas one way or another.

Spatial data and spatial narratives

Scholars have discussed how neogeographic practices might influence the ways spatial data are collected, shared and mapped. Concerns have been raised from the research community regarding the impacts of neogeography on the quality of data and mapping produced, such as those documented in Crampton (2010) and Elwood and Mitchell (2013). Sui (2008) describes neogeography as a manifestation of the increasing wikification of GIS, a process that is linked to the open source software development such as Linux. Some others, however, have called for embracing neogeography (e.g., Goodchild 2007; Batty et al. 2010). Goodchild (2007) suggests VGI can contribute to developing national spatial data infrastructures through the so-called patchworks approach. Batty et al. (2010) consider the possibility of incorporating real-time data through neogeography into GIS such as using spatially referenced social media data to map space and time in new ways. In the field of cartography, for example, Wilmott (2017) calls for more attention to "how the nature of the map is changing with new modes of delivery and user interactivity" (Dodge 2018, 950).

There have been significant efforts examining the data content, quality and validity of neogeography. For example, Haklay (2010) assesses the quality of OpenStreetMap (OSM), a free world map produced by registered users, with Ordnance Survey datasets, an authoritative national mapping agency of the UK. The comparison is focused on London and England. His study shows that OSM can produce high quality data. Researchers also call for reconsidering ways of measuring spatial data validity. Flanagin and Metzger (2008), for example, propose to expand the notion of data credibility from a focus on information accuracy to a framework that pays more attention to people's perceptions. This is because much information provided through neogeography can be more opinion based. Meanwhile, attention has been given to the role of neogeography in mapping narratives. In his useful review of the emergence of new forms of mapping situated in the geoweb era, Caquard (2011) observes the increasing efforts on mapping narratives. He highlights how recent interactive online mapping programs such as Google Maps have facilitated the growth of story maps, as opposed to the grid maps exemplified by reference maps provided by authoritative mapping agencies. Yet, a grid map can also be produced through neogeography such as the case of OSM. These story maps are "new forms of spatial expressions interested in providing different perspectives about places and about stories associated to places" (Caquard 2011, 135). Caquard (2011) argues these new forms of spatial expressions mapping vernacular knowledge and even fiction are important to provide in-depth understanding of places.

Social and political implications of neogeography

Significant efforts have been made to investigate social and political implications of neogeography within geography. Goodchild (2007) argues that spatial data production by volunteers or amateurs allows more input from local people on the area that is being mapped. Another aspect is that, as noted in the preceding subsection, neogeography can facilitate everyday mapping, which might be more difficult through conventional GIS (e.g., Caquard 2011; Gerlach 2010). It has been recognized that neogeography and user-generated geographic content production may be less about aiming to conduct complex spatial analysis, but more about creating and sharing geographic information and narratives.

Such a transformation concerning the ways in which geographic data are produced and shared has brought profound shifts to GIS and related fields, not only regarding types of analysis used and types of data produced, but also regarding meanings and forms of communities, publics, sense of place, and knowledge, as well as relevant institutional arrangements. For example, Foth et al. (2009) suggest that neogeography signals a fourth stage of GIS development, in which everyday views and voices can be shown more easily. Elwood et al. (2012) propose three framings of researching VGI: VGI as a form of spatial data, of evidence, and of spatial practice. Researching VGI as a form of spatial data, the authors investigate a collection of existing VGI initiatives at the time and identify three categories of the purposes of VGI production, which are geovisualization (mapping user-generated information), geoinformation (capturing and manipulating geotagged information), and geosocial (sharing geolocated media data). Investigating VGI as a form of evidence, the authors look at how geographers use VGI as a source of data and discuss issues related to data quality and methodological development to analyze VGI. Examining VGI as a form of social practice, the authors address the role of critical GIS research in shaping the body of work concerning impacts and implications of neogeography. This includes discussing the empowerment potential of neogeography and its constraints, the political-economic relationships and processes that underpin and shape neogeography practices, and questions on users/creators and their experiences.

In particular, there has been a great deal of interest in discussing the empowerment potential of neogeography (e.g., Tulloch 2008; Lin 2012; Elwood and Mitchell 2013; Haklay 2013). Some studies suggest that neogeography is liberating and empowering, as people can create their own maps more easily. Yet, this may underpin a form of deprofessionalization (Crampton 2009). Researchers have examined how neogeography might facilitate public participation and enable agency of marginalized actors (e.g., Tulloch 2008; Elwood and Mitchell 2013). Tulloch (2008) suggests that VGI practices are characterized by personal datasets. As such, unlike many public participation GIS initiatives, VGI initiatives are less about employing public datasets. Elwood and Mitchell (2013) note that there have been two dimensions of addressing the empowerment/marginalization dialectics regarding neogeography in relation to public participation and civic engagement. One dimension focuses on how neogeography might provide greater access to data and technologies to facilitate public participation. In this perspective, neogeography might contribute to public participation and civic engagement through existing mechanisms. For example, there have been discussions on eliciting public input and engagement through neogeographic practices and other related approaches. For a second dimension, studies attempt to explore how neogeography might provide alternative forms of participation or constitute new spaces of contestation and resistance.

Related to the discussion of civic engagement and empowerment is how neogeography might enable and advance certain types of spatial knowledge production. Through interrogating five VGI initiatives, Elwood and Leszczynski (2013) examine how neogeography practices have led to a new knowledge politics, which is characterized by exploratory and experiential knowledge production. In such a form of knowledge production, data credibility is addressed through witnessing, peer verification, and transparency. Gerlach (2010) suggests that OSM cultivates a different kind of politics. Such a politics is not conventional conceptions of counter-mapping or indigenous mapping; rather, it moves towards what Gerlach (2010) calls vernacular mapping.

Neogeographic practices have variable outcomes with respect to the empowerment/marginalization dialect (e.g., Crutcher and Zook 2009; Caquard 2014; Stephens 2013; Elwood and Mitchell 2013). Neogeography including projects based on open source mappings and crowdsourcing might enable community empowerment with greater control over territory and resources. Warf and Sui (2010) argue that that neogeography can provide new epistemological openings for knowledge production with a conversational view emphasizing more on practice, performance, and the speech acts. Meanwhile, many of these neogeographic practices may not produce much change as initially envisioned (Caquard 2011) or perpetuate existing inequalities (e.g., Crutcher and Zook 2009; Stephens 2013). In addition, neogeographic practices may generate new forms of exclusion and surveillance (Kitchin et al. 2017). For example, OSM, while widely recognized as providing an alternative source for open data, its recent collaboration with Microsoft might reflect a greater influence from powerful corporate players in open source mapping practices (Caquard 2014; Leszczynski 2012, 2014). Leszczynski (2014) examines how the discourse of the "neo" in neogeography is used to legitimize certain kinds of technology-society relationships. In doing so, she discusses how this framing of neogeography can "depoliticize spatial media to lay the groundwork for their social naturalization by presenting them as inevitable outcomes of technological progress" (Leszczynski 2014, 62–63).

A number of studies have discussed the political-economic relationships underpinning and reflected by neogeography (e.g., Leszczynski 2012; Caquard 2014; Thatcher et al. 2016; Lin 2018b). The focus in this area is investigating the roles of state and private-section actors in shaping neogeography (Elwood and Mitchell 2013; Caquard 2014). Leszczynski (2012) situates the emergence of the geoweb within the broader restructuring of the state. She argues that "the state is 'rolling back' from public aspects of the cartographic project, market regimes of governance are simultaneously 'rolling out', subsuming the mapping enterprise to the imperatives of technoscientific capitalism" (Leszczynski 2012, 72; see also Caquard 2014).

Subjectivities and subjective experiences

As noted by Wilson and Graham (2013), another area of research is to investigate subject formation associated with neogeography practices (e.g., Schuurman 2004; Wilson 2011; Lin 2013; Elwood and Mitchell 2013). Attention has also been given to understanding user motivations in these practices (e.g., Budhathoki et al. 2008; Lin 2011). For example, Lin (2011) identifies four groups of OSM mappers: those from the business sector, government sector, NGO/third sector, and a loosely coupled individuals.

Through investigating a project engaging youth in neogeography practices, Elwood and Mitchell (2013) suggest that neogeography can serve not only as sites for political engagement, but also as sites of political subject formation. They argue for a broadening conceptualization of neogeography practices that include visual spatial tactics. This means a wider range of knowledge-making practices and knowledgeable subjects can be incorporated. The authors also point to the implications of such an expanded conceptualization for education and citizenship. They argue that neogeography has potential to be used for popular education and progressive pedagogies, moving from a tendency of using neogeography platforms as "a way to teach 'pre-GIS' digital and spatial skills, principles of cartography, or conventional forms of spatial reasoning in more accessible and lower-cost ways" (Elwood and Mitchell 2013, 287). The other implication addresses the role of neogeography in fostering critical spatial citizenship, a notion raised in Gryl and Jekel (2012) referring to "critically reflective uses of geographic information and spatial media to contribute to societal deliberations and decision-making in active and influential ways" (Elwood and Mitchell 2013, 288). Elwood and Mitchell (2013) argue that visual spatial tactics can be an avenue for critical spatial citizenship. This is particularly important considering challenges posed by persistent inequalities and digital divides, which might exclude, or limit room of critical spatial citizenship performed through practices of "strategy," "constituted through the spaces and practices of hegemonic actors/institutions and forms of knowledge" (Elwood and Mitchell 2013, 276).

Ramos (2016, 148) notes that researchers need to "look into how technologies are enabling more sophisticated, nuanced versions of the subject." While not focusing on neogeography, Kinsley (2018) explores subjectivities in relation to digital geographies. He examines the notion of subjectivities through three aspects: subjectivities as a figure of "the subject," subject positions, and subjective experience. Kinsley (2018, 160) notes that particular subject-positions emerging from data practices derived from "surveillant, data-derived representations of people as individuals" can perpetuate existing inequalities, but there is also potential to challenge existing inequalities.

Several studies have called for attention to the role of emotion and affect in mapping (e.g., Kwan 2007; Young and Gilmore 2013) and neogeographic practices (e.g., Gerlach 2014; Lin 2015; Dodge 2018). For example, Gerlach (2014, 35) argues that "alongside an interrogation of representational valence, there is also a speculative concern for the micropolitical lines, contours and legends of maps and mappings," the latter requiring a focus on "the affective processes of maps and mappings coming into being and the generation of the virtual, unqualified political potential." Lin (2015) examines a sound map project based on Google Maps in China, showing how these sound recordings uploaded "provide a repertoire of emotions, memories, and stories embedded in particular places and moments." Dodge (2018, 952) discusses so-called "playful mapping," through an example of collected work by a group of Anglo Dutch cartographic and media researchers. Such playful mapping is "infused with notions from gamification and ludic theory" (Dodge 2018, 952).

Conclusion

Neogeographic practices continue to grow, part of the user-generated geographic information production facilitated by Web 2.0 technologies and location aware devices which can be considered as "spatial big data" (Sui et al. 2012). As briefly reviewed above, this phenomenon has drawn much attention from researchers as well as practitioners as it has brought profound changes to the ways in which geographic information might be collected, curated, and analyzed. Early debates from geography attempt to examine what neogeography might imply for the role of GIS professionals and academic geographers. More discussions subsequently emerged examine data characteristics, actors involved, and social and political implications associated with neogeography. Research has recognized the high level of heterogeneity of neogeographic practices, high speed of such data production and dissemination, and variable outcomes concerning questions of power and equality.

Neogeography tends to be about personalized spatial data produced by laypersons linked to everyday experiences. Research has sought to understand the growth of such everyday mapping (e.g., Caquard 2011; Gerlach 2014; Lin 2015). Yet, neogeography can also be carried out in response to events and incidents that are of public interest such as relief needs mapping. As such, neogeography may constitute an important force for facilitating public participation and civic engagement concerning marginalized groups and actors given the relatively lower technical and financial barriers compared to conventional GIS technologies employment. Neogeography can also be enacted in a way that facilitates political subject formation. Meanwhile, cautions have been raised regarding how neogeography might perpetuate existing inequalities and produce new forms of exclusion, such as considering the role of persisting digital divides in different geographic and social contexts. There have been efforts examining political-economic conditions of neogeographic practices. Researchers have pointed out trends revealing "the replacement of the state as the main reference for the collection and dissemination of cartographic data, by a combination of private interest and individually volunteered contributions" (Caquard 2014, 141; Leszczynski 2012). While it can be liberating and empowering for individuals to produce their own maps and share geographic information more easily, such information might be exploited by large companies and state agencies (Dodge 2018).

Much remains to be explored both conceptually and empirically. For example, Dodge (2018) notes that more research is needed to understand ways in which people are using online mapping through the aspect of improving design. More political-economic accounts are needed (Thatcher et al. 2016; Lin 2018). Also, with increasing reflections on the growing control of large internet corporation actors and their associated business models such as Facebook, Twitter, and Google in relation to the ways information is disseminated and shared, as well as concerns of massive surveillance by powerful state actors, continuous efforts are needed to interrogate various forms of and the ever-evolving neogeography practices, situated in various social and geographic contexts.

References

Batty, M., A. Hudson-Smith, R. Milton, and A. Crooks. 2010. Map mashups, Web 2.0 and the GIS revolution. *Annals of Geographical Information Science* 16(1):1–13.

Budhathoki, N., B. Bruce, and Z. Nedovic-Budic. 2008. Reconceptualizing the role of the user of spatial data infrastructures. *Geojournal* 72(3):149–160.

Caquard, S. 2011. Cartography I: Mapping narrative cartography. *Progress in Human Geography* 37(1):135–144.

Caquard, S. 2014. Cartography II: Collective cartographies in the social media era. *Progress in Human Geography* 38(1):141–150.

Crampton, J. 2009. Cartography: maps 2.0. *Progress in Human Geography* 33(1):91–100.

Crampton, J. 2010. *Mapping: A Critical Introduction to Cartography and GIS*. Oxford, New York: Wiley-Blackwell.

Crutcher, M., and M. Zook. 2009. Placemarks and waterlines: Racialized cyberscapes in post-Katrina Google Earth. *Geoforum* 40(4):523–534.

Dodge, M. 2018. Mapping II: News media mapping, new mediated geovisualities, mapping and verticality. *Progress in Human Geography* 42(6):949–958.

Eisnor, D. 2006. What is neogeography anyway? *Platial.com*. http://platial.typepad.com/news/2006/05/what_is_neogeog.html.

Eisnor, D, 2010. A letter to our mappers'. *Platial.com*. https://platial.typepad.com/news/2010/03/a-letter-to-our-mappers.html.

Elwood, S., and A. Leszczynski. 2013. New spatial media, new knowledge politics. *Transactions of the Institute of British Geographers* 38(4):544–559.

Elwood, S., and K. Mitchell. 2013. Another politics is possible: Neogeographies, visual spatial tactics, and political formation. *Cartographica* 48(4):275–292.

Elwood, S., M. Goodchild, and D. Sui. 2012. Researching volunteered geographic information: Spatial data, geographic research, and new social practice. *Annals of the Association of American Geographers* 102(3):571–590.

Flanagin, J., and M. Metzger. 2008. The credibility of volunteered geographic information. *Geojournal* 72(3/4):137–148.

Foth, M., B. Bajracharya, R. Brown, and G. Hearn. 2009. The second life of urban planning? Using neogeography tools for community engagement. *Journal of Location Based Services* 3(2):97–117.

Gerlach, J. 2010. Vernacular mapping and the ethics of what comes next. *Cartographica* 45:165–168.

Gerlach, J. 2014. Lines, contours and legends: coordinates for vernacular mapping. *Progress in Human Geography* 38:22–39.

Goodchild, M. 2007. Citizens as sensors: The world of volunteered geography. *Geojournal* 69(4):211–221.

Gryl, I., and T. Jekel. 2012. Re-centering geoinformation in secondary education: Toward a spatial citizenship approach. *Cartographica* 47(1):18–28.

Haklay, M. 2010. How good is volunteered geographical information? A comparative study of OpenStreetMap and Ordnance Survey Datasets. *Environment and Planning B: Urban Analytics and City Science* 37(4):682–703.

Haklay, M. 2013. Neogeography and the delusion of democritisation. *Environment and Planning A* 45:55–69.

Haklay, M., A. Singleton, and C. Parker. 2008. Web mapping 2.0: The neogeography of the geoweb. *Geography Compass* 2(6):2011–2039.

Hudson-Smith, A., and A. Crooks. 2008. The Renaissance of Geographic Information: Neogeography, Gaming and Second Life. *Working Papers Series.* www.academia.edu/9765535/The_Renaissance_of_Geographic_Information_Neogeography_Gaming_and_Second_Life.

Kinsley, S. 2018. Subject/ivities. In J. Ash, R. Kitchin, and A. Leszczynski (eds.) *Digital Geographies.* pp. 153–170. Thousand Oaks, CA: Sage.

Kitchin, R., T. Lauriault, and M. Wilson. 2017. *Understanding Spatial Media.* London: Sage.

Kwan, M.P. 2007. Affecting geospatial technologies: Toward a feminist politics of emotion. *Professional Geographer* 59(1):22–34.

Leszczynski, A. 2012. Situating the geoweb in political economy. *Progress in Human Geography* 36(1):72–89.

Leszczynski, A. 2014. On the neo in neogeography. *Annals of the Association of American Geographers* 104(1):60–79.

Lin, W. 2012. When Web 2.0 meets public participation GIS (PPGIS): volunteered geographic information and spaces of participatory mapping in China. In D. Sui, S. Elwood, and M. Goodchild (eds.) *Crowdsourcing Geographic Knowledge: Volunteered Geographic Information (VGI) in Theory and Practice.* pp. 83–103. Thousand Oaks, CA: Sage.

Lin, W. 2013. Situating performative neogeography: Tracing, mapping, and performing "Everyone's East Lake." *Environment and Planning A* 45:37–54.

Lin, W. 2015. The hearing, the mapping, and the Web: Investigating emerging online sound mapping practices. *Landscape and Urban Planning* 142:187–197.

Lin, W. 2018a. Neogeography. In B. Warf (ed.) *The Encyclopedia of the Internet.* Thousand Oaks, CA: Sage.

Lin, W. 2018b. Volunteered geographic information constructions in a contested terrain: A case of OpenStreetMap in China. *Geoforum* 89:73–82.

Lin, Y.-W. 2011. A qualitative enquiry into OpenStreetMap making. *New Review Hypermedia Multimedia* 17(1):53–71.

Ramos, R. 2016. Driving screens: Space, time, and embodiment in the use of waze. In C. Travis, and A. von Lünen (eds.) *The Digital Arts and Humanities Neogeography, Social Media and Big Data Integrations and Applications.* London: Springer.

Scharl, A., and K. Tochtermann. 2007. *The Geospatial Web: How Geobrowsers, Social Software and the Web 2.0 Are Shaping the Network Society.* London: Springer.

Schuurman, N. 2004. Databases and bodies – a cyborg update. *Environment and Planning A* 36:1337–1340.

Stephens, M. 2013. Gender and the geoweb: divisions in the production of user-generated cartographic information. *GeoJournal* 78:981–996.

Sui, D. 2008. The wikification of GIS and its consequences: Or Angelina Jolie's new tattoo and the future of GIS. *Computers, Environment and Urban Systems* 32:1–5.

Sui, D., M. Goodchild, and S. Elwood. 2012. VGI, the exaflood, and the growing digital divide. In D. Sui, S. Elwood, and M. Goodchild (eds.) *Crowdsourcing Geographic Knowledge: Volunteered Geographic Information (VGI) in Theory and Practice.* pp. 1–12. London: Springer.

Thatcher, J., L. Bergmann, B. Ricker, R. Rose-Redwood, D. O'Sullivan, T.J. Barnes, L.R. Barnesmoore, L.B. Imaoka, R. Burns, J. Cinnamon, C.M. Dalton, C. Davis, S. Dunn, F. Harvey, J. Jung, E. Kersten, L. Knigge, N. Lally, W. Lin, D. Mahmoudi, M. Martin, W. Payne, A. Sheikh, T. Shelton, E. Sheppard, C.W. Strother, A. Tarr, M.W. Wilson, and J.C. Young. 2016. Revisiting critical GIS. *Environment and Planning A* 48(5):815–824.

Turner, A. 2006. *Introduction to Neogeography.* Sebastopol, California: O'Reilly Media.

Tulloch, D. 2008. Is VGI participation? From vernal pools to video games. *GeoJournal* 72(3–4):161–171.

Warf, B., and D. Sui. 2010. From GIS to neogeography: ontological implications and theories of truth. *Annals of Geographical Information Science* 16(4):197–209.

Wilmott, C. 2017. In-between mobile maps and media: Movement. *Television & New Media* 18(4):320–335.

Wilson, M. 2011. "Training the eye": formation of the geocoding subject. *Social and Cultural Geography* 12:357–376.

Wilson, M., and M. Graham. 2013. Situating neogeography. *Environment and Planning A* 45(1):3–9.

Wired. 2006. Map Mashups Get Personal. www.wired.com/2006/03/map-mashups-get-personal/.

Young, J., and M. Gilmore. 2013. The spatial politics of affect and emotion in participatory GIS. *Annals of the Association of American Geographers* 103 (4):808–823.

19 Ethnographic research and the internet

Tyler Sonnichsen

The internet has played a significant role in almost all ethnographic research for the past three decades. However, for a litany of reasons, not all social scientists have been eager to acknowledge or embrace this change, no matter how fundamental. Various online research methods (ORM) remain unheralded in many qualitative academic corners due to latent stigmas of pedestrianism and unprofessionalism associated with sites like Facebook, Instagram, Snapchat, and whichever new platforms are popular by the time this Handbook is published. However, as this chapter will posit, the branches of ORM, social media being one crucial one, are all necessary to understand the perpetually changing dynamics of qualitative research.

Most obviously, the reorientation of human networks that the web has enabled vis-à-vis ethnographic data and human informants would have been inconceivable three decades ago. The proliferation of websites, weblogs (blogs), chat rooms, user groups, message boards, and other formats of virtual community has profoundly changed methodologies and general practices of ethnography. Though every researcher has their own set of habits regarding online presence and social media use, to ignore the internet's heavily pervasive role in the development of qualitative data would be irresponsible and necessarily exclude key reflexive components of research.

Two major contributions that the internet has made to ethnographic research have both involved increasing and speeding up access for researchers. The first is the expansion of access to informants. The second has been an expansion in publically accessible, extant ethnographic data. Both of these contributions have fundamentally altered the dynamics of ethnographic research, yet both recall traditional ontological and epistemological theses which have existed for centuries. Similar in ways through which the internet disrupted extant models of music circulation and consumption, the internet has disrupted traditional patterns of human–subject interaction and research.

Understandably, the majority of seminal literature on ethnographic research was published prior to the proliferation of the internet, and so any researcher's ontological and methodological foundations would be heavily informed by classic, pre-internet models. Fittingly, researchers today gather this ethnographic literature from all eras using online conduits like Academic

Search Premiere, Google Scholar, and their University library search engines. Even with a 20th-century ontological foundation, ethnographic research cannot exist outside of the web's superstructure.

This is why social scientists have addressed the internet's pivotal role in ethnographic research over the past 20 years. Even research on spaces and places without internet access (vanishing as they are) is still mediated online. For example, the initial gatekeepers in university-sanctioned human subject research, Institutional Review Boards (IRB), have been predominantly restructured as all-digital interfaces in order to reduce clutter and enhance approval processes. Even a researcher whose data collection relies entirely upon handwritten notes in a hypothetical remote community with no electricity or indoor plumbing would need to file their credentials digitally.

Similarly, university depositories for completed thesis and dissertation manuscripts have adopted digital portals; many libraries are no longer even printing hard copies of these documents. Online portals such as TRACE (Tennessee Research and Creative Exchange) at the University of Tennessee enable institutions to globally expand public access to research composed under their aegis. Prior to the internet, these documents might have only been accessible to members of (or those with physical access to) that institution's library. According to information technologists like Lynch (2003), this represents a fundamental millennial shift in the role of the university from passive to active research collaborator. Ethnographic researchers, often working at the behest of their respective institutions, must adapt to these changes and seek to understand them.

The beginnings of internet research in academia

Much of the earliest published academic work about "electronic mail" centered on concerns around its implementation and social context (Chaum 1981), but little evidence supports any ethnographers predicting its spread to their realm. The first academic work on internet-mediated communication, at least within the humanities, appeared in the early 1990s. By the middle of that decade, email had grown into the chief modern form of asynchronous (not in real-time) communication. Some qualitative and analytical work appeared on the technology's ascent and subsequent mutation into a multi-purpose platform (Whittaker and Sidner 1996). Traditional physical formats of correspondence, still widely in use by older members of the academy, became derided as "snail mail" within popular culture (no doubt abetted by Microsoft's public relations division). Though this was not the first appearance of the term "snail mail," it was the first time that using the Postal Service was categorically less efficient than any alternative.

In 1998, Annette Markham published *Life Online: Researching Real Experience in Virtual Space* (Walnut Creek, CA: Alta Mira), which a reviewer in *Qualitative Research* later called "a bold move in the exponentially increasing field of Internet studies" (Williams 2001). Though generalized conceptions of qualitative research had long existed when the internet started creeping into academic usage, *Qualitative Research* arrived in 2001 and *The International Journal of Qualitative Methods* only first published in 2002. Though several early volumes included articles that mentioned internet-based research, only a handful of articles in the journal's first decade explicitly centered online methods and related epistemological concerns. The rationales for this vary from researcher to researcher, from discipline to discipline – anthropology (Wittel 2000), sociology (McClelland 2002; Murthy 2008), ethnomusicology (Lysloff 2003), and other humanities fields. The Occam's Razor explanation, however, may involve the necessity to give these online platforms some time to breathe for observation, combined with the usually glacial pace of academic publication. Work extolling the potential of collaborating/corroborating internet and offline sources began to emerge (e.g., Seymour 2001; Sade-Beck 2004) in the nascent years of the 21st century.

Though social media was not a thoroughly vernacular term until a few years into the new millennium, some ethnographic literature addressed its seemingly inevitable consumption of quotidian processes. Any conversation about the proliferation of social media, however, must be couched in a historical and technological context. In order to understand how Facebook or Twitter shifted certain paradigms in ethnographic research, we must first understand the dynamics of firmly 20th-century platforms like email.

As a medium through which to conduct ethnographic research, email presented multiple advantages and disadvantages. The advantages, perhaps most evidently at first, was the lower expense both monetarily and resource-wise. Email required the payment of a monthly connection fee, though individual university email accounts were included in the cost of tuition for students and facilities fees for faculty. For unaffiliated researchers and just those who preferred to use private communications, email servers like Hotmail, Yahoo, and Lycos standardized the availability of free accounts, a model upon which Google built, post-millennium. Once registered, the researcher could email any other account in the world for no additional fee. From an economic standpoint, this eliminated long-distance phone bills as well as postage costs. Email also made possible near-instantaneous delivery of messages, unimaginable to those who began conducting their research using the post. Prior to the availability of email and other online conduits, mass mailings had been common practice, costing the university (if the researcher was so fortunate) much more money on postage for only slightly higher rates of return on investment.

The disadvantages of email-mediated ethnographic research, however, were inherent in the mechanics. Though it was unquestionably faster and cheaper than using the postal service or telephone for interview correspondence (especially internationally), it could not avoid the latent "clunkiness" of internet-mediated communication (Sproull and Kiester 1986). This presented a viable gambit for qualitative research, as the investigator's inability to "be there" in person or in voice over the phone made it impossible to discern non-representational data (expressions, pauses, etc.) from the informant. Additionally, like mail correspondence, email gave the informants time and opportunities to amend their answers, fundamentally shifting the non-representational dynamics found in personal interviews. This generated a new set of ethical questions in the collection and reportage of ethnographic data (Madge 2010). Ultimately, email, online surveys, and other forms of asynchronous interviewing may have fundamentally shifted some elements of qualitative research. This does not mean, however, that these complications are unanticipated, or even necessarily all new.

Internet images, video, and ethnography

Cultural geographers have been using video as a mechanism for data collection for as long as video technology has been publically available (Garrett 2011). Gillian Rose, in addition to being among the most prominent feminist geographers, emerged as the foremost authority on visual methods in geography over the past two decades (Rose 1996; Rose 2003; Rose 2012). The greater arc of visual methods encompasses an infinite field of images both static (photos, paintings, maps) and moving (streaming digital videos of various file formats). To make sense of visual images, Rose posits that the researcher must address "the site of audiencing," which encompasses the compositionality of the image, the technological site of the image, and the social context in which it is presented. She wrote in *Visual Methodologies* that "visual images are always practiced in particular ways, and different practices are often associated with different kinds of images in different kinds of spaces" (Rose 2012, 31).

Particularly in the 21st century, the advent of online image searching has reoriented the process of image cultivation from relatively passive to predominantly active. For ethnographic purposes, any images with identifiable people are tagged with their name as well as other characteristics of the picture: race, gender, height, and other physical descriptors. By 2011, Google had amassed such an index of images that users could cross-reference against millions of others using a mathematical algorithm (https://search.googleblog.com/2011/06/search-by-text-voice-or-image.html).

In other words, the process to obtain, index, and/or verify images, which once might have taken some researcher weeks to months, is possible in the span of seconds. However, with heightened accessibility come heightened ethical concerns. A vast majority of pictures that result from an image search are not labeled for noncommercial re-use, free from necessary permissions and questions of public versus private ownership. Google and other image search engines provide a search option that can eliminate images not approved for re-use from search results, though one could argue that this mechanism exists to protect the corporation from copyright violations. Speaking of Google, one search engine becoming the dominant paradigm (often embedded as a default in the DNA of web browsers) poses ethical questions about how images are

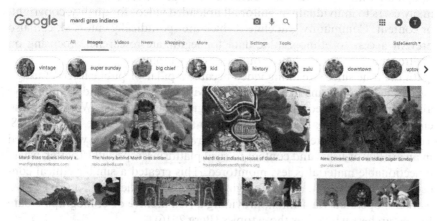

Figure 19.1 Google search results for "Mardi Gras Indians." Screen capture.

curated. A search for the term "Mardi Gras Indians" (Figure 19.1) obviously returns images based upon relevance to the search terms, but also prioritizes images based upon usage data.

One could raise similar concerns about the spate of streaming videos online, subject to a handful of dominant platforms with sufficiently powerful servers. YouTube is the powerhouse of these conduits, owned by Google and functioning almost as a metonym for (free) streaming video. YouTube, Vimeo, and DailyMotion all launched in the mid-2000s, and social scientists were quick to approach their "meanings" for the humanities. Geographers like Robyn Longhurst (2009) and Perry Carter (2015) posed bigger questions about how methods for communicating place-experiences were changing with mass-publishing audio-visual media, and what this may mean for ethnography.

As YouTube and Vimeo both expanded their respective storage capacities and video length allowances, the gates opened for a vast expansion of easily and instantaneously accessible footage of cultural events and materials. These included, but were not limited to, full live musical performances, feature-length films both independently produced and pirated off of copyright holders, and expansive raw footage from events that had been unreleased in pre-production form. These all offer arrays of ethnographic representations for researchers who could not "be there," whether because they lacked funding, access, or were not born when the events depicted took place. In this way, streaming video archives have been a boon to historically oriented researchers.

Unfortunately, as YouTube expands within the control of Google and associated corporate interests, the "public" nature or ownership of these videos can be compromised. The ownership of specific videos and audio records germane to ethnographic research has evaded the control of the original producers, as streaming videos are easy to download, re-edit, and re-upload with a contrary or censored message. These services all have too many users to individually monitor all uploaded videos for quality, copyright, or content. Community guidelines, which are sporadically enforced, can also prevent access to ethnographic data for legitimate researchers focusing on controversial topics like hate groups or pornography.

In the time since streaming video became a pillar of public internet usage, social media platforms like Facebook, Twitter, Instagram, and Tumblr have all incorporated video options, and blogging platforms like Wordpress also enable users to upload and stream video at a premium price. In late 2018, Tumblr removed material deemed pornographic from the site, which resulted in an exodus of users and content to rival platforms, especially Twitter, where objectionable material is less monitored. This created a similar ethical concern, considering how many marginalized communities (e.g., LGBT+, sex workers) operated via Tumblr, providing a rich archive of ethnographic data for researchers who cover those topics (Berg 2016).

Though streaming video sites and social media platforms have expanded the already vast archive of ethnographic data, internet video has enabled researchers to transcend field work. Skype, Oovoo, and other video-chatting services have widened the accessible pool of ethnographic data while decreasing the distance necessary to travel in order to conduct synchronous ethnographic interviews. Video communication enables the ethnographer to note facial expressions, bodily comportments, and other nuances of the personal interview process which often evade capture via written or audio communication.

Though Skype interactions still pale in sensory connection to in-person interviews, they offer many benefits for ethnographic research. Most notably, video interviewing vastly decreases the potential costs of in-person interviews internationally, making these interactions more accessible for researchers who lack the funding or time to visit their informants in person yet still desire face-to-face connections. Though they all have purchase and subscription options, video services like Skype normally offer their programs for free download as well as audio-only options, which have supplanted a long-time need for expensive international long-distance phone calls.

Video-call interviews may work as the closest surrogate for in-person interaction between the researcher and informant, but they can create new obstacles or concerns for either party. First, Institutional Review Boards might not address data gathering that occurs in the interstices of impersonal, digital data collection and in-person qualitative conversations. In many cases, interviews conducted via video are recorded yet are not subject to release form regulations. Second, the webcam rarely presents a peripheral view for the informant, so it may be possible for terms of trust and intimacy to be broken during the interview without their knowledge. A lack of legal recourse or at least explicitly defined expectations could result in a loss of trust or a compromised relationship between the interviewer and informant.

Relatively little literature addressing this interception of video-chat with qualitative methods appeared until well into the 21st century. Over this decade, a multitude of social scientists have addressed these strengths and weaknesses of Skype in greater detail, including Paul Hanna (2012), Hannah Deakin and Kelly Wakefield (2014), and geographers Gail Adams-Hutcheson and Robyn Longhurst (2017).

Smartphones, social media, and access in ethnographic research

Another disruption in the proliferation of ethnographic data and access to ethnographic sources has been the surging popularity of handheld computing devices (smartphones). In the 2000s, companies like Blackberry recognized the potential for transitioning mobile phones into handheld devices that could access email, take photographs, and transfer files in addition to placing calls and sending SMS (short message service text messages). Companies like Apple and Samsung have engaged in a digital and mechanical arms race over the past decade with products like the iPhone and Galaxy, respectively. Ethnographic researchers, particularly those from middle-class backgrounds in developed countries, carry these with them at most times.

When America Online and other public search engines like Yahoo launched in the 1990s, mobile phones were becoming smaller, more inexpensive, and increasingly ubiquitous. Within a decade, a majority of those engaged in civic life in the developed world owned cell phones, which made people more instantly accessible devoid of attachment to place. However, in order to contact an informant, one still needed to know their phone number, and needed to rely upon access to individuals who could provide contact information. This dynamic was similar to email in the early internet era, as there was no accessible database of private phone numbers or email addresses. Smart phones, in tandem with social media platforms, have made individuals virtually accessible at all times through a multiplicity of means.

If a researcher has wireless internet access, they can contact any informant through their social media account anywhere in the world. Many chain restaurants like McDonald's and Starbucks provide Wi-Fi for free without purchase, and an increasing number of public meeting spaces have Wi-Fi available to paying customers. Gillen (2012) investigated the role of gift-giving in qualitative research, but little has been done otherwise on consumption as a means to an end in the humanities, whether engaging informants through laptops, smartphones, or buying food and drink in a personal interview situation.

Considering how the internet (as private consumers know it) has only been prevalent for fewer than three decades, social science is still charting its impacts on music, media, and the general cultural pale. Many have called attention to how internet access is still at a premium for most in the developing world, which has exacerbated inequalities of development. Though internet connection speeds have made the exchange of audio files and video files almost instantaneous in developed countries, much of the Global South is lagging behind. The same could be said for all questions of technological access, especially smartphones.

The exponentially widening pool of accessible ethnographic data must be understood epistemologically. For every new conduit or forum through which the researcher can gather data, they must consider the limitations and specific dynamics of each. Much of the foundational literature in human geography and media studies made more sense when prominent social scientists (e.g., Sauer 1925; McLuhan 1964) published in the era before email, Facebook, and Twitter. While these technologies can easily blur the lines between professional and personal relationships and alter the dynamic of researcher and informant, much human research like this author's own dissertation (Sonnichsen 2017) might have been impossible to complete without it.

Social activities always occur in a place, so when place in the classic Euclidian sense is compromised, so then must the researcher's positionality. Though it may not have been their directive upon founding the site in 2004, Facebook has unwittingly created a user-generated public archive of cultural ethnography. Individual users' private accounts notwithstanding, users still frequently contribute anecdotes and opinions (of debatable veracity or congruity) onto Facebook-linked news sites and comment sections. Within Facebook itself, many topics worthy of academic research like music scene histories are categorically archived through inclusive user groups where people can upload old images, show flyers, and share reflections on contextual places and events.

To some, social media provides a technological extension of the artistic and consumer possibilities offered by human networks. Few would question whether social media has redefined mechanisms through which music has been created, distributed, and appreciated:

> The creative and expressive potentiality of the artistic productions and the ways of musical use have undergone significant changes with the introduction of digital technologies and social media. For example, for many consumers search for music, inquire, possibly by probing the opinions of other lovers, taste it (with the "pre-listen to") and to buy it, have become interconnected operations within a continuous, constant and collaborative process...In general, music has always played a leading role in the diffusion of communication technologies and locates in social media its raison d'etre as it was born to be consumed, exchanged and shared.
>
> (Prattichizzo 2015, 318)

Though Facebook launched in 2004, it took until this decade to draw adequate attention from social scientists, most of whom can no longer deny its societal role and value to researchers (Wilson, Gosling, and Graham 2012). For many contemporary researchers, Facebook and other social media platforms (Twitter, Instagram, Snapchat) have occupied a valuable role in the "snowballing" process (see Small 2009; Longhurst 2010) whereby fieldwork interviews connect them to others. In a seemingly infinite amount of individual cases, social media also made many interviews possible as well as follow-up communication and engagement with one's informants.

Ostensibly, Facebook has provided an open-access phone book (or email directory) for the 21st century. In my summer 2015 field work in France, numerous informants recommended a friend of theirs with whom they had not spoken in years, and in the process had lost their phone number or email address. Facebook provided the only direct means to contact certain sources both in France and in the United States; even entering their name, record label, or musical project into Google would often produce a Facebook or Instagram link. Asking through an extensive network on the chance that one person may have a way to contact a desired subject via email or phone, though necessary in the past, may be a tremendous waste of time today.

Social media also enable a cloak of anonymity. Many of my informants have maintained Facebook pages under pseudonyms. Several times in the course of my snowballing interviews did people write down their friend's name with "on Facebook as" their friends' aliases. Reasons for this can vary, though they often involve a desire for privacy or separation of vocational and personal life. Two of my informants worked for their country's national train company and wished to keep their online personae separate from their professional personae. Gender and intimidation also play a crucial role. Several of the women I have interviewed use aliases on Facebook in equal measure to avoid professional interference and avoid online harassment. The most frequent reason stated across my recorded interviews was simply "so people can't find me unless I know them."

Similar to YouTube, Facebook has been compiling and archiving (however unintentionally) a living history of cultural geographies. For example, one user in that group devoted to the 80's Washington DC punk scene uploaded a photograph of a flyer from a Dead Kennedys Show at the Lansburgh Cultural Center on Sunday, June 5, 1982 with local openers Scream, No Trend, and Void. A comment string ensued with various members of the group volunteering their memories of that show, including details on how the DC Fire Marshall shut the show down and Dead Kennedys singer Jello Biafra led the crowd in a campfire-style sing-along (Figure 19.2).

Facebook and other social media platforms have become so inextricable from daily life that they must be understood as inextricable from qualitative inquiry as well. As Deborah Thien wrote in her critique of unrealistic objectivity, "the very designation 'social scientist' insists upon a distanced and distancing investigation of the social" (Thien 2011, 313). One can only speculate on how a similar story about underground circulation might be written in another three decades, but social media will undoubtedly play as pivotal a role as the internet at large. Some recent research on ethnography in social media

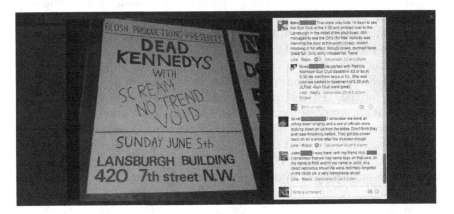

Figure 19.2 Sample of comments in a public Facebook group about 1980s punk and hardcore.

Source: Author: Screen capture.

has taken care to address the inherent messiness of navigating new formats, especially one with such an open-access and paradigm-shifting orientation. This may allow researchers "to follow ethnographically the (dis)continuities between the experienced realities of face-to-face and social media movement and socialities" (Postill and Pink 2012, 123). As publically shared data becomes more interactive at the user level and easier to obtain, researchers and Institutional Review Boards must adapt perpetually in order to anticipate new obstacles in the field like these.

The internet, emotion/affect, and fandoms

While the internet has accelerated many research ontologies, one might argue that some research topics have come into their own due to the online availability of thicker data on human emotions, affect, and fandoms. Two key examples would be the field of emotional geographies (Thien 2005) as well as that of the largely intertwined study on affect (Pile 2010). Both subjects are pivotal in the progression of ethnographic research.

Emotion had never been completely minimized within ethnography, anthropology, and human geography, but explicitly emotion-centered subfields did not rear their heads until well into the internet era. As geographers Kay Anderson and Susan Smith (2001, 7) put it around the turn of the century, engagement and acceptance of emotion were a crucial component in the growth and modernization of their discipline:

> The gendered basis of knowledge production is probably a key reason
> why the emotions have been banished from social science and most other

critical commentary for so long. This marginalization of emotion has been part of a gender politics of research in which detachment, objectivity and rationality have been valued, and implicitly masculinized, while engagement, subjectivity, passion and desire have been devalued, and frequently feminized.

Around that time, geographers like Nichola Wood were spearheading initiatives for understanding emotion's pivotal role in understanding music performance (Wood 2002). Over the course of the 2000s, emotion-centered research became one of the hallmarks of research in cultural geography (Thien 2005). One of the newer and more challenging facets of ethnography, non-representational theory (Lorimer 2008; Thrift 2008), hinges upon the admission that not all crucial data can be (re)presented in concrete terms. It follows, then, that a proliferation of platforms where potential informants can express their loves, hates, and experiences more ephemerally (e.g., Instagram and Snapchat) dig deeper than any formalized focus group or email survey could.

Before the internet, fandom research was limited to fanzines and specifically conducted focus groups and surveys. With the advent of the web, blogging sites democratized fandoms, profoundly democratizing virtual communities and spaces where people could get unite and discuss their favorite foods, movies, music, books, celebrities, and more. Because fandoms necessarily elicit a modicum of emotion and affect on behalf of the participant (and in many cases, researcher), they can provide a valuable arena through which to investigate both.

Because social media took most of the 2000s to gain a hegemonic foothold, social sciences and humanities scholars have been catching up over the course of the 2010s. Video games, for example, which are increasingly tied into internet culture and online connectivity, have been eliciting scholarly attention from various corners of the humanities for most of this time (Taylor 2006). For the few years prior to this writing, a number of researchers have approached these phenomena head on, investigating the relationship between online science fiction fandom communities (Tindall and Hutchins 2016) as well as that of sports fandom and management (Stavros et al. 2014).

Ethics in online ethnographic research

Given the profound expansion of open access to innately personal and emotional ethnographic data in the past two decades, ethnographic social media research has presented debatable ethical quandaries. By entering their opinions, thoughts, and memories into the public record, Myspace, Facebook, and Twitter have made millions of users into unwitting informants for ethnographers. However, depending on the subject matter and adherence of the commenters to the specific network through which the researcher is searching, these informants may never become aware of their contributions. Such is the fluidity of the public record and malleable data on social media. Debates still rage on what constitutes "public" versus "private" on the internet, and the researcher and their informants may not agree on what is fair game. Whether to anonymize these informants is a decision for the researcher; the Institutional Review Board cannot easily mitigate these situations, as online ethnographic research is not conducted in-person. Considering how the investigator did not elicit this data personally, culpability for any unfortunate consequences of their words being published cannot rest completely on one actor's shoulders.

Of course, there are ethical concerns regarding the constantly expanding and updating archive (Postill and Pink 2012). This could include the researcher's presence online and relationship with their subjects (Whiteman 2007), as well as the ability for subjects to (mis)represent data on a variety of levels. This requires more effort out of researchers to triangulate their sources, a process which "offers the opportunity for a thorough understanding of a research topic from multiple data sources, and also invited validation of one source of data by another" (Thien 2009, 75). Most ethnographic researchers should already be triangulating their data to the best of their ability. In some cases, like my oral histories of street violence that took place over 30 years ago, triangulation may be difficult. Alternative data sources on specific incidents and topics may be elusive, exceedingly difficult to contact, or simply not exist. As pervasive as the internet may be, and as often as publicly edited forums debunk certain myths, some urban legends proliferate because they cannot be proven false. Though ethnographers cannot materialize triangulating data out of thin air, the internet's open access nature, like that of oral histories, necessitates information to be prefaced with these qualifiers.

That being said, the internet (especially user-generated content on social media platforms) can open up a wide array of possibilities for triangulation. Over two decades ago, websites, message boards, and email perpetually sped up the process of fact-checking information with available trustworthy sources. Email's usefulness, coupled with searchable public directories of users on Facebook, Twitter, Instagram, and more, continues to provide the most pertinent conduit for these verifications. In many situations, Google, inasmuch as it may front-load search results, can step in and direct the researcher to potential triangulating sources as well.

Ultimately, the internet's role in ethnographic research will only become more pervasive as the internet itself becomes more omnipresent in the life and daily interactions of researchers and their informants. As one beautifully-titled article in *Qualitative Methods* characterized the paradigm shift from an ethnographic perspective, the internet transitioned "from data archive to ethical labyrinth" (Carusi and Jirotka 2009). The rate at which the internet is spreading to incorporate more facets of daily life in more places around the globe is much faster than the rate which technologies have affected the dynamics of ethnographic research in the past. Therefore, social scientists have a prescient and growing responsibility to remain knowledgeable and vigilant about challenges that these changes present.

References

Adams-Hutcheson, G., and R. Longhurst. 2017. "At least in person there would have been a cup of tea": Interviewing via Skype. *Area* 49(2):148–155.

Anderson, K., and S. Smith. 2001. Emotional geographies. *Transactions of the Institute of British Geographers* 26(1):7–10.

Berg, H. 2016. "A scene is just a marketing tool": Alternative income streams in porn's gig economy. *Porn Studies* 3(2):160–174.

Carter, P. 2015. Virtual ethnography: Placing emotional geographies via YouTube. In S. Hanna, A. Potter, E. Modlin, P. Carter, and D. Butler (eds.) *Social Memory and Heritage Tourism Methodologies.* pp. 48–67. London: Routledge.

Carusi, A., and M. Jirotka. 2009. From data archive to ethical labyrinth. *Qualitative Research* 9(3):285–298.

Chaum, D. 1981. Untraceable electronic mail, return addresses, and digital pseudonyms. *Communications of the ACM* 24(2):84–90.

Deakin, H., and K. Wakefield. 2014. Skype interviewing: Reflections of two Ph.D. researchers. *Qualitative Research* 14(5):603–616.

Garrett, B. 2011. Videographic geographies: Using digital video for geographic research. *Progress in Human Geography* 35(4):521–541.

Gillen, J. 2012. Investing in the field: Positionalities in money and gift exchange in Vietnam. *Geoforum* 43(6):1163–1170.

Hanna, P. 2012. Using internet technologies (such as Skype) as a research medium: A research note. *Qualitative Research* 12(2):239–242.

Hutchins, A., and N. Tindall. 2016. *Public Relations and Participatory Culture: Fandom, Social Media and Community Engagement.* London: Routledge.

Longhurst, R. 2009. YouTube: A new space for birth? *Feminist Review* 93:46–63.

Longhurst, R. 2010. Semi-structured interviews and focus groups. In N. Clifford, S. French, and G. Valentine (eds.) *Key Methods in Geography.* pp. 103–115. Thousand Oaks, CA: Sage.

Lorimer, H. 2008. Cultural geography: Non-representational conditions and concerns. *Progress in Human Geography* 32(4):551–559.

Lynch, C. 2003. Institutional repositories: Essential infrastructure for scholarship in the digital age. *Porrtal: Libraries and the Academy* 3(2):327–336.

Lysloff, R. 2003. Musical community on the internet: An on-line ethnography. *Cultural Anthropology* 18(2):233–263.

Madge, C. 2010. Internet mediated research. In N. Clifford, S. French, and G. Valentine (eds.) *Key Methods in Geography.* pp. 173–188. Thousand Oaks, CA: Sage.

McLelland, M. 2002. Virtual ethnography: Using the internet to study gay culture in Japan. *Sexualities* 5(4):387–406.

McLuhan, M. 1964. *Understanding Media: The Extensions of Man.* New York: Signet.

Murthy, D. 2008. Digital ethnography: An examination of the use of new technologies for social research. *Sociology* 42(5):837–855.

Pile, S. 2010. Emotions and affect in recent human geography. *Transactions of the Institute of British Geographers* 35(1):5–20.

Postill, J., and S. Pink. 2012. Social media ethnography: The digital researcher in a messy web. *Media International Australia* 145(1):123–134.

Prattichizzo, G. 2015. Social media is the new punk. User experience, social music and DIY culture. In P. Guerra and T. Moreira Porto (eds.). *Keep It Simple, Make It Fast: An Approach to Underground Music Scenes, vol. 1.* Universidade do Porto – Faculdade de Letras.

Rose, G. 1996. Teaching visualised geographies: Towards a methodology for the interpretation of visual materials. *Journal of Geography in Higher Education* 20(3): 281–294.

Rose, G. 2003. On the need to ask how, exactly, is geography "visual"? *Antipode* 35(2):212–221.

Rose, G. 2012. *Visual Methodologies: An Introduction to Researching with Visual Materials.* 3rd ed. Thousand Oaks, CA: Sage.

Sade-Beck, L. 2004. Internet ethnography: Online and offline. *International Journal of Qualitative Methods* 3(2):45–51.

Sauer, C. 1925. The morphology of landscape. *University of California Publications in Geography* 2(2):19–53.

Seymour, W. 2001. In the flesh or online? Exploring qualitative research methodologies. *Qualitative Research* 1(2):147–168.

Small, M. 2009. "How many cases do I need?" On science and the logic of case selection in field-based research. *Ethnography* 10(1):5–38.

Sonnichsen, T. 2017. Capitals of Punk: Paris, DC, and the Circulation of Urban Counternarratives. Knoxville, TN: University of Tennessee TRACE. https://bit.ly/3aJ3LSW.

Sproull, L., and S. Kiesler. 1986. Reducing social context cues: Electronic mail in organizational communication. *Management Science* 32(11):1492–1512.

Stavros, C., M. Meng, K. Westberg, and F. Farrelly. 2014. Understanding fan motivation for interacting on social media. *Sport Management Review* 17(4):455–469.

Taylor, T. 2006. *Play between Worlds: Exploring Online Game Culture.* Cambridge, MA: MIT Press.

Thien, D. 2005. After or beyond feeling? A consideration of affect and emotion in geography. *Area* 37(4):450–454.

Thien, D. 2009. Feminist methodologies. In R. Kitchin and N. Thrift (eds.) *International Encyclopedia of Human Geography.* pp. 71–78. Oxford: Elsevier.

Thien, D. 2011. Emotional life. In . V. Del Casino, M. Thomas, P. Cloke and R. Panelli (eds.) *A Companion to Social Geography.* pp. 309–325. London: Wiley-Blackwell.

Thrift, N. 2008. *Non-Representational Theory: Space, Politics, Affect.* London: Routledge.

Whiteman, E. 2007. "Just chatting": Research ethics and cyberspace. *International Journal of Qualitative Methods* 6(2):95–105.

Whittaker, S., and C. Sidner. 1996. Email Overload: Exploring Personal Information Management of Email. Paper presented at the Proceedings of the SIGCHI Conference on Human Factors in Computing Systems Boston.

Williams, M. 2001. Book review: Life online: Researching real experience in virtual space. *Qualitative Research* 1(3):426–427.

Wilson, R., S. Gosling, and L. Graham. 2012. A review of Facebook research in the social sciences. *Perspectives on Psychological Science* 7(3):203–220.

Wittel, A. 2000. Ethnography on the Move: From field to net to Internet. Forum Qualitative Sozialforschung/Forum: Qualitative Social Research 1(1). http://www.qualitative-research.net/index.php/fqs/article/view/1131/2517&sa=U&ei=a01.

Wood, N. 2002. "Once more with feeling": Putting emotion into geographies of music. In L. Bondi (ed.) *Subjectivities, Knowledges, and Feminist Geographies: The Subjects and Ethics of Social Research.* Lanham, MD: Rowman and Littlefield.

20 Cyber-spatial cartographies of digital diasporas

Michel S. Laguerre

The phenomenon of digital diasporas has been catalyzed through the use of information technology by both first-generation and long-established immigrant communities in tandem with cross-border virtual interactions between diasporic sites abroad, as well as between each diasporic enclave and the homeland. These sites, deriving from a common homeland and interconnected through online communication, constitute a virtual geographic arena, which can be operationalized to reflect and map the expansion of a nation beyond its designated territorial boundaries (Laguerre 2005, 2010; Srinivasan 2006; Brouwer 2006; Androutsopoulos 2006). The cross-border online activities linking different sites of the cosmonation to each other – diaspora to diaspora and diaspora to homeland – constitute the inclusive cyber-spatial domain of a digital diaspora's public sphere (Parham 2005; Bernal 2005; Royston 2014). To scale a digital diaspora's public sphere is to recognize that it is a component of a cosmonational public sphere, which covers an expansive digital geography (Hanafi 2005;).

This chapter identifies and discusses key aspects of scholarship on the geographies of digital diasporas, addressing issues pertaining to their deployment and everyday practices. It further stresses variables that define digital diaspora communities and explains the social engineering of cosmonational digital social practices. Additionally, this essay considers issues of online interactions within the diasporic public sphere, political facets of digital diasporas, how race and gender are reconfigured online, and how social media contributes to the identitary performances of digital diasporas. Lastly, the chapter concludes with some remarks on critical cyber-cartographical aspects of digital diasporas.

Digital diasporas

The term *digital diaspora* derives from the traditional concept of *diaspora*, which was used at first to refer to the geographical dispersion of the Jewish people, and has since been expanded to include immigrant individuals and neighborhoods of any origin living outside of their physical homelands (Safran 1991; Sheffer 2003; Dufoix 2008; Tölölyan 1991; Boyarin and

Boyarin 1993; Cohen 1997; Gruen 2002; Brubaker 2005; Clifford 1994). This broadened application of the term emphasizes the "dispersion" dimension of diaspora above any particular form of emigration or mode of relation of the group to the homeland, and has been applied to pinpoint the immigrant identities of diverse neighborhood communities (Ben-Rafael 2010, Ben-Rafael and Sternberg 2009).

Likewise, the term "digital diaspora" invokes a plurality of ethnicity-based cyber-spatial groups each emanating from a certain homeland while dispersed in various hostland sites, and forming a cross-border multisite digital nation through internet connections and communications (Bahri 2001; Hiller and Franz 2004; Bravo 2013). For this reason, the cyber-spatial cartography of each digital diaspora is unique in its geographical configuration of variables. The digital mode of interactions used by these groups not only shapes the trajectory, speed, and intensity of diasporic communication, but also enables cross-border practices by establishing links between the homeland and diasporic sites as well as sustains relations among diasporic enclaves.

To define the cyber-spatial cartographies of digital diasporas, one must examine the mapping of various layers of their cross-border cyber-relations: those between the homeland and diaspora sites, between diasporic enclaves, and within each diasporic site. Each digital diaspora encourages participation in a global public sphere where actors in multiple sites of the cosmonation engage in discussions on themes of common interest (Bernal 2005, 2014). Likewise, the cross-border architecture of each digital diaspora is distinct from that of others based on the demographic composition of its public sphere, the varied geographical locations of its constituent enclaves, the network governance of its operation, the frequency and intensity of communication between dispersed sites, the hierarchical positioning of sites, and the cultural content of the information exchanged.

Digital diaspora expresses its singularity through its virtual identity, which affords cyber- communities with characteristics distinguishable from pre-internet era diasporic practices (Chan 2005; Ding 2007; Ignacio 2005). Many factors should be taken into consideration when ascertaining what the use of cyberspace has added to the traditional diaspora model, notably: the geographical expansion of diasporic activities that cyberspace facilitates, the virtual cross-border networking it engenders, the transnational public sphere it enables, the intensity and speed of communications it amplifies, the hybrid integration of the digital and traditional models of diasporic engagement that it makes possible, the one-to-one and one-to-many cross-border communications among members of the cosmonation that it encourages, the collaboration through shared online platforms that it helps materialize, and the digital archive of past conversations – accessible anytime, anywhere, and through any medium – that it makes available to diasporic internauts.

Given the rate of technological progression to date, the integration of the internet into daily life is becoming so fundamental to diaspora individuals and communities that the distinction between physical and digital diasporas will eventually become obsolete. Many immigrant groups have already reached this point, while others are closely following behind. In comparison, the traditional diaspora is characterized partly by the lack of access to and minimal use of the internet. For this reason, it is appropriate to revisit the question and adopt the distinction between the *pre-internet traditional diasporas* and the *contemporary diasporas*. The difference in modus operandi is between the immigrants who have engaged with their diaspora identities before the information technology revolution and those who make use of the internet, cellphone services, and other digital media (e.g., SMS, WhatsApp, Facebook, Twitter, Instagram, Snapchat, etc.) to facilitate communication with family, friends, and compatriots left behind in the homeland, as well as with distant compatriot diasporic individuals or communities.

Digital diasporic community

A digital diaspora community does not occupy a traditional physical place, but it is instead defined as a globally-interconnected virtual site through which one remains connected to the ancestral homeland and other diaspora enclaves, develops, and maintains a hybrid identity, and uses the online environment to nurture cosmonational solidarity. It is foremost a discursive space where mutual concerns, including events in all the sites of the cosmonation, are discussed and debated. This suggests that cosmonational identity, thanks to the dispersion of the citizenry, is developed and consolidated (Pendery 2008; McAuliffe 2007; Sahoo 2006; Graham and Khosravi 2002).

However, communities in cyberspace do not exclusively fall into the domain of digital diaspora, as they may be found by non-diasporic groups as well. Some communities are defined by their shared interests, which are fostered online. In other words, nationality is not the only criterion one uses to form virtual communities.

At first, studies of virtual communities interpreted such entities as either a prolongation of or as separate formations of physical communities (Healy 1997; Hofstede 1997; Rheingold 1993; Skrbis 1998). During this time, questions concerning relationship between virtual and physical spaces were just beginning to surface in academic journals. The concept of virtual community was and is not exclusive to any group, as it was conceived to be made of diverse social formations ranging from interest groups, formal associations, professional organizations, ethnic communities, and so on. Part of the confusion in early discussions of virtuality rests with the geographical metaphor used to characterize its identity and forms of operation. For some scholars, it is important to maintain the separation between virtuality and reality as belonging to different domains, while others theorize the relations in terms of a continuum – virtual reality versus real virtuality (Castells 1996). From the standpoint of the latter, the virtual is seen as being embedded in the real and inevitably entangled in both our experiences and explanations of our daily life.

The phenomenon of digital diasporic communities comes to the fore with the use of the internet, being identified as a virtual manifestation and extension of a pre-existing physical community, while distinguishing itself through the performance of its virtual modalities (Laguerre 2010). The digitalization of communities enhances the capacity to nurture a hybrid identity among diasporans abroad, as this new development increases individual awareness of the societal conditions of the cosmonation; at first, this was useful in identifying two types of community building, the traditional and the virtual (Rheingold 1993). However, with the evolution of practices whereby the internet is used by the majority of the population as an extension of reality, "digital" as a prefix to diaspora will gradually become redundant. The advent of this hybrid ecosystem insinuates a new phase in the evolution and life cycle of diasporas.

The study of digital diasporic communities has raised concerns pertaining to what constitutes the boundaries of any given community and its public sphere, including one's ability to prevent encroachment by those considered outsiders, the hybridization of culture and values, the social implications of transnational migration, and both the sustainability and resilience of a community.

As such, participants of digital diasporic communities maintain contact with one another by contributing to an online space of reverence that acknowledges the ancestral homeland, maintaining relationships of shared diasporic experiences, and engaging in virtual collaboration (Brinkerhoff 2004, 2006, 2009). Furthermore, the common bonds created by physical separation from and cultural attachment to the cherished homeland are reasserted.

As digital diasporas consist of hybrid communities, one must recognize their internal differences – heterogeneity within the homogeneity (Elkins 1997; Lee 2009; Poster 1998). Fragmentation within any diaspora may occur due to differences between immigrant generations (such as first versus second generation), language barriers, countries of origin (whether born in the homeland or in diaspora), religion (as group members may subscribe to different faiths), and income levels. For example, some participants may be fluent in the homeland's national language, while others are more fluent in the language of the host-land. Thus, the demographic composition of a digital diaspora is as varied as that of the physical diaspora. In addition, one must also point out that the traditional diaspora is characterized by its multiple-place basis, while the digital diaspora is characterized by its multisite organizational form of operation.

Despite this multiplicity of backgrounds, digital diasporas form cyberspatial communities when immigrant netizens participate in discussions of diasporic or cosmonational interest spearheaded by compatriots at home or abroad (Bernal 2014). In this context, interactions online can affect the ethnic neighborhood experience offline in small or large ways and allow dispersed diaspora communities to interface virtually with each other without having to travel abroad. Moreover, resources for diasporans are made available through any websites where information can be accessed – where location of service providers are made available, where diaspora entrepreneurs can advertise their businesses, and where community problems and successes are discussed.

Cosmonational digital social practices

It has always been a priority of emigrant communities to maintain ties with family and compatriots living in other countries, as well as in the homeland itself (Blanc et al. 1995; Levitt 2001). Despite living abroad, immigrants have forged new digital paths to belonging, socialization, and embeddedness with the evolving multisite culture of the cosmonation. This expansion of cross-border belonging is distinct for its lack of face-to-face interaction and it requires new forms of operation that account for complications involving distance, mobility, time zones, and electronically mediated communications (Alonso and Oiarzabal 2010). Additionally, the space of a nation expands based on the interactions and imbrications of the diaspora in homeland affairs. In the social realm, emigrant communities use the internet to engage in discursive practices that revisit the history of the homeland, government policies under debate, and concerns of identity, culture, and national attachment that have bound the diaspora to the homeland (Olorunnisola 2000).

Cosmonational digital practices have materialized with the development of new social media (e.g., internet, Facebook, Twitter) forums that facilitate socialization of individuals into the multisite hybrid culture of the cosmonation. By contrast, traditional forms of mass media (ethnic radio, ethnic television, and ethnic newspapers) mostly aim to assimilate their subscribers into the culture of the homeland nation (Laguerre 2016). With the internet, one passes from the nation, enclosed in a territory, to the cosmonation, spread in multiple extraterritorial sites. The cross-border interactions the internet facilitates sustain the reality of the cosmonation, rendering the digital community functional, and the digital diasporic public sphere possible. Thus, the use of the internet, along with increased international migration has led to a decentralization and re-orientation of our conception of a nation by expanding its diasporic tentacles while simultaneously contributing to its formatting as a cosmonation.

Because expatriates undergo different types of resettlement, diasporas evolve in many ways, especially in consideration of the internet. Consequently, such geographical dispersion spurs individuals to develop singular ties with varied sites, and individuals in diaspora may be relatively speaking physically confined to a specific location, whether for economic, religious, political, linguistic, historical, demographic or geographical reasons (Laguerre 2016). Bridging this gap by enabling conversations across fragmented enclaves, the use of information technology by the diaspora is a fundamental process for the formation and unification of the cosmonation, the digital diaspora being one of its defining components.

Politics and foreign policy

In the political realm, diasporans have employed the internet for a variety of reasons: to raise funds for political campaigns, to rescue homeland communities in distress (i.e., from natural disasters and political instability), and to defray expenses incurred in the implementation of village projects undertaken under the aegis of diasporic hometown associations (Laguerre 2016, 2013; Adamson 2015).

At the same time, the role of digital diasporas in politics is multifaceted, and may be used for various ends: to strengthen the power of a national leader, undermine the popularity of a government, support the consolidation of peace accords, contribute to sectarian operational activities and oppositions, and spur political action – either by encouraging rebellion and resistance or advancing the causes of the status quo.

Such diasporic entanglement in homeland conflict transnationalizes the issues, as expatriates become key players to the extent that solutions may require the diaspora's acquiescence to the terms of an agreement, further complicating the application of such a resolution. This transnational entanglement can take different forms (Jelin and Kaufman 2002; Kumat 2012; Sheffer et al. 1986); for example, the use of a common digital platform affords migrants and homeland residents an opportunity to collaborate during times of conflict or change, as seen in the upheavals in Syria, Egypt, and Tunisia during the first decade of the 21st century.

The internet allows for the construction of new virtual sites that facilitate political discourse on a common set of questions: whether to engage in restructuring homeland politics; to endorse candidates and contribute to the expansion of electoral support; to engage in opposition politics or to sustain government policies; or to publish analyses online to broaden accessibility and enlighten the democratic visions of the homeland electorate (Kadende-Kaiser 2006; Navarrete and Huerta 2006; Siddiquee and Kagan 2006). The internet has been used by diasporans to accomplish all of the above.

The role of digital diasporas in protest movements – in spearheading antigovernment rallies and contributing to the upheaval of various regimes – has been documented in the cases of the Tiananmen Square event in China, the fall of the Mubarak regime in Egypt, and the Arab Spring revolution in Tunisia. In these cases, social media was accused for encouraging protests, as activists utilized it to provide tactical advice, undertake virtual fundraising campaigns, and circulate live updates on the movements' developments. However, existing regimes and their supporters use the same tools to denounce protesters and sympathizers with alarming threats and attempts to interfere in their organizations.

Furthermore, the internet has enabled greater engagement between the diaspora and the homeland state in the political arena (Laguerre 2013; Tekwani 2003; Tynes 2007). Digital diasporas are often informed about the everyday activities of the homeland government as well as its current situational and systemic undertakings thanks to accessible online information. Likewise, the government has access to information about diaspora organizations and is aware of diasporic political activities both on the ground abroad and inside the homeland's territory. Thus, the digital diaspora's intervention in hostland politics reappears in yet another configuration of variables. Here, virtual political activism may be undertaken for the election of a member of the diaspora in their new country of residence. Publicity for such a candidate is done online to grow the electoral basis, raise funds, urge compatriots to vote, and encourage people to show up for rallies, political meetings, and at the booths on election day. The aim of this effort is to ensure the electoral success of a political candidate who is a diaspora-compatriot.

Digital cosmonational public sphere

The internet enables diasporans, however dispersed they may be, to maintain an ongoing cosmonational conversation through multiple modes of interaction (Parham 2005; Radhakrishnan 2008). In the digital public sphere, virtual space is used to discuss matters of common concern among a diaspora community and the homeland, whether it is exclusively regarding the diaspora, the homeland, or both (Mayes 2009). Many actors can participate in this space of debate. For instance, cosmonation-based, internet-analyst compatriots, who provide content and play a greater role in distinguishing nuances and presenting the details of issues, should be distinguished from ordinary participants of diaspora, who tend to consume and react to information in circulation among the group rather than provide it.

Digital public sphere scholarship addresses issues of individual agency, institutional organization, and hubs of information, in addition to the open manner in which data is shared (Brouwer 2004). A digital archive is maintained for possible recuperation of information previously posted online. What distinguishes this digital public sphere from Jurgen Habermas' offline public sphere is its virtual rendering.

The digital diasporic public sphere is an essential component of each diaspora community today; considering the role it plays in forging and maintaining the unity of the group and its capacity to reproduce itself (Yang 2003; Richman and Rey 2009; Norris 2004). This is exemplified by the Eritrean diaspora, which uses digital forums to discuss solutions to ending the civil war, the Chinese diaspora, often voicing opinions on the direction of the country under its present leadership, the Haitian diaspora for fundraising to alleviate the plight of the victims of the 2010 earthquake, and the Iranian diaspora to support or oppose the government's foreign policy orientation.

Nonetheless, the diasporic public sphere is consistently fed by the input of local diasporans on various subjects (Kok and Rogers 2016; Kim and Ball-Rokeach 2010). While this can be a source of liberation, it can also further the isolation of ethnic immigrants because of the singularity of learned information, the types of transactions executed, and issues of common interest discussed. This niche (perhaps echo-chamber) of diasporic forums can be attributed to their specific focus and orientation, which chiefly attract those already inclined to participate in such undertakings; likewise, the demographic base of such forums, often specific to the different countries, regions, or cities of residence of the diaspora, impact and perhaps limit the participant perspective found on such sites.

Each diaspora creates its own distinct sector of the cosmonational public sphere, which emerges as an interactive cross-border virtual commons. No government has control over such a space because of the distinct transfrontier parameters of its globality; thus, its composition is broader and more diverse than that of traditional diasporic forms, as are the ideas exchanged there.

Race and digital diasporas

Digital diaspora does not operate in a strictly neutral zone, but in a virtual space where race, ethnicity, and national origin continue to be of significance; nonetheless, these factors manifest themselves differently through this virtual medium, as forms of online communication can be policed by norms of nondiscrimination, netiquette, practices of silencing, and managerial advocacy for transparent language. Thus, ICTs provide platforms for discussions of racial issues, including ways to remove oppressive content and overcome discriminatory practices, either through strategic production of text, biometric profiling, big data analytics, or algorithmic applications (Everett 2009; Byrne 2008).

Many groups of different orientation have utilized the internet to achieve a variety of goals. Some spread racist rhetoric to recruit followers while others aim to combat forms of virtual discrimination, use social media to interact with members of their ethnic group, attract clientele to their businesses or prospective members to their churches, to teach or learn the language of the homeland, provide services to their diasporic compatriots, or seek service providers among members of their ethnic community.

Race factors in the study of digital diasporas in regards to issues of identity – its hybrid content, transnational orientation, and virtual manifestation (Panagakos 2003; Kissau and Hunger 2010; Mitra 2005, 2006; Nakamura 2002). Ethnic groups virtually interact among themselves to consolidate their enclaves, and maintain ongoing communication with the homeland and other diasporic sites, which comprise the cosmonational ensemble (Dahan and Sheffer 2001). Numerous diaspora enclaves in North America and the European Union digitally manifest a vision of cross-border solidarity with compatriots dispersed across many lands. When considering assimilation theory, the global aspect of racial or ethnic solidarity has been less prominent, but is becoming more visible as it intersects with transnationalist theory, and even more so with cosmonationalist theory.

Lack of access to the internet is another prominent complication in the discussion of digital diasporas, as it insinuates a divide between the haves and the have-nots (Kollo and Reid 1998). To elaborate, certain subgroups of the diaspora and distinct ethnic communities may be isolated from the broader and general community due to lack of access to the internet. This inequality in the digital diaspora arena can be identified in terms of first versus second-generation of migration, or the inability of immigrants of color to assert themselves within the mainstream discourse due to the digital divide (Mele 1999).

However, as the internet becomes more and more accessible in diasporic communities, the discourse on the digital divide has receded and is no longer prominent in academic debate (DeHart 2004). Access to the internet is now available not only through computers, but also through other digital devices, including the ubiquitous cellphone. A central concern, however, is how race is taken into account in the discourse on digital diaspora. Much research has been undertaken to analyze manifestations of race online as different from its deployment offline. In addition to studying how internet service providers prevent their sites from being used as canvasses to spread virtual forms of discrimination, studies have examined the dimensions of virtual belonging and how ethnicity is a catalyst that forms virtual diasporic communities who lobby on behalf of their homeland.

New social media

Discussions of the relationship between diasporas and new social media platforms range from substantive, systemic inquiries to situational and transient consideration (Lee 2009; Gillespie 2000; Skrbis 1998,). The critical contribution of diasporan media lies in its contrast to that of their homeland media, as the latter often limits freedom of expression. Remote online diasporan journalists deploy this critical edge over homeland journalists due to their physical separation from the homeland, as it provides them freedom to more actively and aggressively criticize shortcomings of the homeland's government, given the relative unlikeliness of potential brutal reactions.

Diasporas make use of a panoply of communication technologies (Richman and Rey 2009), the most utilized being the internet for communication and access to information; internet telephone services to request remittances; cellphones to receive and send texts; blogs to contribute to ongoing discussions; internet television to keep abreast of recent political developments; and internet radio to access contents of informal and local opinions about conditions in the homeland. This variety of media emphasizes the extent to which the diaspora lives at the pulse of homeland political activities.

Social networking platforms are tools used to empower the diaspora community, provide information to facilitate transactions, and to educate and update members on the everyday survival of the group. Through Twitter, sharing takes a more ubiquitous form, as it is done more often because of its intended informal use among friends, acquaintances, and based on one's personal network.

Additional digital tools provide various modes of virtual interaction with compatriots. Diaspora blogs allow individuals to produce and share contents with subscribers, who tend to be members of the same homeland group, precisely because it is often done using the official language of the homeland country (Hess 2007).

Many social media forums that the diaspora create or use are not geared exclusively toward any specific diasporic enclave, but rather address affairs of the cosmonation, that is, the homeland and diaspora sites abroad, irrespective of the host countries they reside in. While the internet plays a key role in consolidating local sites, it also facilitates the interconnection of distant sites to one another in the global integration of the world network and production of the cosmonational public sphere, where issues concerning each site and the network as a whole are made known to every participating member. New forms of communication like the internet and the cellphone unify a digital diaspora and its enclave internally while integrating it into the cosmonation at large by facilitating its access to other media forms. Using these mediums, one can read the homeland newspaper, listen to local radio programs, and view televised homeland or diasporic programs online to stay updated on the cosmonation's site traffics.

Cyber-spatial cartography

The spatial configurations of digital diasporas are part and parcel of the "geography" of the internet; thus, cartography is a useful metaphor expressing the many forms a diaspora may take, based on the settlement locations of immigrant communities and the varied relations between sites. Proximity to other enclaves, the circulation of people, goods, and the exchange of information from one site to another, as well as the hierarchy of sites all determine the nature of these relations. Thus, territorial conditions affect the ways in which digital diasporas represent themselves in online forums (Blunt 2007; Brah 1996; Carter 2005; Dahlman 2004; Dickinson and Bailey 2007; NiLaoire 2003).

Although functioning in cyberspace, the hierarchy of importance within a digital diaspora depends on physical locations (e.g. the oldest neighborhood where the group has settled, the place where wealthier diasporans live, the settlement locations of the most important and visible political actors, or the size of the population in a given location) as well as upon the digital hubs where important diasporan figures interact (the popularity of certain forums, websites, or ICTs used by a great majority of diasporans).

When drawing the cartography of a digital diaspora, in reference to the geographical locations of interactive cyberspace hubs, one confronts the issue of the dispersion of a given immigrant community, which uses cyberspatial networking as a means to re-assemble itself. This cartography is in constant flux when accounting for both geographical residence and digital participation. Sites cannot be identified without the participation of community members as contributors to or users of digital contents. Thus, cartographies of digital diasporas are tools that provide information on the hierarchies of cyber-spatial sites, the relational intensity between sites, the extent of the geographical dispersion of the diaspora group, and the particular role each site plays in terms of its contribution and functionality in the cosmonational landscape.

In reference to connections between members of different sites, cartography speaks of the digital diaspora as a *space of communication*. Here communication is undertaken between and among active members belonging to networking sites with differing levels of relational intensity depending on cyclical periods of the year (peak versus dormant periods, influenced by factors like elections, holidays, or anniversaries), as well as fluctuations in information contents depending on current circumstances (times of conflict, political crisis, or environmental disaster).

In this context, cartography can delineate various layers of transactions between sites (Harris and Harrower 2006; Crampton 2001; Crampton and Krygier 2015). For example, a digital map of economic transactions may differ from a map of religious communications due to the forms such networks take as they attract participation from people from different demographics within the diaspora.

Cartography also unveils the *space of mobility* characterized by exchanges of goods between sites, especially in regards to transnational commerce, migration of people from one site to another (whether it be from a homeland to a diaspora site, or from one diaspora site to another, or even back to the homeland), the exchange of images, and the circulation of ideas. These are some of the many ways in which mobility between sites is expressed in a digital diaspora; and tracing these networks provides fundamental insights into how social exchanges operate in their varied and transnational manifestations.

The role of digital diaspora in providing a discursive space has been identified as a central function of virtual communities, as it exemplifies how such a community brings continuity with physical places. It is a conversational space – a common arena where physical presence is replaced by telepresence. The centrality of this discursive space to community-building has informed its conceptualization as a digital public sphere.

Furthermore, digital diaspora constitutes an augmented space to an extent that interactions are reconfigured in alternative spaces; this additional dimension provides a locus from which to undertake activities that may not be circumstantially possible in physical space. This augmented space is seen as a continuum, a reflection, and an accomplishment; it is a geography in circulation that expresses a reality in motion, somewhat nomadic and flexible, adding to the mobility of its character.

The geography of digital diaspora is also scalar and can be analyzed in considering how a specific hub or site is used to facilitate interactions, whether communication operates on a local, regional, or cosmonational level, or simply in reference to the types of content (professional, political, educational, or recreational) being exchanged. Thus, a plurality of approaches are possible when attempting to map the digital geography of diasporic exchanges.

Acknowledgements

I am grateful to Shelby Call, Francesca Ciacchella, and Ziheng Liu for research assistance in the preparation of this chapter.

References

Adamson, F. 2015. Blurring the lines: Diaspora politics and globalized constituencies. *World Politics Review* (July 14). www.worldpoliticsreview.com/articles/16224/blurring-the-lines-diaspora-politics-and-globalized-constituencies.

Androutsopoulos, J. 2006. Multilingualism, diaspora, and the internet: Codes and identities on German-based diaspora websites. *Journal of Sociolinguistics* 10(4):520–547.

Alonso, A. and P. Oiarzabal (eds.) 2010. *Diasporas in the New Media Age*. Reno, NV: University of Nevada Press.

Bahri, D. 2001. The digital diaspora: South Asians in the new pax electronica. In M. Paranjpe (ed.) *Diaspora: Theories, Histories, Texts*. New Delhi: Indialog.

Ben-Rafael, E. 2010. Diaspora. Sociopedia.isa.

Ben-Rafael, E. and Sternberg, Y. (eds). 2009. *Transnationalism: Diasporas and the Advent of a New (Dis)order*. Leiden, the Netherlands: Koninklijke Brill NV.

Bernal, V. 2005. Eritrea on-line: Diaspora, cyberspace, and the public sphere. *American Ethnologist* 32(4):660–675.

Bernal, V. 2014. *Nation as Network: Diaspora, Cyberspace, and Citizenship*. Chicago: University of Chicago Press.

Blanc, C., Basch, L. and Schiller, N. 1995. Transnationalism, nation-states, and culture. *Current Anthropology 36*(4):683–686.

Blunt, A. 2005. Cultural geography: Cultural geographies of home. *Progress in Human Geography* 29(4):505–515.

Blunt, A. 2007. Cultural geographies of migration: mobility, transnationality, and diaspora. *Progress in Human Geography* 31(5):684–694.

Boyarin, D. and Boyarin, J. 1993. Diaspora: Generation and the ground of Jewish identity. *Critical Inquiry* 19(4):693–725.

Boyle, M. 2001. Towards a (re)theorisation of the historical geography of nationalism in diasporas: The Irish diaspora as an exemplar. *International Journal of Population Geography* 7(6):429–446.

Brah, A. 1996. *Cartographies of Diaspora: Contesting Identities*. New York: Routledge.

Bravo, V. 2013. *Diaspora Online: Identity Politics and Romanian Migrants*. New York: Berghahn Books.

Brinkerhoff, J. 2004. Digital diaspora and international development: Afghan-Americans and the reconstruction of Afghanistan. *Public Administration Development* 24:397–413.

Brinkerhoff, J. 2006. Digital diasporas and conflict prevention: The case of Somalinet. Com. *Review of International Studies* 32:25–47.

Brinkerhoff, J. 2009. *Digital Diasporas: Identity and Transnational Engagement*. New York: Cambridge University Press.

Brouwer, L. 2004. Dutch Muslims on the internet: A new discussion platform. *Journal of Muslim Minority Affairs* 24(1):47–55.

Brouwer, L. 2006. Dutch Moroccan websites: A transnational imagery? *Journal and Ethnic and Migration Studies* 32(7): 1153–1168.

Brubaker, R. 2005. The "diaspora" diaspora. *Ethnic and Racial Studies* 28(1):1–19.

Butler, K. 2001. Defining diaspora, refining a discourse. *Diaspora* 10(1):189–220.

Byrne, D. 2008. The future of (the) "race": Identity, discourse, and the rise of the computer-mediated public sphere. In A. Everett (ed.) *Learning Race and Ethnicity: Youth and Digital Media*. pp. 15–38. Cambridge, MA: MIT Press.

Carter, S. 2005. The geopolitics of diaspora. *Area* 37(1):54–63.

Castells, M. 1996. *The Rise of the Network Society*. Cambridge: Blackwell.

Chan, B. 2005. Imagining the homeland: The internet and diasporic discourse of nationalism. *Journal of Communication Inquiry* 29(4):336–338.

Clifford, J. 1994. Diasporas. *Cultural Anthropology* 9(3):302–338.

Cohen, R. 1997. *Global Diasporas: An Introduction*. Seattle: University of Washington Press.

Crampton, J. 2001. Maps as social constructions: Power, communication and visualization. *Progress in Human Geography* 25:235–252.

Crampton, J. and J. Krygier 2015. An introduction to critical cartography. *ACME: International Journal for Critical Geographies* 4(1):11–33.

Dahan, M. and Sheffer, G. 2001. Ethnic groups and distance-shrinking communication technologies. *Nationalism and Ethnic Politics* 7:85–107.

Dahlman, C. 2004. Diaspora. In J. Duncan, N. Johnson and R. Schein (eds.) *A Companion to Cultural Geography*. pp. 485–498. Oxford: Blackwell.

DeHart, M. 2004. "Hermano entrepreneurs!" Constructing a Latino diaspora across the digital divide. *Diaspora: A Journal of Transnational Studies* 13:253–278.

Dickinson, J. and Bailey, A. 2007. (Re)membering diaspora: Uneven geographies of Indian dual citizenship. *Political Geography* 26(7):757–774.

Ding, S. 2007. Digital diaspora and national image building: A new perspective on Chinese diaspora study in the age of China's rise. *Pacific Affairs* 80(4):627–648.

Dufoix, S. 2008. *Diasporas*. Berkeley: University of California Press.

Elkins, D. 1997. Globalization, telecommunication and virtual ethnic communities. *International Political Science Review* 18(2):139–152.

Everett, Anna. 2009. *Digital Diaspora: A Race for Cyberspace*. New York: SUNY Press.

Gillespie, M. 2000. *Transnational Communications and Diaspora Communities. Ethnic Minorities and the Media*. Buckingham, UK: Open University.

Graham, M. and Khosravi, S. 2002. Reordering public and private in Iranian cyberspace: Identity, politics, and mobilization. *Identities: Global Studies in Power and Culture* 9(2):219–246.

Gruen, E. 2002. *Diaspora: Jews amidst Greeks and Romans*. Cambridge, MA: Harvard University Press.

Hanafi, S. 2005. Reshaping geography: Palestinian community networks in Europe and the new media. *Journal of Ethnic and Migration Studies* 31(3):581–598.

Harris, L. and M. Harrower 2006. Critical interventions and lingering concerns: Critical cartography/GISci, social theory, and alternative possible futures. *ACME: International Journal for Critical Geographies* 4(1):1–10.

Healy, D. 1997. Cyberspace and place: The internet as middle landscape on the electronic frontier. In D. Porter (ed.) *Internet Culture*. pp. 55–71. New York: Routledge.

Hofstede, G. 1997. *Cultures and Organization: Softwares of the Mind*. New York: McGraw Hill.

Hess, A. 2007. In digital remembrance: Vernacular memory and the rhetorical construction of web memorials. *Media, Culture & Society* 29(5):812–830.

Hiller, H. and Franz, T. 2004. New ties, old ties and lost ties: The use of the internet in diaspora. *New Media & Society* 6(6):731–752.

Ignacio, E. 2005. *Building Diaspora: Filipino Community Formation on the Internet*. New Brunswick, NJ: Rutgers University Press.

Jelin, E. and Kaufman, S. 2002. Layers of memories, twenty years after in Argentina. In D. Lorey and W. Beezley (eds.) *Genocide, Collective Violence, and Popular Memory*. pp. 31–52. Wilmington, DE: Scholarly Resources.

Kadende-Kaiser, R. 2006. The transformation of discourse online: Toward a holistic diagnosis of the nature of social inequality in Burundi. In K. Landzelius (ed.) *Native on the Net: Indigenous and Diasporic People in the Virtual Age*. pp. 220–237. London: Routledge.

Kim Y.C. and Ball-Rokeach, S. 2010. New immigrants, the internet, and civic society. In A. Chadwick and P. Howard (eds.) *The Routledge Handbook of Internet Politics*. pp. 275–287. London and New York: Routledge.

Kissau, K., and Hunger, U. 2010. The internet as a means of studying transnationalism and diaspora. In Bauböck, R. and Faist, T. (eds.) *Diaspora and Transnationalism: Concepts, Theories and Methods*. pp. 245–266. Amsterdam: Amsterdam University Press.

Kok, S., and Rogers, R. 2016. Rethinking migration in the digital age: Transglocalization and the Somali diaspora. *Global Networks* 17:23–46.

Kollo, B. and Reid, E. 1998. Dissolution and fragmentation: Problems in on-line communities. In S. Jones (ed.) *Cybersociety 2.0: Revisiting Computer-mediated Communication and Community*. pp. 212–229. Thousand Oaks, CA: Sage.

Kumar, P. 2012. Transnational Tamil narratives: Mapping engagement opportunities on the web. *Social Science Information* 51:4578–4592.

Laguerre, M. 2005. Homeland political crisis, the virtual diasporic public sphere, and diasporic politics. *Journal of Latin American Anthropology* 10(1):206–255.

Laguerre, M. 2010. Digital diaspora: Definition and models. In A. Alonso and P. Oiarzabal (eds.) *Diasporas in the New Media Age: Identity, Politics, and Community*. pp. 49–64. Reno, NV: University of Nevada Press.

Laguerre, M. 2013. *Parliament and Diaspora in Europe*. New York: Palgrave Macmillan and NYU European Studies Series.

Laguerre, M. 2016. *The Multisite Nation: Cross-border Organizations, Transfrontier Infrastructure, and Global Digital Public Sphere*. New York: Palgrave Macmillan.

Laguerre, M. 2017. *The Postdiaspora Condition: Cross-border Social Protection, Transnational Schooling, and Extraterritorial Human Security*. New York: Palgrave Macmillan and NYU European Studies Series.

Laguerre, M. 2018. Diasporas and the internet. In B. Warf (ed.) *Encyclopedia of the Internet*. Thousand Oaks: Sage.

Lee, H. 2009. *Bittersweet Homecomings: Ethnic Identity Construction in the Korean Diaspora*. Ph.D. Dissertation. University of California, Santa Barbara.

Levitt, P. 2001. Transnational migration: Taking stock and future directions. *Global Networks* 1(3):195–216.

Mavroudi, E. 2008. Palestinians in diaspora, empowerment, and informal political space. *Political Geography* 27(1):57–73.

Mays, W. 2009. Unsettled post-revolutionaries in the online public sphere. *Sojourn* 42(1):89.

McAuliffe, C. 2007. A home far away? Religious identity and transnational relations in the Iranian diaspora. *Global Networks* 7(3):307–327.

Mele, C. 1999. Cyberspace and disadvantaged communities: The internet as a tool for collective action. In Smith, M. and P. Kollock (eds.) *Communities in Cyberspace.* pp. 290–310. London: Routledge.

Mitra, A. 2005. Creating immigrant identities in cybernetic space: Examples from a non-resident Indian website. *Media, Culture and Society* 27(3):371–390.

Mitra, A. 2006. Towards finding a cybernetic safe place: Illustrations from people of Indian origin. *New Media and Society* 8(2):251–268.

Nakamura, L. 2002. *Cybertypes: Race, Ethnicity, and Identity on the Internet.* New York: Routledge.

Navarrete, C. and Huerta, E. 2006. Building virtual bridges to home: The use of the internet by transnational communities of immigrants. *International Journal of Communications Law & Policy* (special issue). Autumn:1–20.

NiLaoire, C. 2003. Editorial introduction: Locating geographies of diaspora. *International Journal of Population Geography* 9(4):275–280.

Norris, P. 2004. The bridging and bonding role of online community. In P. Howard and S. Jones (eds.) *Society Online: The Internet in Context.* pp. 31–41. Thousand Oaks, CA: Sage.

Olorunnisola, A. 2000. African media, information providers and emigrants as collaborative nodes in virtual social networks. *African Sociological Review* 4(2):46–71.

Panagakos, A. 2003. Downloading new identities: Ethnicity, technology, and media in the global Greek village. *Identities: Global Studies in Culture and Power* 10(2):201–219.

Parham, A. 2005. Internet, place, and public sphere in diaspora communities. *Diaspora: A Journal of Transnational Studies* 14(2):349–380.

Pendery, D. 2008. Identity development and cultural production in the Chinese diaspora to the United States: New perspectives: 1850–2004. *Asian Ethnicity* 9(3):201–218.

Poster, M. 1998. Virtual ethnicity: Tribal identity in an age of global communications. In S. Jones (ed.) *Cybersociety 2.0: Revisiting Computer-mediated Communication and Community.* pp. 184–211. Thousand Oaks, CA: Sage.

Radhakrishnan, S. 2008. Examining the "global" Indian middle class: Gender and culture in the Silicon Valley/Bangalore circuit. *Journal of Intercultural Studies* 29(1):7–20.

Rheingold, H. 1993. *The Virtual Community: Homesteading on the Electronic Frontier.* Reading, MA: Addison-Wesley.

Richman, K. and Rey, T. 2009. Congregating by cassette: Recording and participation in transnational Haitian religious rituals. *International Journal of Cultural Studies* 12(2):149–166.

Royston, R. 2014. *Re-Assembling Ghana: Diaspora and Innovation in the African Mediascape.* Ph.D. Dissertation. University of California Berkeley.

Safran, W. 1991. Diasporas in modern societies: Myths of homeland and return. *Diaspora* 1(1):83–99.

Sahoo . A. 2006. Issues of identity in the Indian diaspora: A transnational perspective. *Perspectives on Global Development and Technology* 5:81–98.

Sheffer, G. (ed.) 1986. *Modern Diasporas in International Politics*. London and Sydney: Croom Helm.

Sheffer, G. 2003. *Diaspora Politics at Home and Abroad*. Cambridge: Cambridge University Press.

Skrbis, Z. 1998. Making it tradeable: Videotapes, cultural technologies and diasporas. *Cultural Studies* 12(2):265–273.

Siddiquee, A. and Kagan, C. 2006. The internet, empowerment, and identity: An exploration of participation by refugee women in a community internet project (CIP) in the UK. *Journal of Community and Applied Social Psychology* 16:189–206.

Srinivasan, R. 2006. Indigenous, ethnic and cultural articulations of new media. *International Journal of Cultural Studies* 9(4):497–518.

Staeheli, L., Ledwith, V., Ormond, M., Reed, K., Sumpter, A., and Trudeau, D. 2002. Immigration, the internet, and spaces of politics. *Political Geography* 21(1):989–1012.

Tekwani, S. 2003. The Tamil diaspora, Tamil militancy, and the internet. In K.C. Ho, R. Kluver, and K.C. Yang (eds.) *Asia.com: Asia Encounters the Internet*. London: Routledge.

Tölölyan, K. 1991. The nation-state and its others: In lieu of a preface. *Diaspora* 1(1):3–7.

Trandafoiu, R. 2013. *Diaspora Online: Identity Politics and Romanian Migrants*. Oxford, UK: Berghahn.

Tynes, R. 2007. Nation-building and the diaspora on Leonenet: A case of Sierra Leone incCyberspace. *New Media and Society* 9(3):497–518.

Yang, G. 2003. The internet and the rise of a transnational Chinese cultural sphere. *Media, Culture & Society* 25(4):469–490.

21 Wearable internet for wellness and health

Interdigital territories of new technology

Monica Murero

Wearable technologies are increasingly attracting the attention of academics, industry researchers, the military, health care providers, governments, and millions of consumers all over the world intrigued by affordable opportunities of tracking, quantifying, augmenting and altering the body with wearable accessories. WTs are designed to digitally gather and exchange their user's information by collecting body's electrical, mechanical and biochemical data in the space and over time. Examples include internet connected – or *interdigital* – wearable accessories (Murero 2018) such as wristbands, watches, rings, intelligent patches, tattoos and fabrics, necklaces, virtual reality headsets, exoskeletons, smart hearing and speaking assistants, glasses, and even medical care devices for remote health care monitoring.

In 2016, the overall worldwide number of connected wearable devices amounted to 325 million (Statista 2019). It is expected that within the wearable fitness segment alone, the number of people using at least one wearable device – bracelets, bands, smart socks, excluding smartwatches – will be more than 330 million by 2023, compared to 300 million in 2017 (Statista 2019b). Wearable technologies' overall worldwide revenues sales will reach $34 billion USD and double in three years, up to $72 billion USD by 2022 (Statista 2019a). The wristband segment alone accounted for US$13,615 million in 2019, where most revenues will be produced in China (US$4.5 billion in 2019). Wearable technology is a rapidly growing sector, particularly in the medical field.

Thanks to high-tech micro-components (especially micro sensors reacting to the body's humidity, temperature, movements, electromagnetic fields, and more) wearable devices are able to gather an impressive quantity of physical, environmental, motional, biological, and behavioral information during everyday life. Thanks to the internet, even during sleep time vendors' platforms analyze and than feed back body-data to multiple interlocutors, including the user, whose body generated those data in the first place. The commodification of highly sensitive information is happening for multiple purposes, from self-tracking to custom marketing, generating new threats to privacy but also new opportunities.

In this chapter, I introduce the notion of wearable technologies; analyze the current typologies of wearable devices and their basic technical components. Moreover, I discuss how wearable technologies may contribute to expanding physical and mental *territories* for accessing and providing wellbeing and health care. For example, thanks to wireless systems of mobile care, based on medical wearable technology, patients can be monitored outside the hospital and receive immediate assistance in case *real time* abnormalities are detected, no matter where they are. Then I will discuss how wearable technology data analytics, in the context of artificial intelligence and big data learning, may generate new opportunities but also unprecedented risks for privacy, especially in light of future scenarios, where new biosensors and body implantable devices will impose a reflection on the need for prompt regulation, protection, and security of highly sensitive data.

Defining wearable technologies

Wearable technologies are multimedia-capable computerized devices that can be worn inside, on, or around the body as accessories with digitalized interactive functionalities and wireless connectivity (Murero 2018). Wearable devices communicate via Bluetooth with companion apparel – usually a smartphone or a tablet – that provide internet access. Thousands of downloadable companion apps may elaborate body records received from wearable technologies, offering elaborated statistics to their users in an unobtrusive or minimally invasive manner, that is, notifications and alerts.

According to their characteristics and price, WT devices and their companion applications offer unprecedented functionalities to the non-medical population for health care and body wellness monitoring; these may include: trackers of vital signs, motion, sleeping patterns, weight, body temperature and humidity, heart beat, blood pressure, glucose levels, emotions sensing, body chemical, magnetic and electric activity. Whether for health care or wellness related uses, wearable devices and their companion apps may offer a combination of:

- Personal health and wellbeing monitoring, statistics and alerts;
- Real-time information sharing;
- Physical activity tracking;
- Analysis of body performances;
- Location-based services (software that analyzes the geographical position of mobile devices);
- Environmental and navigation tools (weather, map, directions, commercial suggestions, and more);
- Notifications and assistantship. Some wearable technologies offer Web searches via voice commands;
- Productivity (calendar appointments on the go, and more) and access to stored information.

Thanks to Bluetooth technology and the internet communication, data gathered from wearable technologies might feedback single consumer's *sousveillance* systems (self-tracking), but also multiple interlocutors' platforms, algorithms and artificial intelligence (AI) systems. For the first time in human history, gathering and exchanging very large amounts of sensible body records ubiquitously, and over extended periods of time through the internet are becoming a common reality among the wearable technology customers, generating new opportunities but also threats, which I discuss later.

A brief history of wearable technologies

The practice of tracking, quantifying, augmenting and altering the body with wearable accessories has been diffused since the beginning of the human history. According to Brown (1999), human and animal tracking can be considered as the very first applied science practiced by early hunters. Tattoos and body piercing, wearable analogic accessories like animal and human parts, clothing and jewelry have been offering early examples of body alteration and enhancement for communicating social meanings and status in any field, from fashion to military strategy. For centuries, meaningful symbols in the form of accessories worn on the body have been used to attract partners, scare enemies, and reinforce sports fans' identity in many different cultures.

The recent introduction of digitalized wearable technology is the result of several inventions developed in human history. During the Ming Dynasty (1368–1644), a Chinese mathematician, Cheng Dawei invented the abacus, an early analogic precursor of the modern systems of computation. The abacus is an analogic device that performs calculations by moving circular beads along a few rows, representing values and numbers – still in use today, particularly in Asia.

Wearable computers

Throughout history, many inventions contributed to the creation of the current systems of computations that led to the development of wearable technologies. For example, during the Second World War, Alan Turing invented a computational system to decrypt the German "Enigma" machine in use by the Nazi military to secretly exchange strategic orders and information. Turing's machine made a large contribution to the history of computing – and to the military of Europe once the "Enigma" secret code was revealed to Allies forces. During the 1960s and 1970s the early *wearable com*puters appeared in the USA. In 1961, two mathematicians, Edward Thorpe and Claude Shannon, created the first wearable computer – in a shoe – and applied it to a less-than-noble cause: roulette gambling. Thanks to portable computing capabilities and a micro switcher the "cheating shoe" was able to calculate in which of the eight segments of the roulette the rolling ball would have fall. The gambling advice was then communicated via radio waves in form of eight music tones audible thanks to a tiny loudspeaker worn by the two gamblers in one ear canal. The computer capabilities were quick enough to allow gamblers make an instant bet, while the ball was still rolling at the table roulette and the croupier was still accepting wagers.

The personal computer was introduced in the 1970s and rapidly diffused all over the world during the 1980s. In the 1980s, the state of Nevada introduced a new law to prohibit computerized devices of any kind in the casinos.

Self-quantification in the 18th century

In the 18th century, daily quantification of self-actions and thoughts in the form of a handwritten personal diary was a common practice. For example, Benjamin Franklin, a US politician and inventor, used to quantify himself by keeping track of how he spent his daily time by using a very detailed diary written on an analogic support – paper – to be shared with others. His daily quantifications were than retrospectively used to assess if his actions were or not coherent with his stated values. At the end of the 18th century, the first attempts to measure time on mobile supports (wearable pocket watch) started circulating, thanks to miniaturized analogic technology. One hundred years later, the first wearable analogic watches strapped to the wrist became very popular in Europe and the USA, but only after digitalization; during the 1970s, the first wearable digital watches appeared in Western countries and in Asia.

During the 1990s, one of the first wearable computerized analogic systems for heart beat self-tracking during sports and fitness consisted of an elastic band equipped with a small transmitter positioned around the torso. The transmitter was able to communicate wireless with a wristwatch – Pulsar – or a treadmill machine during training.

Wearable cameras in the 20th century

With the invention of photography and its popularization at the beginning of the 20th century, the early examples of wearable analog cameras appeared in Europe thanks to pigeons aerial pictures. Julius Neubronner (1907/1908), a German apothecary with a passion for professional photography, invented the predecessor of the current GoPro camera that was strapped to homing pigeons. The camera was able to take only one aerial picture during the pigeon's flight thanks to a timer mechanism. During the 1970s analog cameras become very cheap and popular (e.g., the Polaroid), and only twenty years later, during the 1990s digital cameras become available to consumers. Shortly after, in 2004, the first *wearable GoPro* device appeared on the market, a camera/video camera for hands-free live recording of self and group activity, adjustable on several wearable supports, from head helmets to floating handle grip.

The new territories of Wearable Technology

Ultimate wearable devices for wellbeing

Today, popular wearable technologies targeted at consumers' wellness like smart wristbands – Fitbit, Xiaomi, Nike and more – are designed to gather body parameters such as heart beats per minute, movement, sleep and walking habits, running performances, and more. There are thousands of apps that elaborate WT data for wellness purposes, such as measuring and alerting water intake, suggesting relaxing music for stress relief, helping reach personal goals, sensing emotions, offering home exercises, dieting, meditation, breathing mindfulness, and more. In psychological professional practice, recent examples of wearable technology applications include the use of intelligent virtual reality headsets for treating anxiety disorders, fear of flying in airplanes, agoraphobia – fear of open spaces – and many other psychological conditions.

Wearable medical instruments

While wearable technology for leisure and body self-tracking are becoming more and more popular, the most innovative prototypes in wearable technology are currently appearing in the medical field. The management of chronic diseases like diabetes, asthma, allergy, heart disease and even phobias is experiencing original solutions both for hospitalized and home patients that may help reduce errors and improve the timely administration of treatments – especially drugs. New wearable robots (exoskeletons) are helping more seniors and even quadriplegics stand up, do physical therapy, and even perform assisted walking. Medical wearable accessories for enhancing bodily abilities and supporting those with disabilities appeared during the Middle Ages. Thanks to recent innovations, digitalized wearable hearing aids not only compensate for severe degrees of hearing loss but can also connect wireless to multiple home apparatuses, including the smartphone, TV, and home assistive technology. Hearing-impaired people could count on the first hearing aid – the trumpet horn – since the seventeenth century, but only further technology improvements like the invention of a system for transmitting audio signals (telegraph, radio, telephone, and television) and the diffusion of microprocessors led to interdigital wearable hearing aids.

In the 13th century, Italians invented eyeglasses, which became rapidly popular among jewelers, artists, monks, and scholars. Today pioneer prosthetic devices implantable on and within the body are rapidly evolving. For example, in 2011 a bionic eye called Argus II was invented to restore partial sight to the blind by implanting an artificial retina. The system included a video camera that recorded the vision field by tracking the direction of the head and sent the data to a companion device worn in the belt that converted the video images into electrical signals. The signal was wirelessly transmitted to the prosthesis embedded in the bionic eye, creating an artificial *smart retina* that restored, in part, the user's vision (Futuristic News n.d.). Although Google Glass (2013) was not invented to enhance defective vision, it was the first wearable device computer combining hands-free internet access operated by voice with an optical head-mounted display and a virtual reality capability. For example, in the operating room (OR), Google Glass has the ability to capture a doctor's "vision" and voice during a surgical procedure and remotely projecting them to an audience of students and peers over the internet for educational purposes. The audience is virtually "in the OR" without the risks of contaminating the sterile surgical field, and could also benefit from the first operator's vision of the surgical field and the capabilities of Google Glasses to enhance details. The WATCH Society (Wearable Technology in Healthcare Society) incorporates surgeons using Google Glass in the operating room; they foster the use of smart technology in healthcare, educate other doctors, and create new forms of collaborations.

In the medical field, wearable technology is developing revolutionizing opportunities, such as smart contact lenses that check body parameters like pressure and glucose levels, wireless EEG brain headsets for quadriplegic environmental control (domotics), and smart prostheses (hands, legs, eyes) offering wireless systems of control exploiting the brain electromagnetic field. Innovative wearable devices are able to geo-localize and audio-guide visually impaired individuals in space. People with severe speech impairments will be able to communicate by a *donated* voice system thanks to Voiceitt, an unprecedented speech-recognition technology whose algorithm translates unintelligible speech in real time, supported by a companion device and app.

In the next few years, improved wearable technologies using systems for checking glucoses levels during sport for the active community, and medication reminders are expected to improve the everyday lives of diabetics. Fetal heart monitoring system for pregnant women, baby wireless-enabled monitors, smart thermometers scales and changing pad, and even sensors for tracking diapers are all offering unexplored territories of access to health and wellbeing This is possible, in part, due to sophisticated wearable technology, including reliable biosensors, faster Bluetooth and internet connectivity for mobile devices.

Bodily implanted interdigital devices

An extended definition of wearable technology (Murero 2018) should include not only smart accessories that people wear on or around the body, but also computerized devices implantable "inside" the body by health care professionals. Implantable devices for medical purposes start appearing during the 20th century, when the first pacemaker was invented to assist regular heart beats in 1958. During the 1970s and 1980s, the mass production of microchips led to the creation of smaller and lighter computerized systems implanted within the body that improved pacemakers' features and performances.

BAN and WBAN

Pioneer applications show that an implantable biosensor (for example, into the heart, liver, pancreas, under the skin) can help monitoring health to unprecedented extent and provide new insights. When multiple biosensors are implanted within the body they can create a Body Area Network (BAN), a pioneer system of multimedia-capable electronic devices, consisting of one or more miniaturized implanted apparatus. When the BAN is equipped with wireless technology it is called a WBAN. Current examples include internet connected pacemakers and wireless cardiac re-synchronizers implanted in the body. Inside the body, WBAN technology is finding early applications, particularly in preventive telemedicine and mhealth – health care provided via smartphone and mobile technology.

A rapidly evolving field in future years, BAN AND WBAN applications include continuous vital signs and single organ monitoring for chronic diseases and even anxiety disorders or depression. A single organ implant equipped with wireless technology – for example, a mechanical heart – BAN and WBAN biosensor systems can offer life-saving technology to patients who do not need immediate hospitalization or acute care. They can be remotely monitored during everyday activities, ubiquitously and over time, thanks to an internet connection, a companion device and an app.

How does Wearable Technology work?

In order to explain how wearable technologies are able to track and communicate mechanical, electrical and biological signs of the human body and generate innovative system of remote health and wellbeing monitoring, it is important to introducing the reader to a few basic *technical* concepts. Technological aspects can help clarify current and future benefits and threats that may occur with the use of wearable technology, as I discuss later.

Miniaturized components

High-tech miniaturized components (such as sensors, accelerometer, microcontrollers, temperature and water resistance trackers, and more) are crucial to wearable technology functioning. Wearable technology is a very rapidly evolving field. More and more sophisticated and reliable micro-components are able to perceive and react to a body's signs like movement, temperature, and humidity.

Sensors and the reliability of measurement

WT sensors are particularly important because their performance and precision affect wearable technology reliability. Sensors are rapidly growing not only in terms of quality and technical performances, but also in the extension of market demand. Hayward and Chansin (2016) estimate that by 2026 the market of sensors for wearable technology will reach US$6.1 billion, not only in the area of wellness but also in the medical wearable field. Currently, there are different types of sensors enabling wearable technology, such as:

- Sensors reacting to stretch and pressure;
- Inertial Measurement Units (IMUs): (accelerometers, barometers magnetometer and gyroscopes);
- Temperature and audio sensors;
- Optical sensors: (optical heart rate trackers, cameras, Photoplethysmography – PPG);
- Wearable electrodes (brain helmets)

Wearable technology sensors are designed to collect properties of the body they track. For example, sensors embedded in smart watches are able to track heartbeat, approximate the number of steps walked or sleep habits by elaborating an accelerometer's movements of the wrist over time, combined with geo-localization.

Continuous and intermittent monitoring

Wearable devices are designed to monitoring vital signs and body activity continuously or intermittently at one's request. Wearable technologies may offer intermittent or continuous monitoring of vital signs by using different supports such as sensors-equipped bracelets, skin adhesive patches, and more. In a recent pilot study conducted in a hospital, Downey and colleagues (2018) analyzed wearable technology chest adhesive patch performances for monitoring complex information like vital signs and alerting nurses about crucial information. Notifications included the correct timing of administration of antibiotics in post-operative patients with sepsis. Out-of-control body system infection is the third leading cause of death in the USA (Bartels et al. 2013). Wearable technology offered nurses a system for continuously monitoring vital signs of critical subjects, and receiving notifications or reminders; continuous monitoring proved better results in terms of patient's management, overall less mortality and recurrence of sepsis, versus intermittent technology – activated on demand, by manual push.

Bluetooth and the wearable internet

Bluetooth technology allows wireless communication to happen among two or more devices that are closely located in space. For example, an iPhone could exchange its music and pictures wirelessly to a second device when they are connected via Bluetooth nearby. One or more wearable technology apparels communicate via Bluetooth wireless technology to a main companion device such as a smartphone, tablet, or a medical device collecting data. The companion device allows for data synchronization and provides internet connectivity to any Bluetooth connected object. This is a crucial passage to comprehend wearable technology and the term "wearable internet." The companion device can share its internet connection with the wearable device. For example, a smart bracelet (e.g., Fitbit) can access the internet only if it is connected to a smartphone via Bluetooth, and only if the smartphone has internet access. If the companion device is not connected to the internet, the bracelet cannot access the internet. However, the wearable technology and the companion device remain connected via Bluetooth as long as the battery of the two devices is sufficiently charged. Internet synchronization will occur when the connection become again available.

In 2000, the first Bluetooth headset was sold on the market. Since 2015, wearable technology has been offering enriched features, thanks to faster Bluetooth and internet wireless connectivity combined. For example, early applications of the Bluetooth technology together with internet connection in cloth design led to the invention of a pioneer smart fabric equipped with sensors, for long distance transmit ion of the sensation of "touching," the Hugshirt (2014). In wearable technology, the rapid diffusion and improvements of wireless internet connection, and Bluetooth technology opened up unprecedented opportunities to mobile computerized systems, like continuous mobile monitoring systems.

While some wearable technologies are able to directly access to the internet, at the current stage (early 2019) the large majority of wearable technologies communicate via Bluetooth with the companion device. For example, a smartphone can share its internet connection with a wristband, if the two devices are located in the close space proximity. The smartphone becomes the "companion" device, offering internet access to one or more Bluetooth-connected devices.

From simple to complex features

There are hundreds of wearable devices available on the market at the moment, both for professional (hospitals, organizations) and non-professional usage (self-tracking). wearable technology can offer a large variety of services. Depending on thehardware (micro-components, type connectivity, capacity of elaboration) and software characteristics (customizable app) combined, wearable technology can offer customers simple, multiple or complex features (Murero 2018).

- **Simple features**: Wearable devices providing simple features have limited capacity and are able to gather reduced set of data within a little storage capacity. Examples of wearable technology offering simple features include wristband jewelry, rings and patches, step counters, wireless weight scale, glucose monitors, blood pressure measuring devices, smart fabrics, GoPro Camera for continuous live recording. Simple wearable technologies rely on the companion device and app for elaborating gathered information. Companion apps may offer extended services, from activity statistics to alerts and notifications. Information can be stored on cloud computing and shared with other devices or even via social media.

- **Extended multiple features**: This is the case of wearable devices offering multiple services and features, like the case of smart watches (e.g., iWatch), home-care medical devices, and WBAN apparels. These devices provide a multiple range of services, from simple (standalone) to extended features, when they are connected to a companion device that is able to access to the internet. Wearable devices offering multiple features can gather and exchange large quantity of body records and offer personalized services. For example, the iWatch is able to receive and perform phone calls, take pictures, visualize social media notifications, access the calendar, talk to Siri assistive technology and more, according to the features offered by a multitude of companion apps that can be downloaded and customized.
- **Complex features** (or Wearable Interdigital Devices, WID). Wearable Interdigital Devices (Murero 2018) have the same multiple features previously observed, and in addition they are able to directly connect to the internet. Complex wearable technology offer multiple features and do not need a companion device to function and go online, thanks to an integrated micro sim card (for example, the Samsung Gear S2, IWatch series 4).

Companion apps

Thanks to its micro-components, the wearable device is specialized in acquiring and exchanging information, but does not perform body data analysis. The companion device elaborates information thanks to specific apps. Usually producers of wearable technology offer "companion apps." All apps require installation and a registration to the platform providing the service. Wearable technology companion apps are strategically designed to provide information by communicating to their platforms over the internet, and then provide feedback to the user. For example, a wearable technology bracelet may gather wrist movements during sleep. The companion device – smartphone – and app receive data from the bracelet, also according to the permissions allowed during the registration process. The companion app synchronizes with the wristband and communicates with its platform via an internet service provider (ISP). Data on wrist movements, sleep habits, and heartbeats could be continuously gathered during the night without the user's awareness. Body sleep information is then compared to other people's performances on the platform's base, and than elaborated through proprietary algorithms and big data analytics. Sleeping performances – quantity, quality – then appear in the companion app. People may receive notifications aimed at improving their sleeping habits in the form of alerts – for example a message appears both on the smartphone and on the bracelet when it is time to go to sleep. If sleeping statistics are gathered over time, graphical performances may appear as well.

Companion apps receive, elaborate and synchronize wearable technology information. This is mainly because companion devices and apps are designed to offer extended and easy-to-use features. For example, in the majority of cases wearable devices do not have a display or their dimensions are too small for comfortable reading. Companion software apps may be built-in the wearable device, but they are usually downloaded via the internet. There are thousands of apps that offer the possibility of elaborating bodily information gathered thanks to wearable technology micro-components both for Android and IOS-based systems.

The wearable internet: controversial issues

Wellness coaching: attitudinal and behavioral challenges

Based on different types of features, software applications, and price ranges, wearable technologies may offer high quality medical devices for remote health monitoring or simple body activity tracking such as number of steps walked, heartbeat counters, GPS localization, time measurement, and more. The recent development of miniaturized internet-connected technologies embedded in wearable digitalized accessories and gadgets has opened up unprecedented possibilities and applications that might influence to some degree the user's attitudes and behaviors. Monitoring vital signs of the body through comfortable supports such as super-tech textiles, skin patches and bands, special eyeglasses, smart jewels connecting to an ad hoc companion app may translate into benefits. For example, data gathered through a smart bracelet during the day – for example, frequency of movement of the wrist, heartbeat, step counter, geo-localization – may be elaborated by the companion apps to approximate how long the consumer has jogged or walked. By gathering personal entries at the time of registration like weight, gender, stature, and more, the wellness app may send notifications to create awareness about the "fitness" behavior and may provide suggestions to change attitudes and health behaviors, as necessary. For example, hundreds of popular apps such as MyFitnessPal, Lose it!, Runkeeper, Yazio, and Melarossa use data from wearable devices to elaborate custom programs of dieting and exercising, and help people stay on track with their goals over time.

Personal sousveillance and the quantified self

Wearable technologies provide data gathered from non-medical wearable devices are sent to a multitude of interlocutors, including the individual who generated data in the first place. The term *susveillance* indicates the recording of a body activity typically by way of small wearable or portable personal technologies (Mann et al. 2001). The self-tracking or quantified self is an emerging field. This is due, in part, to the easy access to unprecedented bodily records, geo-localized and gathered over time but also organized per day, week, and months on the display.

In the current scenario, wearable technology app users may receive analytics about his or her performance through ad hoc apps by means of notifications, suggestions, and statistics. Companion apps allow checking and tracking large quantities of personal information, from running activity to sleeping habits. For example, the possibility of tracking and recording movement in the space, calories burnt, food intake, and sleeping habits is designed to providing visual statistics and feedbacks about fitness performances over time, for different purposes.

The Quantified Self is a movement of individuals and groups engaged in self-tracking (available on display on a smart device) providing physical, environmental, biological, and/or behavioral information and statistics over time (Murero 2018). The popularization and affordability of wearable technologies has increased the Quantified Self phenomenon, with several implications. Wearing a smart device can reinforce or even change attitudes and behaviors. On the one hand, self-care knowledge and apps notifications in the form of suggestions may engage the individuals to better care and self-motivation to achieve specific goals. When shared via social media, Quantified Self records may influence others' attitudes and behaviors such as competition, adoption, avoidance, and more.

Expanding the physical and the mental territories of health care and wellbeing

Complex remote medicine-based services and wearable technologies for mobile monitoring of patients with chronic diseases such as diabetes, cardiovascular heart disease, hypertension, and more are offering challenging developments in that they may extend the physical territories of health care distribution over space and time. For example, a pacemaker normally sends an electrical stimulus to the myocardium if the system detects significant abnormalities in the heart rate. Moreover, if a pacemaker has access to the internet through a companion device and a dedicated app, it may be able to exchange crucial information about normal and abnormal heart functioning to a remote facility that alerts immediate medical care, if needed. In most cases, caregivers do not need to do anything because the system is smart enough to automatically communicate and monitor from a distance.

An interesting development of the interdigital home care systems based on wearable technology devices is that the medical facility may be located virtually anywhere in the world. Once records are sent over the internet, and the need for care is detected by the system's algorithms, local assistance can be immediately arranged by directly contacting the patient to assess a more precise diagnosis, or by sending medical staff in the location where the wearable device is geo-located.

Internet-distributed systems of medical knowledge may offer new insights to remote medical care management. For example, traditional space and time barriers can be overcame because people may receive medical monitoring even during every day's outdoor life, while freely moving in an interdigital connected environment. As far as a reliable internet connection is available and medical data are continuously exchanged with the point of remote care, the latter may be located even at thousands miles of distance from the patient.

Constant remote monitoring of health conditions no matter where a person is located is a complex process. New forms of distributed real-time medical care that act in mobile space have no precedent in human history and would not be possible without several sophisticated conditions, including the mobile access to the internet and the availability of reliable technological instruments, particularly in rural and developing countries. Moreover, a biosensor's quality affects the reliability of biological data reading, and the significance of information exchanged.

Wearable technologies could expand the physical and the mental territories for accessing wellbeing and health care. Ongoing research (Murero et al. 2019) shows that wearing an interdigital medical device offering constant health monitoring outside the hospital may have important psychological implications for the users' attitudes. Wearing a medical device can affect the perception of safety and quality of care, the levels of anxiety and stress, the attitudes and perceived comfort towards the wearable device, and improve the process of decision-making when it comes to health care, and may potentially improve the overall quality of life.

Medical wearable devices: organ and body data

Wearable medical devices and in-body (WBAN) implanted technology collect unprecedented data about body parameters. Thanks to high-tech miniaturized components, wearable technologies are able to track unprecedented information such as individual locations, sleep habits, heartbeat, communication patterns, daily activities, body biological performances, accurate information about diseases, disabilities, and more. Thousands of companion devices and "ad hoc" apps gather, exchange, and analyze billions of records collected through WT, ubiquitously and continuously.

While wearable technology may offer unprecedented benefits, major concerns are raising. Collecting and analyzing highly sensitive information ubiquitously and continuously bring about new concerns in the current debate regarding privacy, security, and pervasiveness of new technology. Current academic discussion (Murero 2014; Pasquale 2015; Neff and Nafus 2016; Murero 2018) shows that the use of body-log data is fostering technology pervasiveness into daily life. In fact, thanks to the diffusion of wearable technologies, multiple subjects can gather and sell highly sensitive information, anytime and anywhere, to different stakeholders. Evidence also shows that wearable technology users, whose data are commercialized, have no clear awareness of the process. The phenomenon is part of a recent economy where personal data plays a growing role in commodification and marketing.

WT and artificial intelligence

Vast datasets of sensitive patient information are expected to feed artificial intelligence projects aimed at medical preventive care (Murero et al. ongoing). For example, artificial intelligence findings may support the medical decision-making process. Artificial intelligence analytics using several inputs including (but not limited to) information from distributed health care systems and patient's data coming from wearable and WBAN technology may improve protocols of medical intervention over time, improve automatized diagnostics algorithms for early detection of disease and continuous machine learning. Moreover, AI findings may direct local health prevention programs and rationalize scarce resources.

Thanks to artificial intelligence techniques, combining multiple sources of information, new threats to the use of sensitive body-log data emerge as well. This recent phenomenon is particularly threatening when Big Data analytics combined with wearable technologies ultra sensitive information may show a pretty accurate and powerful picture based on an enormous quantity of personal information, from social media comments to web browsing navigation and online purchases.

The need for regulation

The use of data collected from wearable technologies raises several questions in terms of civil liberties because the policy to protect emerging technology is limited or does not yet exist in the majority of countries where innovative devices are becoming popular. There are significant variations in what wearable technology vendors and third parties do with body data, often without the involved person's knowledge; evidence show that terms of service are largely ignored and hard to read. Consumers have very limited options to access and safeguard their privacy, besides not buying a WT. However, even implanted medical devices that a subject may be forced to wear due to life saving situations raise several questions.

Multiple interlocutors accessing wearable technology data may include internet service providers, vendors and manufacturers of the device, cloud service generators, third parties mediators, business oriented companies, app developers, local administrators, and governments, and even other internet or Bluetooth connected smart objects. Multiple subjects interested in personal data commodification may acquire health datasets for business related purposes, despite the fact that body data are highly sensible and even without the wearable technology users' explicit permission. The use of big data and artificial intelligence techniques in the current context of personal data treatment, combined with the lack of adequate regulation of rapidly emerging technology are, a matter of serious concerns that need immediate attention and intervention from policy makers.

Why patients cannot access their own data

Another issue that needs urgent regulation regards the access to information when a patient receives one or more implanted medical devices. For example, data from cardiac devices implanted in the body are as important to individuals with heart disease as blood glucose data are important for self-monitoring and self-administering of insulin in diabetics. While diabetics' self-access data, it is almost impossible for people with cardiac conditions to visualize their own records when carrying an implanted wireless pacemaker or a cardioverter defibrillator (ICD). This is because the producer of the implantable device, in the majority of cases, refuses to share this information. There are several reasons for this. One argument is that some people may not be able to cope with information that might generate anxiety or stress. On the other hand, access to heart device notifications, alerts and other information would allow self-care patients for corrective action to be taken, fostering autonomy and empowerment, particularly when the implanted device is not interdigitally checked. In this scenario, currently, in the large majority of cases the patient himself, or a caregiver, must call the hospital in case suspicious symptoms emerge.

Threats to personal privacy and security are a matter of concern particularly due to the lack of adequate regulation because the policy to protect sensible data is limited or does not exist in the majority of countries where emerging technologies like wearable technologies are becoming popular (Murero 2018). For example, the recent use of quantified self-tracking information by multiple stakeholders is not yet regulated (Swan 2013; Murero 2014). All these emerging developments need careful consideration and strong regulation starting from emerging WBAN prototypes that soon will reach the market, because wireless life-saving medical apparatus that are implanted inside the body and generate data can not be self-removed or turned off if the patient refuses to share their bodily organ information. The same criteria that national laws impose for analogic health data records protection should be enforced and extended to digital data handling.

Conclusions

Wearable technology for altering, modifying or enhancing the body abilities is rapidly developing. For example, interactive wearable virtual reality headsets are helping improve personal wellness and psychological conditions like phobias and depression. Non-intrusive wearable technology solutions are expected to facilitate the daily self-management of chronic diseases like diabetes. Innovative wearable products may support directional assistance enhancing ear, vision, and speech impairments (e.g., Voiceitt). New generations of wearable technology are expected to connect directly to the net and offer extended capabilities to stakeholders.

The recent diffusion of wearable technology is creating new territories of health care management and wellness. For example, thanks to innovative wearable technologies, health conditions can be remotely monitored anywhere and anytime, as far as an internet connection is available and the battery of the mobile apparels involved in the communication process are sufficiently charged. For example, wearable devices tracking body activity anywhere and anytime are extending the traditional locations where medical conditions are checked, such as a doctor's office or hospital. Nowadays a wearable technology can continuously check a body's activity, anywhere a person goes, and even during eating or working time. Providing health care outside the hospital may have very important implications in terms of mental stress-relief for patients and their families, who can comfortably receive continuous care, if needed, without taking particular actions. This phenomenon is unprecedented and may contribute to enlarge the boundaries of access to health care and wellbeing thanks to a combination of sophisticated and multiple factors. These include, but are not limited to, reliable technology, personal motivation, and professional expertise. In the next years, more and more innovative smart medical centers offering remote monitoring through the internet may multiply and contribute to overcome current barriers in health care education and management. Innovative systems of continuous health monitoring in chronic patients who do not need immediate care can avoid the burden and costs of unnecessary hospitalization. Thanks to the diffusion of the internet and wireless technology, extended geographies for health care access, particularly in developing countries and rural areas may further expand where professional specialties are still lacking, if adequately supported by policy makers, international organizations and responsible vendors.

In the next few years artificial intelligence may offer unprecedented knowledge by analyzing multiple and vast personal information. Artificial intelligence results may improve the effectiveness of health prevention campaigns and help expand resources for self-care and wellbeing. However, the risks of technology pervasiveness and commodification of super-sensitive body data, combined with the lack of adequate regulation and even ethical guidelines for wearable technology emerging innovations show more than a concern. The rapid evolution of wearable technologies shows that upcoming scenarios need urgent regulation but they may also create unprecedented opportunities for self-caring and new scientific knowledge.

References

Bartels, K., J. Karhausen, E. Clambey, A. Grenz, and H. Eltzschig. 2013. Perioperative organ injury. *Anesthesiology* 119(6):1474–1489.

Brown, T. 1999. *The Science and Art of Tracking*. New York: Berkley Books.

Donovan, T., J. O'Donoghue, C. Sreenan, D. Sammon, P. O'Reilly, and K. O'Connor. 2009. A Context Aware Wireless Body Area Network (BAN). Pervasive Computing Technologies for Healthcare, 2009. *PervasiveHealth 2009*. 3rd International Conference on. IEEE.

Downey, C., R. Randell, J. Brown, and D. Jayne. 2018. Continuous versus intermittent vital signs monitoring using a wearable, wireless patch in patients admitted to surgical wards: Pilot cluster randomized controlled trial. *Journal of Medical Internet Research* (12):e10802. www.jmir.org/2018/12/e10802.

Franklin, B. n.d. The Electric Ben Franklin. www.ushistory.org/franklin/autobiography/page41.htm.

Futuristic News. n.d. Argus II artificial retina will restore partial sight to the blind. http://futuristicnews.com/argus-ii-artificial-retina-will-restore-partial-sight-to-the-blind/.

Hayward, J., and G. Chansin. 2016. Wearable Sensors 2016–2026: Market Forecasts, Technologies, Players Shedding Light on the Components Enabling Wearable Technology. www.idtechex.com/research/reports/wearable-sensors-2016-2026-market-forecasts-technologies-players-000470.asp.

Mann, S., J. Nolan, and B. Wellman. 2001. Sousveillance: Inventing and using wearable computing devices for data collection in surveillance environments. *Surveillance & Society* 1(3):331–355.

Murero, M. 2014. *Comunicazione Post-Digitale: Teoria Interdigitale e Mobilità Interconnessa*. Webster, Padova: Libreriauniversitaria.it.

Murero, M. 2018. Wearable technologies. In B. Warf (ed.) *Encyclopedia of the Internet*. Thousand Oaks, CA: Sage.

Murero, M., S. Albairak, G. D'Ancona, and H. Ince. 2019. *Smart@heart Artificial Intelligence for Assistive Distributed Knowledge. Ongoing Funded Project*. TU Universitat, Berlin, Germany and Unina, Italy.

Neff, G., and D. Nafus. 2016. *Self-Tracking*. Cambridge, MA: MIT Press.

Pasquale F. 2015. *The Black Box Society*. Cambridge, MA: Harvard University Press.

Statista. 2016. *Statista Wearable Report, 2016–2020*. Statista Digital Market Outlook www.statista.com/statistics/490231/wearable-devices-worldwide-by-region.

Statista. 2019. *Statista Wearable Report, 2016–2020*. Statista Digital Market Outlook. www.statista.com/topics/1556/wearable-technology/.

Statista. 2019a. Wearable Technology – Statistics & Facts. www.statista.com/topics/1556/wearable-technology/.

Statista. 2019b. Wearables worldwide. www.statista.com/outlook/319/100/wearables/worldwide#market-revenue.

Swan M. 2013. The quantified self: Fundamental disruption in Big Data science and biological discovery. *Big Data* 1(2):85–99.

22 The Internet of Things

Anurag Agarwal and Bhuvan Unhelkar

The "Internet of Things," or the IoT, refers to an infrastructure of ubiquitous, context-aware devices, or "things," capable of communicating over the internet, interacting with each other through messages and sensors, to accomplish some objective(s) (Agarwal and Unhelkar 2017). Objectives could be personal, commercial, or industrial in nature. Traditionally, the internet is thought of as a network of "computers" communicating with each other. The IoT is an extension of that idea, in which besides just computers, any network-enabled device can be connected: devices such as cars, phones, appliances, watches, security cameras, and so on. The applications of the IoT range from those for individuals, such as remote health monitoring or navigating using a GPS system, to large industrial uses such as smart cities, smart farming, weather monitoring, etc.

The Internet of Things consists of two fundamental concepts: the internet, signifying connectivity and communication, and things, representing a variety of objects or devices capable of communicating and interacting over the internet. Advances in wireless communications and networking technologies, together with the shrinking size and cost of microcomputers, sensors, and actuators are driving the development of several IoT applications. Affordable miniaturized portable or wearable devices (pens, watches, buttons, and even clothing) are only adding to the popularity and pervasiveness of the IoT in our daily lives. This has enabled hitherto unimaginable innovative potential of the internet connectivity to be reflected in applications such as road navigation, health monitoring, contactless purchasing, crime monitoring, and even grocery shopping. This is so because the varied interconnected devices are themselves becoming smarter, collecting large amount of context-sensitive data automatically through their sensors without the intervention of their users. Once this data is transmitted over the internet, it can be used to undertake analytics for a group of people depending on their common areas of interest.

The Internet of Things is evolving rapidly, starting with the convergence of many technologies that have been developing over the last few decades, such as RFIDs, sensors, sensor networks, mobile computing, real-time analytics, machine learning, control systems, the internet platform, broadband communication, cellular and wireless connectivity and the cloud. Traditionally, devices equipped with RFID and NFC chips, could only communicate locally due to their limited near-field communication capabilities. Advances in telecommunications and the internet technologies have resulted in the ability of these very same devices to now connect globally and interact with each other on the cloud – ushering in the era of the IoT. Devices interacting in the IoT infrastructure are referred to as the IoT devices or smart objects. Each device is identified by a unique IP address. Theoretically, any device such as a refrigerator, an appliance, a car, a thermostat or a drone, when equipped with sensing, processing and networking capability can act as an IoT device. Some well-known examples of IoT devices include self-driving vehicles, smart phones, and smart watches. The number of active IoT devices in the world is estimated to reach over 50 billion by 2020, up from an estimated 25 billion in 2015 (Figure 22.1).

For IoT devices to be useful, they have to be able to sense the context (i.e., their operating environment) through appropriate sensors, process the sensory data and become context aware and thereby build ambient intelligence and communicate with other IoT devices or humans towards some objective (Agarwal et al. 2016). IoT devices can be programmed to take certain corrective actions with the help of actuators based on the ambient intelligence gathered through sensors. For example, a home security camera can be programmed to text the owner or call law enforcement if it senses an intruder. Or a smoke detector can be programmed to call the fire station automatically instead of simply setting off the alarm. They can also be programmed to learn from the environment in which they are operating and take appropriate actions.

This chapter starts by defining the IoT and outlining its origins and history. This discussion is followed by a brief outline of the technology of the IoT and describing what constitutes the Industrial IoT (IIoT). The applications of the IoT are outlined with specific focus on geospatial applications. Challenges in the use of the IoT in practice are then highlighted followed by the conclusion and future directions.

Definition of the IoT

Because of the broad nature of the IoT, there is no single universally accepted, all-encompassing definition of the IoT. We provide here definitions from some well-known and reliable sources, The Internet of Things Global Standards Initiative (IoT-GSI 2012), which defines the IoT as a global infrastructure for the information society, enabling advanced services by interconnecting (physical and virtual) things based on existing and evolving interoperable information and communication technologies. Gartner defines the IoT as the network of physical objects that contain embedded technology to communicate and sense or interact with their internal states or the external environment (Gartner). Techopedia describes the IoT as a world where just about anything can be connected and communicate in an intelligent fashion. It describes the IoT as one big information system. Wikipedia defines the IoT as the network of devices, vehicles, and home appliances that contain electronics, software, actuators, and connectivity which allows these things to connect, interact and exchange data. While there is no universal definition for the IoT, the core concept is that everyday objects can be equipped with identifying, sensing, networking, and processing capabilities that will allow them to communicate with one another and with other devices and services over the internet to achieve some useful objective (Whitmore, Agarwal, and Xu 2015). Some other terms that are also used to describe the concept of the IoT include ubiquitous computing, invisible computing and pervasive computing.

Origins and history

The basic idea behind the IoT is rooted in the concepts of telemetry and telecommand. Telemetry is an automated communications process by which measurements and other data are collected at remote or inaccessible points and transmitted to receiving equipment for monitoring. An example of telemetry is a radiosonde, which is a helium balloon carrying sensors for atmospheric readings. Radiosodes measure atmospheric data such as temperature, pressure, and humidity and transmit them to a weather station on the ground. Satellites are also essentially telemetry devices.

Telecommand is a command sent remotely to control a device. Both telemetry and telecommand become possible through machine-to-machine (M2M) interaction, where the machines are connected wirelessly. A garage opener or a TV remote are simple examples of telecommand devices that send command to a receiving device. Through telecommand, the receiving device is controlled remotely. As another example, through telecommand, NASA can control a rover sent to Mars; the rover in turn acts as a telemetry device for data about Mars.

M2M communication precedes the internet. So some form of telemetry and telecommand has existed ever since M2M communication has existed, which dates back to the 1960s and 1970s. As the sensor technology and wireless communication technology advanced, telemetry and telecommand concepts are being applied to more and more devices. When communication happens through TCP-IP, we get the present day IoT.

The conceptual idea behind the IoT – of adding sensors and intelligence to basic objects, connected through the internet – has been under discussion since the 1980s. In 1989, a scientist named John Romkey created the first internet device, a toaster that could be turned on and off over the internet. The toaster was a device on TCP/IP network and could be controlled from another computer in the network. In 1998, Mark Weiser, the Chief Scientist at Xerox PARC, constructed a water fountain outside his office that mimicked the volume and price trends of the stock market. Basically, the fountain was in constant communication with the stock market computers. In 1999, the term "Internet of Things" was coined by Kevin Ashton, then-executive director of the Auto-ID Center (Ashton 2009). In 2000, LG announced its first plans for "connected appliances" – the Internet Refrigerator.

In 2005, the UN's International Telecommunications Union (ITU) published its first report on the IoT and had this to say:

> A new dimension has been added to the world of information and communication technologies: from anytime, anyplace connectivity for anyone, we will now have connectivity for anything. Connections will multiply and create an entirely new dynamic network of networks – an Internet of Things.

In 2008 the first European IoT Conference was held. A group of 50 companies launched the IPSO Alliance to promote the use of Internet Protocol (IP) in networks of "smart objects" and to enable the Internet of Things. In 2010, the number of internet-connected devices (12.5 billion) surpassed the number of human beings (7 billion) on the planet. In 2011, IPV6 protocol was launched which allowed 2^{128} unique addresses, enough to provide an IPV6 address to every atom in every human on the earth. The IoT is considered the third wave in the development of internet-based information systems. In the first wave (1990s), the internet connected about 1 billion users. In the second wave (2000s), the internet connected another 2 billion users. In the third wave (2010s) the IoT has the potential to connect as many as 50 billion objects by 2020. As mobile technology develops, mobile crowd sensing becomes possible through mobile IoT (Liu et al. 2018). In future, the size of IoT-enabled devices is expected to shrink to the nano scale, giving rise to what is termed the Internet of Nano-Things (Akyildiz and Jornet 2010).

Year	Number of connected devices (in billions)
1992	0.001
2002	0.5
2009	*IoT Inception*
2012	8.7
2013	11.2
2014	14.4
2015	18.2
2016	22.9
2017	28.4
2018	32.8
2019	42.1
2020	50.1

Figure 22.1 Number of objects connected to the Internet of Things.
Source: www.mesh-net.co.uk/what-is-the-internet-of-things-iot/.

The technology

The underlying IoT technologies can be broadly divided into hardware, software and communications architecture. In terms of hardware for the IoT, the critical elements include the internet communication platform on which everything else resides, RFIDs (Radio Frequency Identification Devices), NFC (Near Field Communication) devices, sensors, and local sensor networks. The RFID technology is used to uniquely identify objects using RFID tags that include an Electronic Product Code or EPC. This RFID technology has been used for decades for purposes of tracking big-ticket items such as vehicles and livestock. As the cost of RFID tagging has declined over the years, RFID tags are being used to tag even low-value items such as items at a grocery store or a retail outlet. Such RFID tagging facilitates tracking the inventories in a retail outlet; it can also facilitate rapid scanning of checkout items. The benefits of RFID, which were limited to close proximity, can be extended globally by replacing the EPC with a unique IP address and letting it communicate over the internet.

The NFC technologies, which are rooted in RFID technologies, are a set of short-range wireless technologies, typically effective within a 10 cm range. One NFC device is considered the active device (acting as the initiator) and the other is the passive (acting as the target). The initiator device generates a radio frequency that powers the target device. NFC tags contain data that are usually read-only, but may also be writable. Another critical hardware component for IoT devices is sensors, used to sense and monitor the relevant aspects of the environment. Some examples of sensors include cameras, thermometers, barometers, motion detectors, distance sensors, speed sensors, heart rate sensors, blood pressure readers, transducers etc. When multiple sensors are used together, they form a sensor network. Sensor networks may also contain gateways that collect data from the sensors and pass it on to a server. The most critical hardware component for the IoT is the internet communication network, without which no communication is possible.

For all the hardware pieces to work together, new software applications are required in order to support the devices and their corresponding interoperability. The software to power the IoT can be broadly classified as middleware and searching/browsing software. The IoT middleware sits between the IoT hardware and data and the applications that developers create to exploit the IoT to accomplish the intended objective (Bandyopadhyay et al. 2011). The IoT also relies on searching/browsing software for maximum effectiveness. The IoT browsers are different from the traditional web browsers because while the web search engines are designed to display and index relatively stable web content, the data in an IoT environment is very dynamic and massive as the data is most likely generated by a device in motion. The IoT search engines need to be capable of searching high velocity, rapidly changing information generated by IoT-enabled objects.

To make the IoT function effectively, well-designed architectures are needed to represent, organize, and structure the IoT. Architecture for the IoT can be further classified as hardware/network architecture, software architecture and process architecture. Hardware/network architectures can be peer-to-peer or autonomic. Software architectures are necessary to provide access to and enable the sharing of services offered by IoT devices. Service-oriented architectures (SOA) and representational state transfer (REST) model have been proposed for the IoT because of their focus on services and flexibility.

The biggest challenge to the IoT is M2M communication that needs communication over long ranges with low power level usage. This issue is solved by a new wireless technology known as LoRa (Long Range). LoRa is long range, low power consumption technology that is used for building IoT networks worldwide. Public and private networks using this technology can provide coverage that is greater in range compared to that of existing cellular networks. LoRa technology uses LoRaWAN protocol, which is developed by LoRa alliance. It uses unlicensed radio spectrum in ISM band 868 MHz–915 MHz for communication between sensors (nodes) and gateways connected to a network server.

Sourcing of data with IoT devices can include a wide variety – such as "crowd generated" machine sensor data, audio and video data and collaborative meta-data from one or more IoT devices. This generation of data by machine sensors results in high volume data on a continuous, streaming basis. This propensity of IoT and machine sensors to significantly increase data workloads needs to be complimented by corresponding technologies for storage, retrieval and processing (Unhelkar 2018). As a result, process architectures become necessary to structure and optimize the usage of data according to needs of the corresponding business processes.

Thus, the potential of the devices is brought to fruition only when they are connected together through back-end cloud technologies. The value of an IoT device to its user is apparent when such a device is "alive." For example, GPS car navigation is of greater value if its back end collaborates with weather, traffic, and sporting events databases. The value of a pacemaker lies not only in its support to the heart, but also, potentially, its ability to alert relevant services in case of an emergency. And the wearer of a smart-watch (e.g., Fitbit) is encouraged to perform better through the encouragement of a group of friends all collaborating on a back-end cloud. It is thus imperative that IoT devices be innovatively integrated with the back-end cloud where data is stored, shared, and analyzed.

The Industrial Internet of Things (IIoT)

The Industrial Internet of Things (IIoT) extends the technologies of the IoT and applies them to industrial processes and digital transformations (Jeschke 2017). A key requirement of IIoT is that the devices are able to operate with intelligence based on feedback loops. Machine Learning (ML) algorithms are embedded in the devices as well as in the backdrop (typically on the Cloud) to enable the processes driving the devices to "learn" from rapid as well as detailed data analytics. Such analytics – depending on their granularity – are able to provide immediate feedback to the IIoT devices to enable them to simulate intelligence and undertake decision with minimal or nil human intervention. Thus, the IIoT is characterized by the following:

- Big Data: this characteristic of IIoT deals with the sensing, ingestion and storage of vast amounts of data, most of which is on the Cloud. By storing data on the Cloud, it is possible to undertake Analytics on the Cloud across a range of devices, situations, and times.
- Machine Learning: this characteristic of the IIoT framework enables detailed as well as rapid Analytics of the Big Data on the Cloud to produce insights. These insights generated by ML are continuously being updated to enable increasingly refined (intelligent) decision making.
- Automated Communication: this characteristic of the IIoT framework/ ecosystem enables machine-to-machine communication in an automated manner to enable the IIoT devices to organize their activities, correct their actions and suggest improved actions – with minimal or nil human intervention.

These characteristics of the IIoT differentiate item from the typical personal IoT devices such as smart watches and cell phones. While PIoT also have the possibility of the above characteristics, they are not mandatory requirements in personal usage. Furthermore, the individual user's processes are not necessarily based on the aforementioned characteristics, whereas the IIoT has these as mandatory characteristics.

Applications

The IoT is a rich technology for many practical applications at both individual and industry level. While many applications are already implemented in practice, many more are likely to emerge in the future as the possibilities are endless. Some areas of applications include smart infrastructure, smart homes, retail, healthcare, law enforcement, energy and mining, supply chains and logistics, manufacturing, military, and social applications. We can classify the applications as consumer oriented, commercial, industrial and infrastructure ones.

Consumer applications include devices such as Apple's iWatch for personal fitness and health. Another application of the IoT is smart homes that are equipped with sensors to optimize the use of electricity and gas and reduce waste. Smart homes can also allow remote access to control the thermostats, appliances, and shades. Smart home technology that is IoT-based can also be used to monitor the residents, particularly elderly or disabled residents and call for response team or other help if necessary. Smart assistant devices such as Amazon Echo, Google Home, Apple's HomePod, and Samsung's SmartThings Hub are also consumer-oriented applications.

For example, the IoT-enabled smart infrastructure enables the vision of self-driven vehicles to become a reality. Self-driven vehicles use IoT-driven GPS technology to navigate to their destinations. Through sensors, they are able to stop at traffic lights, change lanes, and navigate traffic with the help of context awareness and ambient intelligence. On a broader scale, IoT-driven technologies can be employed to make cities more efficient. For example, in smart cities, IoT-enabled emergency vehicles and public transportation buses can interact with IoT-enabled traffic lights to allow them to reach their destinations faster and serve citizens better. Taxi hailing services have been made possible through the IoT infrastructure.

Retail sector can capitalize on IoT applications such as "Fast Retail" checkout optimizers (integrated with RFID chips), shopper analytics, supply-chain visibility and optimizing service provider staff's processes. Telecommunications and Information sector can use IoT-enabled (video, mobile, social) applications that assist in operations optimization, equipping next-generation worker with current statuses, tower management through remote chips and tools, and even optimized service fleet management.

Health care is another domain where extended care, remote patient monitoring, staff mobility, and security are made possible through the application of IoT devices and the IoT infrastructure. Good health care often depends on timely detection of a current or impending health issues and taking timely action to address the issues. Often, the patient living alone is unable to call the emergency number on account of disability due to a health condition. With appropriate sensing and actuating IoT devices, many health crises can be preempted by calling for help in a timely manner. For example, devices that can monitor heart rates, or falls or stroke symptoms or any other detectable medical condition can call for emergency care when necessary. In that sense, monitoring and decision making can be shifted from the humans to machines. Also, data from sensors can be continuously transmitted from the patient to the doctor's office, or family members and other interested parties who can take appropriate actions in a timely manner. With advances in wearable sensors, such as smart watches and even sensors embedded in fabric, a much closer health monitoring is possible. Smart watches can send data to personal trainers who can recommend modifications to one's life style to improve their health.

In supply chain management, RFID and sensor networks have long played a role in improving their effectiveness through continuous monitoring of flow of assets through the supply chain. The ubiquity and pervasiveness of the IoT will enable the use of these technologies across organizations and geographic boundaries. So, theoretically, it will be possible to locate any asset anywhere in the world, as long as it is tagged with a unique identifier and is the IoT enabled. Traditionally, assets could only be located while they were at a warehouse or a store, and not during transit. With the IoT, assets can even be tracked during transit. Theoretically, every piece of mail can be IoT-enabled and tracked without having to explicitly enter its current location. Inventory levels at retail shelves and warehouses can be monitored and manufacturers can get an idea of how fast certain items are moving through the supply chain. Manufacturers can plan their manufacturing activities more accurately, warehouses and retail can place their orders more timely and in appropriate quantities, resulting in less surplus or shortage throughout the supply chain (Balaji and Roy 2017).

The IoT can also be used in the preemptive and timely maintenance of equipment. The sensing technology has advanced enough that impending failures of equipment can be detected prior to actual failure. This information, when transmitted to the appropriate persons, can allow preemptive maintenance and saving losses resulting from failed equipment.

The IoT can be very useful in law enforcement. Stolen cars, for example, can be traced with pin-point accuracy as each car will have an IP address and can be tracked remotely. Another practical IoT application in law enforcement is the Breath Alcohol Analyzer (BAA). This BAA is an IoT-based device used to monitor the breath-alcohol level of drivers. By periodic sampling of breaths while driving a vehicle, the driver and the law enforcement agencies can be alerted to the danger on the road. Additionally, smoothness of the drive itself can be evaluated using appropriate sensors, such as telematics device and vehicle's curb camera.

In the domain of physical security and safety, IoT-based surveillance of public places will enhance the overall physical security. The technology can also be used to detect counterfeit goods and spare parts used in airline or automotive industry.

There are a number of potential social applications. With each mobile phone acting as an IoT device that can transmit its location information, individuals can be informed when they are in proximity to friends, social events, or other activities of interest. Smart watches that collect data on fitness can be shared with a community of friends for mutual encouragement. Further, IoT-enabled phones may connect directly to other mobile phones and share contact information when certain predefined friendship or dating profiles are matched.

Spatial and geographic implications

When IoT technologies are specifically focused on the three-dimensional mapping of a point in space and then carry out its subsequent analysis, it is considered to be spatial analytics. The data point in space is the IoT device itself or it could be an entity being located and moved by an IoT device. A large number of applications in day-to-day use are spatial in nature: for example, the "Google maps" and "Waze" applications utilize the position of a vehicle to which they are attached (usually through a smart phone) to analyze, in real time, the movement of the vehicle. The real-time nature of most of these applications also implies time management within these applications. Thus, all such applications incorporate time together with the geospatial locations. As a result, these spatial applications are specifically able to optimize the performance across a geographical space. Such applications are, therefore, specifically called geospatial applications.

Sensors on IoT devices detect the location of the device followed by tracking its movement in time. These IoT devices can be put together in a group (temporarily, such as when a group of vehicles are traveling on a road, or on a more permanent basis, such as a group of emergency vehicles belonging to a county of city). The data generated by these devices in real-time is a source of substantial information which, when put together on the Cloud and analyzed, can reveal patterns that would not be possible to identify without the streaming data from these IoT devices.

The potential applications of geospatial analytics are limitless:

- Traffic management in busy cities and express ways (such as Google maps and Waze mentioned above)
- Emergency vehicle management and movement (such as ambulances and police vehicles)
- Sports applications – training as well as execution of sports through precision in location-specific applications (such as timing a large group of runners in a marathon)
- Deliveries of goods and services (such as Amazon's drops and pizza deliveries)
- Optimizing travel (such as Uber and Lyft).

Combining big data and machine learning provides opportunities to identify patterns and make sensible predictions in each of the above example applications.

Challenges

The IoT faces many challenges, both technical and social. Technical challenges include interoperability, standardization, and security challenges. Social challenges include privacy, legal/accountability and general challenges (Ma 2011). Following are some of the risks and challenges in applying the IoT and cloud in the business space:

- Security. The more devices connected to the network, the more vulnerable they are to being hacked. Thus the popularity and growth of the IoT is confounding the industry users and the researchers in terms of security of devices. Additional challenges from a security standpoint arise due to the fact that for many of these consumers the IoT devices are not easily upgradable for their operating systems and software versions (Hancke et al. 2010). This lack of currency in the operations of these IoT devices leads to a major gap in security because the devices themselves are not upgraded corresponding to the threats faced by them. Furthermore, the IoT devices communicate through wireless network and this communication needs to be made secure through encryption. Basic IoT devices may not be advanced enough to support robust encryption. The encryption technology needs to advance to be more efficient and less energy consuming. Besides encryption challenges, identity management is also a challenge. Ensuring that smart objects are who they say they are is critical to the success of the IoT. The possibility of identity theft, which is a direct consequence of weak identify management, which can result from weak security, is a big challenge in the IoT. When critical processes depend on machine-to-machine interaction, compromising a device's identity can potentially lead to very undesirable outcomes (Roman et al. 2011). The IoT is also a fertile ground for hackers. With so many devices sending signals around the world, preventing people from tapping onto this communication with malicious intent is going to be difficult

- Standardization and interoperability. There are too many competing standards, making the IoT devices incompatible with each other. Google, Microsoft, Intel, Apple, and Samsung are all pushing their own versions, and there is no consensus in the industry on how to centralize around common standard(s). Without compatibility between devices from different manufacturers, the IoT domain finds it challenging to capitalize on the data collection, its analysis and provisioning of actionable insights. The IoT brings together a host of heterogeneous devices and technologies that interact with each other globally. Interoperability of these devices is at the core of the success of the IoT. To achieve interoperability, standards must be agreed upon and be acceptable across organizations and various geographical regions. Companies and governments must come together to agree on standards allowing interoperability. It is clear that an emerging idea is to consider the IoT standardization as an integral part of the future internet definition and standardization process (Bandyopadhyay and Sen 2011).

- Manageability. One analyst firm predicts that "the IoT market will grow from an installed base of 15.4 billion devices in 2015 to 30.7 billion devices in 2020 and 75.4 billion in 2025." How are these devices going to be managed? The 2.3 billion smartphones in use today – a number that will only grow – have spawned the global service provider industry (Verizon, AT&T, Vodafone, Bharti, China Mobile, etc.). The scaling of the smartphone coupled with smart devices is resulting in an industry that is extremely challenging to monitor, regulate, and standardize.

- Network optimization. The IoT is characterized by a large amount of control traffic (the device connecting to the network, authenticating itself, going to sleep when not active, paging periodically to announce its presence, etc.) and very little data traffic (a few bytes to a few Kbytes per data cycle). In contrast, current networks are meant to handle a small amount of control traffic and a large amount of data traffic. This is a challenge from networks and communications viewpoint as there is a need to align current networks with requirements that are opposite – emanating from the emerging the IoT devices.

- Analytics. It is clear that for the IoT to be useful, it has to process of lot of data, data that is not stationary but streaming. the IoT therefore requires real-time or streaming analytics algorithms. Many existing algorithms, such as Map Reduce works on stationary data. Thus, there is a need for analytics to evolve in order to facilitate incorporation of streaming data from the IoT devices to be of immediate value to the users. The question as to what kind of big data analytics will be needed to harvest useful information from the massive flood of data that the IoT devices generate is an interesting challenge. For example, "a Boeing jet generates 10 terabytes of information per engine every 30 minutes of flight, according to Stephen Brobst, the CTO of Teradata. So for a single six-hour, cross-country flight from New York to Los Angeles on a twin-engine Boeing 737 – the plane used by many carriers on this route – the total amount of data generated would be a massive 240 terabytes of data." How and where to store this data and how to enable its sensible analytics are challenges that demand creative approaches to data sourcing, storage, analytics, and display.

- Social aspects. There are a number of social challenges as well. How will society evolve when we are being watched or monitored by the IoT devices (e.g., Amazon Alexa) all the time? How will governments use this information to serve, spy on, or prosecute their citizens? Will there be a few mega-corporations controlling the IoT ecosystem, or will there be a more democratic setup of constructively competing smaller players?

- As devices become traceable through the IoT, they increase the threat to personal privacy. Theoretically, one's location, at all times can be known to someone else in public, unless privacy is managed carefully. To protect privacy, it is critical to manage ownership of data collected from smart objects. The data owner must be assured that the data will not be used by any third party without their consent. For example, if health data through a smart watch can be tapped by a life insurance company, it can terminate the policy of those at a higher health risk. To tackle this challenge it is important to have data exchange protocols based on privacy policies. Whenever two objects interact with each other, they must check each other's privacy policies for communication before communicating.

Finally, the IoT will create new legal challenges. Establishing laws governing such a global resource as the IoT is difficult to outline and even more difficult to enforce. Governance cannot be dictated by a single group, but by a group of broad-based stakeholders. In addition, global accountability and enforcement will also be necessary.

Conclusion

This chapter provides a summary of the Internet of Things (IoT) from a technology and applications viewpoint. The ability of the devices to connect over the internet provide significant opportunities for automated data collection, analytics, and provisioning of insights for the users – in making better business decisions as well as improving the quality of life of the individual users. Furthermore, this chapter discusses the various challenges that are presented by the IoT – and an approach to handling those challenges in order to apply the IoT in business.

References

Agarwal, A., and Unhelkar, B. 2017. The Internet of Things. In B. Warf (ed.) *Encyclopedia of the Internet*. Thousand Oaks, CA: Sage.

Agarwal, A., Govindu, R, Ngo, F., and Lodwig, S. 2016. Solving the jigsaw puzzle: An analytics framework for context awareness in the Internet of Things. *The Cutter Journal* 29(4):6–11.

Akyildiz, I., and Jornet, J. 2010. The Internet of Nano-Things. *IEEE Wireless Communications* 17(6):58–63.

Ashton, K. 2009. *The RFID Journal*, www.rfidjournal.com/articles/view?4986.

Atzori, L., Iera, A., and Morabito, G. 2010. The Internet of Things: A survey. *Computer Networks* 54(15):2787–2805.

Balaji, M., and Roy, S. 2017. Value co-creation with Internet of Things technology in the retail industry. *Journal of Marketing Management* 33:1–2, 7–31.

Bandyopadhyay, D., and Sen, J. 2011. Internet of Things: applications and challenges in technology and standardization. *Wireless Personal Communications* 58(1):49–69.

Bandyopadhyay, S., Sengupta, M., Maiti, S., and Dutta, S. 2011. Role of middleware for Internet of Things: A study. *International Journal of Computer Science & Engineering Survey* 2(3):94–105.

Gartner, www.gartner.com/it-glossary/internet-of-things/.

Hancke, G. P., Markantonakis, K., and Mayes, K. 2010. Security challenges for user-oriented RFID applications within the "Internet of Things." *Journal of Internet Technology* 11(3):307–313.

IoT-GSI Internet of Things Global Standards Initiative. 2012. www.itu.int/en/ITU-T/gsi/iot/pages/default.aspx.

Jeschke S., Brecher C., Meisen T., Özdemir D., and Eschert T. 2017. Industrial Internet of Things and cyber manufacturing systems. In Jeschke S., Brecher C., Song H., and Rawat D. (eds.) *Industrial Internet of Things*. Dordrecht: Springer.

Liu, J., Shen, H., Narman, H.S., Chung, W., and Lin, Z. 2018. A survey of mobile crowdsensing techniques: A critical component for the Internet of Things. *ACM Transactions on Cyber-Physical Systems – Special Issue on the Internet of Things: Part 2*, 2(3).

Ma, H. 2011. Internet of Things: objectives and scientific challenges. *Journal of Computer Science and Technology* 26(6):919–924.

Roman, R., Najera, P., and Lopez, J. 2011. Securing the Internet of Things. *IEEE Computer* 44(9):51–58.

Unhelkar, B. 2018. *Big Data Strategies for Agile Business*. Boca Raton, FL: Auerbach Books.

Whitmore, A., Agarwal, A. and Xu, L. 2015. The Internet of Things – A survey of topics and trends. *Information Systems Frontiers* 17(2):261–274.

Index

Printed in the United States
by Baker & Taylor Publisher Services

Printed in the United States
by Baker & Taylor Publisher Services